REAL ANALYSIS

by Nicolas K. Artémiadis

Southern Illinois University Press
Carbondale and Edwardsville

Feffer & Simons, Inc.
London and Amsterdam

Artémiadis, Nicolas K.
Real analysis.

Bibliography: p.
Includes Index.
1. Functions of real variables. I. Title.
QA331.5.A77 515'.8 75-29189
ISBN 0-8093-0727-8

Printed by offset lithography in the United States of America

TO MY BROTHER, *ARTEMIS*

Contents

Preface

This book is primarily designed as a text for the course usually called "Real Variables" in most universities in the United States and which is generally offered for the first-year graduate or senior undergraduate student of mathematics. The text can be covered in a year's course, assuming three one-hour meetings per week, if some sections of Chapters 7 and 10 are omitted. The prerequisite for reading this book is a workable knowledge of advanced calculus.

In developing this text we had several objectives in mind. First, we have tried to include material essential at the first-year graduate level for further study in real analysis and related subjects. Second, we have endeavored to present the material rigorously and in a modern spirit. Finally, we have been guided by the hope that a student with mathematical potential will finish the course with a liking for the subject matter and an interest in pursuing it further.

It is perhaps too easy for the mathematics student of today to lose sight of the roots of classical analysis if he is forced to concentrate all effort on extreme generalization of existing mathematical structures. Some may find the purely formalistic point of view so abusive that they come to a complete rejection of modern mathematics. We have made an effort in this book to bridge the gap between the extreme points by fusing, to a certain extent, the classical material with modern developments. It is our intention to convince the reader of the advisability of staying in the middle of the bridge without losing sight of either end, so that he can realize that there is no gap but only continuity in the evolution of mathematics. Except for Theorem 7 and Propositions 8, 9, and 10 of Chapter 5 (see Artémiadis [2]) the rest of the material is classical and well-known.

Since the book is written as a textbook it includes a Problem section at the end of each chapter. Some of these Problems are easy and involve direct application of the results obtained in the text, while others require various degrees of ingenuity on the part of the student.

I would like to express thanks and to acknowledge indebtedness to my wife for her patience and skill in typing the manuscript, and also to thank my colleague, Professor John M. H. Olmsted, for his close reading of the manuscript and his willingness to oversee the publication of the book itself. And I wish to express my appreciation to Sharon Champion and Rosemary Crocker for their help in the final stages of preparation of this book for publication.

Carbondale, Illinois   Nicolas Artémiadis

REAL ANALYSIS

CHAPTER 1

Set Theory

1.1 INTRODUCTION

The theory of sets began to be developed at the turn of the last century. The ideas and concepts of set theory penetrated deeply into all branches of modern mathematics, and one can now say that most mathematics is based on set theory. The founder of this discipline was the German mathematician G. Cantor (1845-1918).

Set theory has an axiomatic development that we shall not give here. For this we refer the reader to Bourbaki [5] and Kelley [16]. There are however some places in analysis where it is necessary to have precise formulations of some principles of set theory. One of them is known as the *axiom of choice* and is discussed below (see 1.10). We will be rather informal about the use of this axiom. In this chapter any statement not accompanied by a proof or a definition must be considered as an axiom connecting undefined terms.

The concept of *set* is a primitive one and does not lend itself to an accurate definition. The following lines illustrate the concept, but they do not pretend to give a definition.

According to Cantor, *a set* S *is any collection of definite, distinguishable objects of our intuition or our intellect, to be conceived of as a whole. The objects are called the elements or members of* S.

The word "set" is equivalent to "aggregate," "ensemble," "collection," or "class," and refers to certain objects that are combined under a certain criterion or according to a certain rule.

All the elements of any set are to be regarded as distinct from each other. If an object x is an element of a set S, we express this fact by writing $x \in S$. If x does not belong to S we write $x \notin S$. A set is determined by its elements. This means that two sets are equal if and only if they consist of the same elements.

Although sets can be elements of other sets, to avoid contradiction it is convenient to postulate that *no set can be an element of itself*. The theory of sets presented here is that used by mathematicians in their daily work. Unfortunately it is not free of difficulties and it presents contradictions (paradoxes). A cornerstone of Cantor's theory is that one is guided by intuition in deciding which objects are sets and which are not. This is the reason why the name "intuitive set theory" is often used. The faith that individuals have in their intuition seems to be responsible for the paradoxes of intuitive set theory. Generally speaking the principle which asserts that every property determines a set may be regarded as the weak point of the intuitive set theory. Indeed when used without restriction this principle yields contradictions. The following paradox was pointed out by Bertrand Russell (1872-1970). Let A be the set of all sets each of which does not contain itself as an element, and let B be the set of all sets each of which contains itself as an element. We now ask, to which one of the sets A or B does the set A belong? If A belongs to A then A contains itself as an element. This contradicts the definition of A. Therefore $A \notin A$. Suppose now A is an element of B. Since all elements of B are sets each of which contains itself as an element, by definition A cannot be an element of B. Therefore $A \notin B$. In other words A can neither be a member of itself nor fail to be a member of itself.

If A and B are sets such that every element of A belongs to B, then A is said to be a *subset* of B. We express this fact by writing $A \subset B$. Every set is a subset of itself; $A \subset A$. Two sets A and B are equal, written $A = B$, if and only if $A \subset B$ and $B \subset A$.

In this book, in any particular discussion, the sets encountered will be regarded as subsets of some set which we call the *universe* and denote usually by X.

If P is a given property and Y is a set, there is a unique subset A of Y such that, for all elements of A, P is true. We denote this subset by $A = \{x \in Y: P(x)\}$. The set $\emptyset_Y = \{x \in Y: x \neq x\}$, $x \neq x$ being the negation of

$x = x$, is called the *empty subset* of Y; it contains no element. We can in fact prove that $\emptyset_Y$ does not depend on Y. Let $\emptyset_Z$ be the empty subset of the set Z. We have $\emptyset_Y \subset \emptyset_Z$. For if not, then there would exist an element in $\emptyset_Y$ not in $\emptyset_Z$, thus contradicting the fact that $\emptyset_Y$ has no element. Similarly $\emptyset_Z \subset \emptyset_Y$, and so $\emptyset_Y = \emptyset_Z$. This means that all empty sets are equal, and they are denoted $\emptyset$. For any set S we have $\emptyset \subset S$.

The (unique) set whose elements are the objects $x_1, x_2, \ldots, x_n$ is denoted by $\{x_1, x_2, \ldots, x_n\}$. In particular $\{x\}$ is the set whose unique element is the object x. The unique set of all subsets of a given set S is denoted by $\mathcal{P}(S)$ or by 2^S. This last notation is justified by the fact that if S is a set with only n elements then the set 2^S has 2^n elements. (Prove this.) We also have $\emptyset \in \mathcal{P}(S)$, and $S \in \mathcal{P}(S)$.

Corresponding to any two objects x and y is a new object, their *ordered pair* $\langle x,y \rangle$. A rigorous definition and treatment of the notion of the ordered pair is given in Kelley [16]. We call x and y the first and the second coordinate respectively of the ordered pair $\langle x,y \rangle$. Two ordered pairs $\langle x,y \rangle$ and $\langle a,b \rangle$ are *equal* (identical) if and only if $x = a$ and $y = b$. Thus $\langle x,y \rangle = \langle y,x \rangle$ if and only if $x = y$. In a similar way we can consider ordered triplets $\langle x,y,z \rangle$, ordered quadruplets $\langle x,y,z,w \rangle$, etc. For any two sets X and Y, the unique set whose elements are all of the ordered pairs $\langle x,y \rangle$ with $x \in X$ and $y \in Y$, is called the *cartesian product* or *direct product* or *product* of X and Y and it is written $X \times Y$. Similarly $X \times Y \times Z$ is the set of all ordered triplets $\langle x,y,z \rangle$ with $x \in X$, $y \in Y$, and $z \in Z$, etc.

We assume that the reader is familiar with the *principle of finite induction* and the *well-ordering principle of the set* $N = \{1,2,3,\ldots\}$ of natural numbers, as stated below.

The principle of finite induction: If $S \subset N$, $1 \in S$, and if for every $n \in S$ it follows that $n + 1 \in S$, then $S = N$.

The well-ordering principle: If $\emptyset \neq S \subset N$, then S has a smallest element.

1.2 OPERATIONS ON SETS

The *intersection* of two sets A and B, denoted $A \cap B$, is the set of all elements that belong to both A and B. If $A \cap B = \emptyset$ then A and B are said to be *disjoint*.

The *union* of two sets A and B, denoted $A \cup B$, is the set of all elements that belong to at least one of the sets A and B.

The *complement* of a subset A of a set E, with respect to E, is the set of all elements of E that do not belong to A. It is denoted by $\complement_E A$ or $\complement A$ when there is no possibility of confusion.

The *difference* $A - B$ of the sets A and B is the set of all elements of A that do not belong to B. It is easy to see that $A - B = A - (A \cap B) = \complement_A (A \cap B)$. The following propositions are immediate consequences of the above definitions:

$$A - A = \emptyset, \qquad A - \emptyset = A.$$
$$A \cup A = A, \qquad A \cap A = A.$$
$$A \cup B = B \cup A, \qquad A \cap B = B \cap A.$$

The relations $A \subset B$, $A \cup B = B$, and $A \cap B = A$ are equivalent. The statement $A \subset A \cup B$ is equivalent to $A \cap B \subset A$.

$A \cup B \subset C$ if and only if $A \subset C$ and $B \subset C$.

$C \subset A \cap B$ if and only if $C \subset A$ and $C \subset B$.

$A \cup (B \cup C) = (A \cup B) \cup C$ denoted $A \cup B \cup C$.

$A \cap (B \cap C) = (A \cap B) \cap C$ denoted $A \cap B \cap C$.

$A \cup (B \cap C) = (A \cup B) \cap (A \cup C)$

(Distributivity of the union with respect to the intersection).

$A \cap (B \cup C) = (A \cap B) \cup (A \cap C)$

(Distributivity of the intersection with respect to the union).

If A and B are subsets of E, then

$$\complement(\complement A) = A, \quad \complement(A \cup B) = (\complement A) \cap (\complement B),$$

and

$$\complement(A \cap B) = (\complement A) \cup (\complement B).$$

These last two relations are called, after Augustus De Morgan (1806-71), *De Morgan's laws.*

$A \subset B$ if and only if $\complement A \supset \complement B$.

The statements $A \cap B = \emptyset$, $A \subset \complement B$, and $B \subset \complement A$ are equivalent.

The statements $A \cup B = E$, $\complement A \subset B$, and $\complement B \subset A$ are equivalent.

1.3 FUNCTIONS

Definition *Let* X *and* Y *be two given sets. A* function f *from (or on)* X *to (or into, or with values in)* Y *is a rule which assigns to each element* x *of* X *a unique element* $f(x)$ *of* Y.

We often show that f is a function on X into Y by writing $f\colon X \to Y$.

We wish to emphasize here that the characteristic property of a function is that with every element of X there is associated one and only one element of Y. It is of course legitimate to define a function which, to every $x \in X$, associates a subset of Y containing more than one element (a multiple-valued function), but such a definition is useless in practice because it is impossible to define, in a sensible way, algebraic operations on the values of such a function.

A synonym for function is *mapping*, and we shall use both words interchangeably. Let f be a function from X to Y. Then the set X is called the *domain* of f. The set of all elements y in Y for which there is an $x \in X$ such that $f(x) = y$ is called the *range* of f. If the range of f is the whole set Y, then we say that f

is a function from X *onto* Y, or that the function is *surjective*. The subset G of the Cartesian product $X \times Y$ consisting of all elements of the form $\langle x, f(x) \rangle$ is called the *graph* of f. It follows from the definition of a function that if $\langle x_1, f(x_1) \rangle$ and $\langle x_2, f(x_2) \rangle$ are two elements of the graph G of the function f, then $x_1 = x_2$ implies $f(x_1) = f(x_2)$. Conversely, if G is a subset of $X \times Y$ such that for any two elements $\langle x_1, y_1 \rangle$ and $\langle x_2, y_2 \rangle$ of G, $x_1 = x_2$ implies $y_1 = y_2$, then G is the graph of some function f on X into Y; namely, for $x \in X$ the value of this function is the second coordinate of the unique pair whose first coordinate is x. It follows from the above discussion that a function is determined by its graph. This leads to the following definition: *a function on* X *into* Y *is a subset* G *of* $X \times Y$ *such that* $\langle x_1, y_1 \rangle \in G$, $\langle x_2, y_2 \rangle \in G$, $x_1 = y_1$ *imply* $y_1 = y_2$.

Two functions are said to be equal if and only if they have the same graph. This is also equivalent to saying that the two functions f and g are equal if and only if they have the same domain X and for every $x \in X$, $f(x) = g(x)$.

If f is a function on X into Y, and if A is a subset of X, the *image* of A under f is the set $f[A] = \{y \in Y: \exists x \in A$ such that $y = f(x)\}$. The *inverse image* of a subset B of Y is the set $f^{-1}[B] = \{x \in X: f(x) \in B\}$. The set $f^{-1}[\{y\}]$ is also written $f^{-1}(y)$. Throughout the book the symbols $\exists x$ and iff mean "there is an x" (or "for some x") and "if and only if" respectively.

A function $f: X \to Y$ is called *one-to-one* (or *injective*) iff for every y in $f[X]$ the set $f^{-1}(y)$ contains only one element of X. In other words, if $x_1 \in X$, $x_2 \in X$, and $x_1 \neq x_2$, then $f(x_1) \neq f(x_2)$. A function $f: X \to Y$ which is one-to-one and onto Y is called a *one-to-one correspondence* (or *bijective*) between X and Y. If f is a one-to-one correspondence between X and Y then the function $g: Y \to X$, such that, for every $y \in Y$, $g(y)$ equals the unique element of the set $f^{-1}(y)$, is called the *inverse function* of f. It is denoted by f^{-1}.

If $f: X \to Y$ and $g: Y \to Z$, then the function $h: X \to Z$, defined by setting for $x \in X$, $h(x) = g(f(x))$ is called the *composite function* of g and f (in that order) and is written $h = g \circ f$. For any subset A of Z we have $h^{-1}[A] = f^{-1}[g^{-1}(a)]$.

Let $f: X \to Y$ and let A be a subset of X. The function $g: A \to Y$ such that for $x \in A$ $g(x) = f(x)$ is called the *restriction* of f to A and is denoted by f/A.

Characteristic Function of a Set

Let A be a subset of a set E. The characteristic function of A (with respect to E) is defined to be the function χ_A, whose domain is E, such that $\chi_A(x) = 1$ if $x \in A$ and $\chi_A(x) = 0$ if $x \notin A$.

Conversely, every function f defined on E and which assumes only the values 0 or 1 is evidently the characteristic function of the set $f^{-1}(1)$, which is a subset of E.

Sequences *A* finite sequence *from (or in) a set X is a function on the set of the first n natural numbers (that is, the set* $\{i \in N: i \leq n\}$*) into X, for some* $n \in N$.

An infinite sequence *from (or in) a set X, is a function on the set* N *of natural numbers into X.*

If S is a sequence (finite or infinite), then the image S(n) of n is denoted by x_n. We denote a finite and an infinite sequence by $\langle x_i \rangle_{i=1}^{n}$ and $\langle x_i \rangle_{i=1}^{\infty}$ respectively. If no confusion is possible we simply write $\langle x_i \rangle$. The range of the sequence $\langle x_i \rangle$ is denoted by $\{x_i\}$.

One must carefully distinguish a sequence from its range. The range of a sequence can be a set with only one element. Two different sequences can have the same range.

Finite, Infinite, Denumerable, and Countable Sets

A set is called *finite* iff it is empty or if it is the range of a finite sequence. A set is called *infinite* iff it is not finite. A set is called *denumerable* iff it is infinite and it is the range of an infinite sequence. A set is called *countable* iff it is finite or denumerable. From the definition of countable set it follows immediately that the range of a function whose domain is a countable set is also a countable set.

A nonempty finite set $S = \{x_1, x_2, \ldots, x_n\}$ can also be the range of an infinite sequence, defined by setting $x_i = x_n$ for $i > n$.

Families The notion of family is a generalization of the notion of sequence. Let I be any set, whose elements will be called *indices*, and let X be another set. *A family of elements* of X, indexed by I, is a function f on I into X. For $i \in I$ we write $f(i) = x_i$. We denote this family by $\langle x_i \rangle_{i\in I}$. If I is a finite (respectively denumerable) set we say that the family is finite (respectively denumerable). To every family $\langle x_i \rangle_{i\in I}$ of elements of X corresponds the range $\{x_i\}_{i\in I} = f[I] \subset X$ of f, but, as we pointed out in the case of a sequence, the range $f[I]$ itself does not determine the family $\langle x_i \rangle_{i\in I}$. Two different families can have the same range.

Example

If X is a given set and $Y = \mathcal{P}(X)$, then a family $\langle y_i \rangle_{i\in I}$ of elements of Y is called a family of subsets of X.

Let $\mathcal{F} = \langle X_i \rangle_{i\in I}$ be a family of subsets of a set E. Then the *intersection* of the X_i is the set $\bigcap_{i\in I} X_i$ (also written $\bigcap_i X_i$) of all elements x in E such that $x \in X_i$ for every $i \in I$. The *union* of the X_i is the set $\bigcup_{i\in I} X_i$

(also written $\bigcup_i X_i$ or $\bigcup_{F \varepsilon \mathcal{F}} F$ or $\bigcup \mathcal{F}$) of all elements x in E such that $x \varepsilon X_i$ for at least one $i \varepsilon I$. We can easily see that many of the propositions obtained for the union and intersection of two sets can be generalized for the union and intersection of any family of sets. In particular we have two generalized distributive laws

$$\complement\left(\bigcup_{i \varepsilon I} X_i\right) = \bigcap_{i \varepsilon I} \complement X_i$$

$$\left(\bigcup_{i \varepsilon I} X_i\right) \bigcap \left(\bigcup_{j \varepsilon J} Y_j\right) = \bigcup_{\langle i,j \rangle \varepsilon I \times J} \left(X_i \bigcap Y_j\right)$$

as well as the relations that we obtain if the symbols $\bigcap$ and $\bigcup$ are interchanged.

If $f\colon X \to Y$, and if $\langle X_i \rangle_{i \varepsilon I}$ is a family of subsets of X, then $f\left[\bigcup_i X_i\right] = \bigcup_i f\left[X_i\right]$. If $\langle Y_j \rangle_{j \varepsilon J}$ is a family of subsets of Y, then:

$$f^{-1}\left[\bigcup_j Y_j\right] = \bigcup_j f^{-1}\left[Y_j\right]$$

and

$$f^{-1}\left[\bigcap_j Y_j\right] = \bigcap_j f^{-1}\left[Y_j\right].$$

We now define the notion of an *infinite subsequence* of a given infinite sequence S. Let $k: N \to N$ be a mapping from the natural numbers N into N with the property that $n < m$ implies $k(n) < k(m)$. Then the composite mapping $S \circ k$ of S and k is called an infinite subsequence of S. The element $(S \circ k)(n)$ is denoted by S_{k_n}, and the subsequence $S \circ k$ by $\langle S_{k_n} \rangle$.

Example

Let $S = \langle 1/n \rangle$ and let k be defined by $k(n) = 3n$; then $S \circ k = \langle 1/3n \rangle$.

1.4 RELATIONS

Any subset R of the product $A \times B$ of two sets A and B is called a *binary relation* (or *relation*) *between the sets A and B*. If $\langle a,b \rangle \in R$, we say that the relation R is verified by a and b, and we write aRb.

A binary relation on a set A is a binary relation between the set A and the set $B = A$; that is a subset R of $A \times A$. The following is an example of such a relation. If A is a set, consider the set $R = \{\langle x,x \rangle : x \in A\}$, called the *diagonal* of $A \times A$. Writing xRy, is equivalent to saying that x equals y. In other words the equality relation on A is defined by the diagonal of $A \times A$.

A binary relation on a set A is said to be an *equivalence relation*, denoted by $\equiv$ (or by $\sim$) iff it satisfies the following conditions:

Reflexivity: $x \equiv x$ for every $x \in A$.
Symmetry: $x \equiv y$ implies $y \equiv x$.
Transitivity: $x \equiv y$ and $y \equiv z$ imply $x \equiv z$.

Using a geometric language suggested by the case where A is an interval of the real line, we may say that the set R which defines an equivalence relation on A must contain the diagonal of $A \times A$. It must also contain the fourth vertex of every "rectangle" whose two diagonally opposite vertices belong to R and whose third vertex also belongs to the diagonal.

Let $\equiv$ be an equivalence relation on a set A. For every $x \in A$ the set $E_x = \{y \in A: x \equiv y\}$ is called the *equivalence class of* x. Obviously $x \in E_x$. We prove that any two equivalence classes E_{x_1}, E_{x_2} are either disjoint or coincide. In fact suppose $E_{x_1} \cap E_{x_2} \neq \emptyset$ and $z \in E_{x_1} \cap E_{x_2}$. If x is any element of E_{x_1} we have $x \equiv z$; since $z \equiv x_2$ we have $x \equiv x_2$ and $x \in E_{x_2}$. Thus $E_{x_1} \subset E_{x_2}$. Similarly we have $E_{x_2} \subset E_{x_1}$, and so $E_{x_1} = E_{x_2}$.

It is left to the reader to verify that the following relations are equivalence relations.

1) Let A be the set of all positive, negative, or zero integers, and let k be a fixed integer. Put xRy iff $x - y$ is a multiple of k. This relation is usually denoted by $x = y$ (modulo k).

2) Let A be the set $N \times N$. Put $\langle p,q\rangle \, R \, \langle p_1,q_1\rangle$ iff $pq_1 = p_1q$.

3) Let A be the set of all straight lines L in the Euclidean plane. Put L_1RL_2 iff $L_1 \cap L_2 = \emptyset$ or if $L_1 = L_2$.

4) Let A be the set of all pairs of points (M_1,M_2) of the Euclidean plane. Put $(M_1,M_2) \, R \, (P_1,P_2)$ iff there exists a straight line containing the points M_1,M_2,P_1,P_2 and iff on this line we have $\overline{M_1M_2} = \overline{P_1P_2}$.

5) Let A be the complement, in the Euclidean plane, of a fixed straight line L of the plane. Put xRy iff the segment with endpoints x and y does not intersect L.

Let $\equiv$ be an equivalence relation on a set A. Then denote by $A/\equiv$ the collection of the equivalence classes under the equivalence relation $\equiv$, and call this collection the *quotient* of A with respect to $\equiv$.

The *natural mapping* of A onto $A/\equiv$ is the mapping that associates to every $x \in A$ the equivalence class E_x. The inverse image of any element of $A/\equiv$ is an equivalence class in A.

Conversely, let A be any arbitrary set, and let f be a mapping from A onto a set B. Define a binary relation $\equiv$ on A as follows:

$$x_1 \equiv x_2 \quad \text{iff} \quad f(x_1) = f(x_2).$$

Then $\equiv$ is an equivalence relation on A. The

equivalence classes are the inverse images $f^{-1}(y)$ of the elements of B. We easily see that the mapping $\phi: (A/\equiv) \to B$ defined by $\phi(E_x) = f(x)$ is a one-to-one correspondence between the sets $A/\equiv$ and B, and for this reason we identify the sets $A/\equiv$ and B. It follows from the above discussion that the two notions of function and equivalence relation can be reduced one to the other, that is, to a given equivalence relation $\equiv$ on a set A, there corresponds a function (the natural mapping) on A onto $A/\equiv$, and if f is a function on A onto a set B, then there is an equivalence relation R on A, such that f is the corresponding natural mapping to R and such that there is a one-to-one correspondence between A/R and B.

These two notions of function and equivalence relation can also be reduced to the notion of partition of a set. By a *partition* of a set X we mean a pairwise disjoint collection P of nonempty subsets of X such that each member of X is a member of some (and hence, exactly one) element of P.

Next we show how the notion of equivalence relation and the notion of partition can be reduced one to the other. In fact from the definition of equivalence class it follows that given an equivalence relation on a set A the corresponding set of equivalence classes form a partition of A. Conversely if $\{A_i\}_{i \in I}$ is a partition P of A, we associate to P the binary relation $\equiv$ on A defined by $x \equiv y$ iff x and y belong to the same set, say A_i. The relation $\equiv$ is an equivalence relation. In fact the reflexivity and symmetry properties being obvious we only prove the transitivity of $\equiv$. Suppose that $x \equiv y$ and $y \equiv z$. This means that x and y belong to a set A_i and y and z to another set A_j. Since $A_i \cap A_j \neq \emptyset$ (y is in both of them) it follows that $A_i = A_j$, which means that $x \equiv z$.

1.5 PARTIAL ORDERING

A binary relation $\prec$ on a set A is called a *partial ordering* iff the following three conditions are satisfied:

Reflexivity: $x \prec x$ for every x in A.
Antisymmetry: $x \prec y$ and $y \prec x$ imply $x = y$.
Transitivity: $x \prec y$ and $y \prec z$ imply $x \prec z$.

The ordered pair $\langle A, \prec \rangle$ is called a *partially ordered* set, and the set A is said to be partially ordered by $\prec$. A partial ordering $\prec$ on a set A is said to be a *linear ordering* (or ordering) iff for any two elements x, y of A we have either $x \prec y$ or $y \prec x$. In this case the ordered pair $\langle A, \prec \rangle$ is called a *linearly ordered set*.

Examples

1) Let A be the set of positive integers. The relation "$x \prec y$ iff x divides y exactly" is a partial ordering.

2) Let A be the collection of all subsets of a set E. The relation "$x \prec y$ iff $x \subset y$," is a partial ordering.

3) Let A be the set of all rational numbers. The relation "$x \prec y$ iff $x \leq y$," is a linear ordering.

Definitions *Let* A *be a set partially ordered by* $\prec$. *If there is an element* $a \in A$ *such that for every* $x \in A$ *we have* $x \prec a$, *then* a *is called the (unique)* largest element *of* A. *For example, every finite linearly ordered set has a largest element.*

An element $a \in A$ *is said to be* maximal *iff there is no* $x \in A$ $(x \neq a)$ *such that* $a \prec x$.

The largest element of A, if there is one, is also a maximal element and it is unique. If there is no largest element in A then there may exist more than one maximal element.

We define the *smallest element* and a *minimal element* in a similar way.

If $\langle A, \prec \rangle$ *is a partially ordered set and* $B \subset A$, *then an element* x *in* A *is an* upper bound *of* B *iff for all* b *in* B, $b \prec x$. *We define a* lower bound *of* B *in a similar way.*

1.6 COUNTABLE SETS

In section 1.3 we defined a set to be countable iff it is either empty or the range of a sequence. Furthermore, we defined a set to be denumerable, iff is infinite and countable. The following proposition provides another definition of denumerable set equivalent to the given one.

1. Proposition *An infinite set* E *is denumerable iff there is a one-to-one correspondence between* E *and the set* N *of natural numbers.*

Proof If there is a one-to-one correspondence f between N and E then f is an infinite sequence and E is its range. Since E is infinite, it follows that E is denumerable. Next suppose that E is the range of an infinite sequence $\langle x_n \rangle$. We first define the function ϕ from N into N as follows: Put $\phi(1)$ equal to 1, and put $\phi(n + 1)$ equal to the smallest value of m such that $x_m \neq x_i$ for all $i \leq \phi(n)$. Such an m always exists because the set E is by assumption infinite. Define now a mapping $f\colon N \to E$ such that $f(n) = x_{\phi(n)}$. It follows from the way that ϕ is constructed that f is a one-to-one correspondence between N and E. This proves the proposition.

Here are some propositions concerning countable sets.

2. Proposition *Every subset of a countable set is countable.*

Proof Let E be a countable set. If E is empty then there is nothing to prove. If E is not empty, then it is the range of a sequence, say $\langle x_n \rangle$. We have $E = \{x_n\}$. Let A be a subset of E. If A is empty then by definition A is countable. If A is not empty let x be an element of A. Define the mapping $f\colon N \to A$ as follows: $f(n) = x_n$ if $x_n \in A$ and $f(n) = x$ if $x_n \notin A$. Then f is a sequence and A is its range, and hence A is countable.

3. Corollary *Every infinite subset of a denumerable set is denumerable.*

4. Proposition *Every subset of a finite set is finite.*

The proof is similar to the proof of Proposition 2.

5. Proposition *The set of all finite sequences of natural numbers is denumerable.*

Proof Let $\{2,3,5,11,\ldots,p_k,\ldots\}$ be the set of prime numbers, let S be the set of all finite sequences from N, and let S_o be the set of all finite sequences from the set $N_o = N \cup \{0\}$. It is clear that the set of prime numbers is denumerable. Every n in $N - \{1\}$ has a unique factorization of the form $n = 2^{a_1} 3^{a_2} \ldots p_k^{a_k}$ where $a_i \in N_o$ and $a_k > 0$. Define the mapping $f: N \to S_o$ as follows: $f(n) = \langle a_1, a_2, \ldots, a_k \rangle$ if $n \neq 1$ and $f(1) = \langle 0 \rangle$. From the uniqueness of the factorization of every n it follows that f is a one-to-one correspondence between N and f[N] so that f[N] is denumerable by Proposition 1. Since S is an infinite subset of f[N] it follows from Corollary 3 that S is denumerable.

6. Corollary *The set of all finite sequences of a denumerable set is denumerable.*

7. Proposition *The set of all rational numbers is denumerable.*

Proof Let A be the set whose elements are the sequence $\langle 1,1,3 \rangle$ and all the sequences of natural numbers of the form $\langle p,q,1 \rangle$, and $\langle p,q,2 \rangle$. Then A is infinite and as a subset of the set of all finite sequences of natural numbers is denumerable. Define now the mapping f from A to the set of rational numbers as follows:

$$f(\langle p,q,1 \rangle) = \frac{p}{q}, \quad f(\langle p,q,2 \rangle) = -\frac{p}{q}, \quad f(\langle 1,1,3 \rangle) = 0.$$

Since the mapping f is surjective, the set of rational numbers as the range of the countable set A under f is also countable, and since it is infinite it is denumerable.

8. Proposition *The union of a countable collection of countable sets is a countable set.*

Proof Let $\mathcal{F}$ be a countable collection of countable sets. If $\bigcup \mathcal{F}$ is the empty set, it is countable by definition. If $\bigcup \mathcal{F}$ is not empty, then there are nonempty sets in $\mathcal{F}$ and since the empty set contributes nothing to the union of the sets of $\mathcal{F}$ we may suppose that all the elements of $\mathcal{F}$ are nonempty. Also (see section 1.3) we may suppose that $\mathcal{F}$ is the range of an infinite sequence $\langle A_n \rangle_{n=1}^{\infty}$ where each A_n is nonempty and is the range of an infinite sequence $\langle x_{nm} \rangle_{m=1}^{\infty}$. Now the mapping $\langle n,m \rangle \to x_{nm}$ is from the set of all ordered pairs $\langle n,m \rangle$ of natural numbers onto the set $\bigcup_{n=1}^{\infty} A_n$. Since the set of all ordered pairs of natural numbers is countable it follows that $\bigcup_{n=1}^{\infty} A_n$ is also countable.

9. Proposition *Every infinite set S has a denumerable subset.*

Proof Let $x_1 \in S$. Suppose that for every positive integer n, there are distinct elements $x_1, x_2, \ldots, x_n$ of S. Since S is infinite there is an element $x_{n+1} \in S$ distinct from all the elements $x_1, x_2, \ldots, x_n$. By the principle of finite induction, S has a subset $\{x_1, x_2, \ldots, x_n, \ldots\}$ which is in one-to-one correspondence with the natural numbers. Therefore S has a denumerable subset.

1.7 CARDINAL NUMBERS

If A and B are two finite sets and if there is a one-to-one correspondence between them, then we can affirm that A and B have the same number of elements. Of course we can arrive at the same conclusion if we determine the number of elements in A and the number of elements in B and find them equal. But the method of establishing a one-to-one correspondence between A and B in order to compare the number of elements of these sets, has the advantage that it can be applied when the sets to be compared are infinite. The method of one-to-one correspondence was precisely the method used by Cantor in his research to find a process of comparing various "infinite" sets.

The concept of equivalence of two sets, introduced by the following definition, is a generalization of the concept of "having the same number of elements" for finite sets.

Definition *Two sets A and B are said to be* equivalent *or to* have the same power *(written $A \sim B$) iff there is a one-to-one correspondence between them.*

If $A \sim B$ then obviously $B \sim A$.

Definition *With every set A we associate a sym-*

bol called the cardinal number of A, *such that two sets* A *and* B *have the same cardinal number, iff* $A \sim B$. *The cardinal number of* A *is denoted* $\overline{\overline{A}}$.

10. Theorem (Bernstein) *Let* A *and* B *be two sets. If there is a subset* B_1 *of* B *and a subset* A_1 *of* A *such that* $A \sim B_1$, *and* $B \sim A_1$, *then* $A \sim B$.

Proof Let g be a one-to-one correspondence between A and B_1. The restriction of g to A_1 is a one-to-one correspondence between A_1 and a subset B_2 of B_1. Since $A_1 \sim B$, it follows that $B \sim B_2$. We prove that $B \sim B_1$. Let f_0 be a one-to-one correspondence between B and B_2. The restriction of f_0 to B_1, say f_1, is a one-to-one correspondence between B_1 and a subset B_3 of B_2. Similarly the restriction of f_1 to B_2, say f_2, is a one-to-one correspondence between B_2 and a subset B_4 of B_3 and so on. We obtain in this way a sequence $\langle B_n \rangle$ of sets and a sequence $\langle f_n \rangle$ of one-to-one correspondences such that

$$B = B_0 \supset B_1 \supset B_2 \supset \ldots \supset B_n \supset \ldots$$

$$B_0 \sim B_2 \sim \ldots \sim B_{2n} \sim \ldots$$

$$B_1 \sim B_3 \sim \ldots \sim B_{2n+1} \sim \ldots$$

The restriction of f_{n-1} to B_n, say f_n, is a

one-to-one correspondence between B_n and a subset B_{n+2} of B_{n+1}.

Put $C = \bigcup_{n=1}^{\infty} (B_{n-1} - B_n)$, $D = \bigcap_{n=1}^{\infty} B_{n-1}$.

Obviously $C \cup D \subset B$. We also have $B \subset C \cup D$. In fact let x be any element of B. If $x \in C$ then there is nothing to prove. Suppose $x \notin C$. This means that $x \notin (B_{n-1} - B_n)$ for $n = 1,2,3,\ldots,$ which means that $x \in B_n$ for $n = 1,2,3,\ldots,$. Since x is also an element of B_0, it follows that $x \in \bigcap_{n=1}^{\infty} B_{n-1}$ or $x \in D$. This proves that $B = C \cup D$.

Similarly we can prove that

$$B_1 = \left(\bigcup_{n=1}^{\infty} (A_n - A_{n+1})\right) \cup \left(\bigcap_{n=1}^{\infty} A_n\right).$$

The restriction of f_{2n} to $B_{2n} - B_{2n+1}$, say f_{2n}^*, is a one-to-one correspondence between $(B_{2n} - B_{2n+1})$ and $(B_{2n+2} - B_{2n+3})$, that is $(B_{2n} - B_{2n+1}) \sim (B_{2n+2} - B_{2n+3})$ for $n = 0,1,2,\ldots$.

Also, since $\bigcap_{n=0}^{\infty} B_n = \bigcap_{n=1}^{\infty} B_n$, we have $\bigcap_{n=0}^{\infty} B_n \sim \bigcap_{n=1}^{\infty} B_n$. We now define the mapping $F: B \to B_1$ as follows:

$$F(x) = \begin{cases} f^*_{2n}(x) & \text{if } x \in (B_{2n} - B_{2n+1}) \\ & \qquad\qquad (n = 0,1,2,\ldots) \\ x & \text{if } x \notin (B_{2n} - B_{2n+1}) \end{cases}$$

It is easily seen that F is a one-to-one correspondence between B and B_1. Hence $B \sim B_1$. Since by assumption $B_1 \sim A$, it follows that $A \sim B$. The theorem is proved.

Comparison of Cardinal Numbers

Let a and b be two cardinal numbers and let A and B be two sets with cardinal numbers a and b respectively. If A is equivalent to a subset of B we say that a is <u>less than or equal</u> to b and we write $a \prec b$. It is obvious that the definition of the relation $\prec$ does not depend on the choice of the sets A and B. Moreover the relation $\prec$ is reflexive, transitive, and antisymmetric. In fact the reflexivity and transitivity of $\prec$ are immediately seen. We now prove the antisymmetry of $\prec$. Let a and b be two cardinal numbers such that $a \prec b$ and $b \prec a$, and let A and B be two sets such that $\overline{\overline{A}} = a$ and $\overline{\overline{B}} = b$. Since $a \prec b$, there is a subset of B which is equivalent to A. Similarly, since $b \prec a$, there is a subset of A equivalent to B. It follows from Bernstein's theorem that $A \sim B$, or $a = b$. This proves that the relation $\prec$ is antisymmetric.

11. Proposition *Let A be an infinite set and B another set such that $A \subset B$ and $B - A$ is countable. Then $\overline{\overline{A}} = \overline{\overline{B}}$.*

Proof Since A is infinite, by Proposition 9 there is a subset E of A such that $A - E$ is denumerable. We have $B - E = (B - A) \cup (A - E)$. Since $B - A$ is countable and $A - E$ is denumerable, it follows that $B - E$ is denumerable. Therefore the sets $A = E \cup (A - E)$ and $B = E \cup (B - E)$ are equivalent, so that $\overline{\overline{A}} = \overline{\overline{B}}$.

1.8 OPERATIONS ON CARDINAL NUMBERS

If a and b are any cardinal numbers, their *sum* $a + b$ is the cardinal number of the set $A \cup B$, where A and B are two disjoint sets such that $\overline{\overline{A}} = a$, $\overline{\overline{B}} = b$.

The above definition does not depend on the choice of the sets A and B. For if A and A_1 are sets whose cardinal number is a, and B, and B_1 are sets whose cardinal number is b, and if $A \cap B = \emptyset$ and $A_1 \cap B_1 = \emptyset$, then $(A \cup B) \sim (A_1 \cup B_1)$, so that $A \cup B$ and $A_1 \cup B_1$ have the same cardinal number.

The *product* ab of two cardinal numbers a and b is defined to be the cardinal number of the set of all ordered pairs $\langle x,y \rangle$, $x \in A$, $y \in B$, where A is a set with cardinal number a and B a set with cardinal number b. Again this definition is independent of the choice of the sets A and B. The proof is left to the reader.

The denumerable sets all have the same cardinal number, that of the set of natural numbers. The symbol $\aleph_0$ (aleph zero, aleph being the first letter of the Hebrew alphabet) designates this cardinal number.

Since every infinite set has a denumerable subset (Prop. 9) we have $\aleph_0 \prec \overline{\overline{S}}$, for every infinite set S. In other words *the cardinal number* $\aleph_0$ *is the smallest infinite cardinal number*, that is the smallest cardinal number among the cardinal numbers of infinite sets. Some of the properties of *finite cardinal numbers* (that is the cardinal numbers of finite sets) are not true for infinite cardinal numbers. For example, if n is the cardinal number of a nonempty finite set then $n + n \neq n$. However, this is not always true for infinite cardinal numbers. As an example we prove that $\aleph_0 + \aleph_0 = \aleph_0$.

Let A be the set of all even positive integers, and let B be the set of all odd positive integers. Since $A \cap B = \emptyset$ and $\overline{\overline{A}} = \overline{\overline{B}} = \aleph_0$, it follows from the definition of the sum of two cardinal numbers that $\aleph_0 + \aleph_0 = \aleph_0$.

12. Theorem (Cantor) *If* E *is any set, then* $\overline{\overline{E}} \prec \overline{\overline{\mathcal{P}(E)}}$ *and* $\overline{\overline{E}} \neq \overline{\overline{\mathcal{P}(E)}}$.

Proof Since $\mathcal{P}(E)$ contains a subset equivalent to E, we have $\overline{\overline{E}} \prec \overline{\overline{\mathcal{P}(E)}}$. To prove $\overline{\overline{E}} \neq \overline{\overline{\mathcal{P}(E)}}$, suppose the contrary - that $\overline{\overline{E}} = \overline{\overline{\mathcal{P}(E)}}$. Then there exists a one-to-one correspondence between E and $\mathcal{P}(E)$, say ϕ. Let B be set of all elements x in E such that $x \notin \phi(x)$. Then B, as a subset of E, is an element of $\mathcal{P}(E)$. But B cannot be the image of any element in E. For if for some $b \in E$ we have $\phi(b) = B$ then either $b \notin \phi(b)$ which implies $b \in B$, that is, $b \in \phi(b)$, or $b \in \phi(b)$ which implies $b \notin B$, that is, $b \notin \phi(b)$. Both cases lead to the contradiction that ϕ is not

not a one-to-one correspondence between E and $\mathcal{P}(E)$.

The above theorem asserts that *there is an infinite number of distinct infinite cardinal numbers.*

The following statements are immediate consequences of Theorem 12.

For any positive integer n, *it is true that* $n < 2^n$.

If A *is a denumerable set then the set* $\mathcal{P}(A)$ *is not denumerable.*

The cardinal numbers of the sets N, $\mathcal{P}(N)$, $\mathcal{P}(\mathcal{P}(N))$,... *form a strictly increasing sequence.*

Although a rigorous definition of the system $\mathbb{R}$ of real numbers will be given in Chapter 2, we take the liberty at this point to borrow the following property of real numbers. Let x be a real number, $0 \leq x \leq 1$. Then there is a sequence $\langle a_n \rangle$ of integers with $0 \leq a_n < 2$ such that

$x = \sum_{n=1}^{\infty} (a_n/2^n)$ and this sequence is unique except when x is of the form $p/2^n$, in which case there are exactly two such sequences. The sequence $\langle a_n \rangle$ is called the binary expansion of x. We now denote by c the cardinal number of the set of real numbers in the closed interval [0,1].

13. Theorem $\overline{\overline{\mathcal{P}(N)}} = c$.

Proof Let $A \in \mathcal{P}(N)$ and χ_A be the characteristic function of A. Then the mapping $A \to \chi_A$ is a one-to-one correspondence between the set $\mathcal{P}(N)$ and the set of characteristic functions of all elements of $\mathcal{P}(N)$. Since the set of characteristic functions of all elements of $\mathcal{P}(N)$ is equivalent to the set D of the binary expansions of the real numbers in the interval $[0,1]$, we have $\overline{\overline{\mathcal{P}(N)}} = \overline{\overline{D}}$. Now each number in $[0,1]$ has a unique binary expansion, except those numbers of the form $p/2^n$ which have exactly two expansions. Since the set of the numbers $p/2^n$ is denumerable the set of their binary expansions is also denumerable. Let A_r be the set of all numbers in $[0,1]$ having a unique binary expansion, and D_r be the set of their binary expansions. Since $A_r \sim D_r$ we have $\overline{\overline{A_r}} = \overline{\overline{D_r}}$. On the other hand, by Proposition 11, we have $c = \overline{\overline{A_r}}$ and $\overline{\overline{D}} = \overline{\overline{D_r}}$. It follows that $c = \overline{\overline{D}}$ and $\overline{\overline{\mathcal{P}(N)}} = c$.

14. Corollary *The cardinal number of the set of real numbers is* c.

Proof By Proposition 11, we have $[0,1] \sim (0,1)$. The mapping $x \to 2x - 1$ is a one-to-one correspondence between $(0,1)$ and $(-1,1)$, so that $(0,1) \sim (-1,1)$, and the mapping $x \to x/(1-x^2)$ is a one-to-one correspondence between $(-1,1)$ and the set $\mathbb{R}$ of real numbers. It follows that $[0,1] \sim \mathbb{R}$ so that $\overline{\overline{\mathbb{R}}} = c$.

By Theorem 12 we have $\aleph_0 \neq c$. Since $\aleph_0$ is the smallest infinite cardinal number we have $\aleph_0 < c$. Now one may ask the question: Has every infinite subset of the set of real numbers either cardinal number $\aleph_0$ or c? The conjecture that the answer to this question is in the affirmative is called the *Continuum Hypothesis.*

More generally, the conjecture: "If A is any infinite set, then there is no cardinal number a such that $\overline{\overline{A}} \neq a$, $\overline{\overline{\mathcal{P}(A)}} \neq a$, $\overline{\overline{A}} \prec a \prec \overline{\overline{\mathcal{P}(A)}}$," is called the *Generalized Continuum Hypothesis.*

K. Göded has proved that *if the Generalized Continuum Hypothesis is added to the other axioms of set theory no inconsistencies are introduced into the system that were not there previously.*

1.9 ALGEBRA OF SETS

Let X be a set. A nonempty collection $\mathcal{A}$ of subsets of X is called an *algebra of sets* iff the following conditions are satisfied:

(a) $A \in \mathcal{A}$ implies $\complement A \in \mathcal{A}$

(b) $A \in \mathcal{A}$, $B \in \mathcal{A}$ imply $A \cup B \in \mathcal{A}$.

An equivalent definition of algebra of sets can be obtained if condition (b) is replaced by condition (b'): $A \in \mathcal{A}$, $B \in \mathcal{A}$ imply $A \cap B \in \mathcal{A}$. For suppose that (a) and (b) are satisfied and $A \in \mathcal{A}$, $B \in \mathcal{A}$. Then by (a) we have $\complement A \in \mathcal{A}$, $\complement B \in \mathcal{A}$ and by (b) and (a),

$\complement(\complement A \cup \complement B) = A \cap B \in \mathcal{A}$, so that (a), (b') are satisfied. Similarly we can prove that if (a) and (b') are satisfied then (a) and (b) are also satisfied.

It is easily seen that an algebra of sets is closed under the formation of finite unions and of finite intersections. A σ-*algebra* of subsets of

X is an algebra $\mathcal{A}$ such that if $\langle A_n \rangle$ is a sequence from $\mathcal{A}$ then $\bigcup_{n=1}^{\infty} A_n$ is also an element of $\mathcal{A}$.

If $\mathcal{A}$ is a σ-algebra and $\langle A_n \rangle$ a sequence from $\mathcal{A}$ then, since $\complement A_n \in \mathcal{A}$, we have

$$\complement\left(\bigcup_{n=1}^{\infty} \complement A_n\right) = \bigcap_{n=1}^{\infty} A_n \in \mathcal{A}.$$

15. Proposition *If* $\mathcal{E}$ *is any family of subsets of* X *then there exists a* σ*-algebra* $\mathcal{A}$ *such that:*

(i) $\mathcal{E} \subset \mathcal{A}$.

(ii) If $\mathcal{B}$ *is any* σ*-algebra of sets which contains* $\mathcal{E}$ *then* $\mathcal{B} \supset \mathcal{A}$.

Proof Let $\mathcal{F}$ be the family of σ-algebras each of which contains $\mathcal{E}$. $\mathcal{F}$ is not empty because it contains the set $\mathcal{P}(X)$ of all subsets of X. Put $\mathcal{A} = \bigcap_{\mathcal{B} \in \mathcal{F}} \mathcal{B}$. $\mathcal{A}$ is a σ-algebra. For if $A \in \mathcal{A}$ then $A \in \mathcal{B}$ for every $\mathcal{B}$ in $\mathcal{F}$. Since $\mathcal{B}$ is a σ-algebra we have $\complement A \in \mathcal{B}$ for every $\mathcal{B} \in \mathcal{F}$, therefore $\complement A \in \mathcal{A}$. Next, let $\langle A_n \rangle$ be a sequence of elements of $\mathcal{A}$. Then $\langle A_n \rangle$ belongs to $\mathcal{B}$ for every $\mathcal{B} \in \mathcal{F}$. Since $\mathcal{B}$ is a σ-algebra we have $\bigcup_{n=1}^{\infty} A_n \in \mathcal{B}$ for every $\mathcal{B} \in \mathcal{F}$, so that $\bigcup_{n=1}^{\infty} A_n \in \mathcal{A}$.

Now let $\mathcal{B}$ be a σ-algebra which contains $\mathcal{C}$. Then $\mathcal{B} \in \mathcal{F}$, and from the definition of $\mathcal{A}$ it follows that $\mathcal{B} \supset \mathcal{A}$.

The σ-algebra $\mathcal{A}$ considered in Proposition 15, is called the *σ-algebra generated by the family* $\mathcal{C}$, *or the smallest σ-algebra containing* $\mathcal{C}$. Proposition 15 remains true if the words "σ-algebra" are replaced by the word "algebra." The proof is the same.

16. Proposition *Let* $\langle A_n \rangle$ *be any sequence of sets. Let* $\langle B_n \rangle$ *be the sequence defined by*

$$B_1 = A_1, \text{ and for } n > 1, \; B_n = A_n - \bigcup_{k=1}^{n-1} A_k.$$

Then $\bigcup_{n=1}^{\infty} A_n = \bigcup_{n=1}^{\infty} B_n$, *and the sets* B_n *are pairwise disjoint.*

Proof Let $x \in \bigcup_{n=1}^{\infty} A_n$. Then for some n, we have that $x \in A_n$. Let n_0 be the smallest integer such that $x \in A_{n_0}$. We have $x \notin A_n$ for $n = 1,2,\ldots,n_0-1$, which implies that $x \in B_{n_0} \subset \bigcup_{n=1}^{\infty} B_n$. Thus $\bigcup_{n=1}^{\infty} A_n \subset \bigcup_{n=1}^{\infty} B_n$. On the other hand, since $B_n \subset A_n$ for all n, we have that $\bigcup_{n=1}^{\infty} B_n \subset \bigcup_{n=1}^{\infty} A_n$ which implies $\bigcup_{n=1}^{\infty} A_n = \bigcup_{n=1}^{\infty} B_n$. Finally, if $m \neq n$, say

$m < n$, then $B_m \cap B_n = \emptyset$. For if $x \in B_n$, then

$x \notin \bigcup_{k=1}^{n-1} A_k$, $x \notin A_m$ which implies $x \notin B_m$.

1.10 THE AXIOM OF CHOICE AND SOME OF ITS EQUIVALENT FORMULATIONS

A partially ordered set $\langle X, \prec \rangle$ is called *well-ordered* if and only if each nonempty subset of X has a smallest element. A typical example of a well-ordered set is the set N of natural numbers with its usual ordering. In 1904 the German mathematician *Ernst Zermelo* gave a demonstration of the so-called *well-ordering theorem* which asserts that *on any given set* X, *a partial ordering* $\prec$, *can be defined so that* $\langle X, \prec \rangle$ *is well-ordered*. Soon after the publication of the well-ordering theorem, the French mathematician Emile Borel proved that a property of sets, now known as the *axiom of choice* was assumed by Zermelo in the proof of his theorem, and that this property could be deduced from the well-ordering theorem. In other words Borel pointed out that the well-ordering theorem and the axiom of choice are two equivalent properties. The axiom of choice, which has excited vigorous controversy, has many important consequences. The proof of certain important theorems in mathematics depend on assuming this axiom; at least no other way of proving them is known.

Whenever a new axiom is added to a system of axioms, there is a possibility of introducing inconsistencies into the system that were not there previously. In the case of the axiom of choice, this possibility was ruled out by K. Gödel who proved that *if the axioms of set theory, without the axiom of choice, form a consistent system, then this system remains consistent if the axiom of choice is added.* In other words a proposition which is proved by using the axiom of choice can never be disproved unless the other axioms already contain an inconsistency which has nothing to do with the axiom of choice.

Another point to be clarified was whether or not the axiom of choice was really an axiom and not a theorem deriving from the axioms of set theory. This question also was answered recently by P. J. Cohen who established the impossibility of proving the axiom of choice from the other axioms of set theory; therefore it is proper to call it an axiom.

There are several formulations of the axiom of choice. One of them is the following: *Let* S *be a set and* $\mathcal{C}$ *a nonempty collection of nonempty subsets of* S. *Then there exists a mapping* $f: \mathcal{C} \to S$ *such that, for every* $A \in \mathcal{C}$, $f(A) \in A$.

The function f is called a *choice function*; it is somehow a formal mechanism that picks out "simultaneously" an element from any set of the collection $\mathcal{C}$.

Before we state some equivalent formulations of the axiom of choice we need the following:

Definition *Let* $\langle X_i \rangle_{i \in I}$ *be a family of sets indexed by* I. *Then the Cartesian product* $\prod_{i \in I} X_i$ *of the sets of this family is defined to be the set of all functions* $f: I \to \bigcup_{i \in I} X_i$ *such that* $f(i) \in X_i$ *for every* $i \in I$.

This definition agrees with the one given in section 1.1 of the Cartesian product of two sets. To see this, suppose $I = \{1,2\}$. Then by definition of section 1.1, $X_1 \times X_2$ consists of all ordered pairs $\langle x_1,x_2\rangle$ where $x_1 \varepsilon X_1$, $x_2 \varepsilon X_2$. But the mapping $f: I \to X_1 \cup X_2$ defined by $f(1) = x_1$, $f(2) = x_2$ is the element of $\prod_{i\varepsilon I} X_i$ that was denoted by $\langle x_1,x_2\rangle$, and vice versa. So the set of all such mappings is the set of all points of $X_1 \times X_2$ and conversely.

The mapping $f: I \to \bigcup_{i\varepsilon I} X_i$ which is an element of $\prod_{i\varepsilon I} X_i$ is, as in the case of a sequence, denoted by $\langle x_i\rangle_{i\varepsilon I}$ where $x_i = f(i)$ for every $i \varepsilon I$. If $\langle x_i\rangle$ is an element of $\prod_{i\varepsilon I} X_i$ we shall call x_i the ith coordinate of $\langle x_i\rangle_{i\varepsilon I}$. We now give, in the form of a theorem, without proof, some equivalent formulations of the axiom of choice. For a proof the reader is referred to McShane and Botts [19].

17. Theorem *The following statements are equivalent to one another:*

(1) Axiom of choice.

(2) If $\langle A_i\rangle_{i\varepsilon I}$ *is a family of nonempty sets indexed by a nonempty set* I*, then* $\prod_{i\varepsilon I} A_i$ *is*

nonempty.

(3) Hausdorff maximal principle *Every partially ordered set* $\langle X, \prec \rangle$ *includes a maximal linearly ordered subset, that is a subset* S *of* X *which is linearly ordered by* $\prec$ *and has the property: if* $S \subset T \subset X$ *and* T *is linearly ordered by* $\prec$, *then* $S = T$.

(4) Zorn's lemma *Every nonempty partially ordered set in which each linearly ordered subset has an upper bound, contains a maximal element.*

(5) Every set can be well-ordered.

1.11 ORDINAL NUMBERS

In section 1.5, a linearly ordered set was defined to be a set L, together with a linear ordering on it. For simplicity, such a linearly ordered set will be called a *linear order* L. Also, again for simplicity, the same symbol will be used for the orderings of any linear orders L,L',L",..., even though their orderings may be distinct.

Definition *Two linear orders* L *and* L', *are said to be* similar, *iff there is a one-to-one, order-preserving correspondence* f *between them; that is if* a *and* b *are two elements in* L, *with* $a \prec b$, *then* $f(a) \prec f(b)$.

It follows from the above definition that: *If* L *and* L' *are two similar orders and if one of them has a smallest element then the other also has a smallest element.* Indeed, suppose L has a smallest element say a. Then for every $b \in L$ we have $a \prec b$. If f is an order-preserving one-to-one correspondence between L and L' we have

$f(a) \prec f(b)$ which implies that $f(a)$ is the smallest element in L'.

It is clear that *if two linear orders* L *and* L' *are similar, then the sets* L *and* L' *are equivalent. The converse of this statement is not true.* To see this consider the linear orders $L = \{1,2,3,\ldots\}$, $L_1 = \{\ldots,3,2,1\}$. Then L and L_1 as sets are equivalent but as linear orders are not similar, for L has a smallest element while L_1 does not.

Examples

1) Let $L_1 = \{1,2,3,\ldots\}$, $L_2 = \{\ldots,3,-3,-1,0,1,2,3,\ldots\}$. Then L_1 and L_2 are not similar linear orders. If we order the set L_2 in the manner: $L_3 = \{0,1,-1,2,-2,\ldots\}$ then L_1 and L_3 are similar linear orders. Also if we order the set L_1 into: $L_4 = \{\ldots,6,4,2,1,3,5,\ldots\}$ then L_2 and L_4 are similar linear orders.

2) The linearly ordered sets in the intervals $1 \leq x \leq 2$ and $5 \leq y \leq 10$ are similar linear orders. The one-to-one correspondence between them is given by $y = 5x$.

3) The ordered sets $A = \{0 < x \leq 1\}$ and $B = \{0 \leq y < 1\}$ are not similar linear orders, if the linear ordering is the usual ordering of the real number system. But if the linear ordering on B is defined by $y_1 \prec y_2$ iff $y_1 \geq y_2$, and if we consider the usual ordering on A, then A and

B are similar orders. The one-to-one correspondence between them is given by the relation $y = 1 - x$.

18. Proposition *The similarity relation for linear orders is an equivalence relation.*

The proof is left for the reader.

Just as the concept of equivalence of two sets led to the definition of cardinal number, so the concept of similarity leads to the definition of *order-type.*

A class of similar linear orders is called an order-type. The statement that two linear orders have the same order-type means merely that they are similar.

We shall be concerned only with order-types of well-ordered sets, that is, linearly ordered nonempty sets such that every one of their nonempty subsets has a smallest element. Examples of well-ordered sets are the following:

1) Any finite linear order.
2) The linear order $1,2,3,\ldots,n,\ldots,$ of positive integers.
3) The linear order $1,3,5,7,\ldots; 2,4,6,8,\ldots,$.

Examples of linear orders that are not well-ordered sets are the following:

1) The linear order $\ldots,-3,-2,-1,0,1,2,3,\ldots$ of all integers.
2) The rational numbers in their usual order.

Definition *The order-type of a well-ordered set is called an* ordinal number.

For any finite set A, any two linear orderings are similar. In other words, if two finite sets A and B are equivalent in the sense of cardinality, and if L and L' are similar orders for A and B respectively, then L and L' are similar. For it is easy to find a one-to-one order preserving mapping between L and L'. For practical reasons the ordinal number of a finite linear order with n elements is also denoted by n. Thus, the natural numbers attain a double significance.

Definition *The ordinal number of the linear order of positive integers (i.e., the usual order of positive integers by magnitude) is denoted by* ω.

We wish to emphasize here that two equivalent well-ordered sets may not have the same ordinal number. To see this, consider the well-ordered sets given in the Examples 2 and 3 following Proposition 18. The well-ordered set given in Example 2 has ordinal number ω while the well-ordered set given in Example 3 does not have ordinal number ω.

The following propositions are immediate consequences of the definition of well-ordered set:

(i) *Every nonempty subset of a well-ordered set is well-ordered.*

(ii) *Every ordered set that is similar to a well-ordered set, is itself well-ordered.*

Comparison of Ordinal Numbers

Definition *Let* A *be a well-ordered set. Then by an* initial segment Aa *of* A *we mean all* $b \in A$ *such that* $b \prec a$, $b \neq a$.

The following theorem is stated without proof. It is the analogue of Bernstein's theorem for ordinal numbers.

19. Theorem *Let A and B be two well-ordered sets. Then either:*

(a) A and B are similar, or

(b) A is similar to an initial segment of B, or

(c) B is similar to an initial segment of A.

Let α and β be two ordinal numbers and A and B two well-ordered sets with ordinal numbers α and β respectively. If A is similar to an initial segment of B we say that α is *less than or equal* to β and we write $\alpha \prec \beta$. When we say that α is *strictly less than* β, we mean that $\alpha \neq \beta$ and $\alpha \prec \beta$. It is easily seen that the definition of the relation $\prec$ does not depend on the choice of the well-ordered sets A and B used in defining the statement $\alpha \prec \beta$. Moreover the relation $\prec$ is reflexive, transitive, and antisymmetric. The proof is similar to the proof given in the case of cardinal numbers (see p. 25).

20. Proposition *Every set of ordinal numbers is a well-ordered set.*

Proof Suppose the proposition is not true. Then there is a sequence $\langle \alpha_n \rangle$ of distinct ordinal numbers with $\alpha_2 \prec \alpha_1, \alpha_3 \prec \alpha_2, \alpha_4 \prec \alpha_3, \ldots$. Let A be any well-ordered set of ordinal number α_1. Then by Theorem 19, A has initial segments $Aa_1, Aa_2, \ldots, Aa_n, \ldots$ whose ordinal numbers are $\alpha_2, \alpha_3, \ldots, \alpha_{n+1}, \ldots$ respectively. It follows that $\alpha_3 \prec \alpha_2, \alpha_4 \prec \alpha_3, \ldots,$ so that A has a subset which has no smallest element. This contradicts the assumption that A is a well-ordered set.

If k is a cardinal number, let us denote by $S(k)$ the set of all ordinal numbers λ with the property that λ is the ordinal number of a well-ordered set of cardinality k. By Proposition 20, the set $S(k)$ is well-ordered. Therefore it has a smallest element, say $\alpha(k)$. The ordinal number $\alpha(k)$ is called the *initial ordinal* of the cardinal k.

As an example consider the case where $k = \aleph_0$. Then $S(\aleph_0)$ is the set of all ordinal numbers of all well-ordered denumerable sets. The initial ordinal of $\aleph_0$ is ω. For let B be a denumerable well-ordered set with ordinal number β. Let 1^* be the smallest element of B, 2^* the smallest element of $B - \{1^*\},\ldots,$ and for each $n \in N - \{1\}$ let n^* be the smallest element of $B - \{1^*,2^*,3^*,\ldots,(n-1)^*\}$. We obtain, by induction, the well-ordered set $B^* = \{1^*,2^*,3^*,\ldots\}$, which is obviously similar to the well-ordered set $N = \{1,2,3,\ldots\}$. If $B - B^* = \emptyset$ then $\omega = \beta$ and hence $\omega \prec \beta$. If $B - B^* \neq \emptyset$ then let b be the smallest element of $B - B^*$. Then for every $n^* \in B^*$ we have $n^* \prec b$ and $n^* \neq b$. This means that B^* is an initial segment of B and since N is similar to B^* we also have $\omega \prec \beta$.

Other ordinals of the cardinal $\aleph_0$ are, the ordinal number of the well-ordered set $\{1,3,5,\ldots;1'\}$ denoted $\omega + 1$ and the ordinal number of the well-ordered set $\{1,2,3,\ldots;1',2',3',\ldots\}$ denoted $\omega + \omega$.

Thus we proved that ω is the smallest ordinal number among the ordinal numbers of the well-ordered denumerable sets. We express this fact briefly by saying that ω *is the smallest infinite ordinal number* and that it has denumerable cardinal number.

Next we prove that there are ordinal numbers that correspond to nondenumerable sets. We first prove the following lemmas.

21. Lemma *A well-ordered set* A *is not similar to any of its initial segments.*

Proof Let Aa be an initial segment of A similar to A, and f an order-preserving one-to-one correspondence between A and Aa. Put $f(a) = a_1$. We must have $a \neq a_1$ and $a_1 \prec a$. Then $a_2 = f(a_1) \prec f(a) = a_1, a_3 = f(a_2) \prec f(a_1) = a_2, \ldots$. It follows that the sequence $\langle a, a_1, a_2, \ldots, a_n, \ldots \rangle$ of distinct elements in A has no first element, which contradicts that A is a well-ordered set.

An immediate consequence of Lemma 21 is that *no two distinct initial segments of a well-ordered set are similar.*

22. Lemma *Let* α *be an infinite ordinal number. Then the set of ordinal numbers strictly less than* α *is a well-ordered set whose ordinal number is* α.

Proof Let A be a well-ordered set whose ordinal number is α. We prove that there exists a one-to-one order-preserving correspondence between the ordinals strictly less than α and the initial segments of A. In fact for every ordinal $\beta \prec \alpha$, $\beta \neq \alpha$ there is an initial segment $A\beta$ whose ordinal is β. By Lemma 21 the initial segment $A\beta$ is unique. This establishes the one-to-one correspondence sought. Moreover, since A is infinite there is a one-to-one order-preserving correspondence between its elements and its initial segments. This proves the lemma.

Now let W be the set of all ordinal numbers that correspond to countable sets. By Proposition 20, W is a well-ordered set. Let Ω be the ordinal number of W.

23. Theorem *The ordinal number* Ω *is the first nondenumerable ordinal number.*

Proof By Lemma 22, if α is a denumerable ordinal the initial segment $W\alpha$ of W has ordinal number α. Since Ω is the ordinal of W, and since a well-ordered set is never similar to any of its initial segments, we have for every denumerable ordinal number α, that $\alpha \neq \Omega$ and $\alpha \prec \Omega$. This means that Ω is a nondenumerable ordinal. On the other hand if β is any nondenumerable ordinal number and B a well-ordered set whose ordinal number is β then B is not similar to any initial segment of W. This implies that $\Omega \prec \beta$, which means that Ω is the smallest nondenumerable ordinal number.

As we mentioned in a previous section the most important theorem about well-ordered sets is the well-ordering theorem which Cantor had accepted as true but which was first proved rigorously by Zermelo. According to this theorem every set can be well-ordered. The prime importance of the well-ordering theorem lies in the possibility that we can generalize to any well-ordered set the principle of finite induction for natural numbers. This generalized principle, called the *principle of transfinite induction*, is given by the following theorem.

24. Theorem *If* S *is the set of all ordinals less than a given ordinal* α, *and* A *is a subset of* S *such that*

(i) *The first ordinal* I *belongs to* A,

(ii) *For every ordinal* $\beta \prec \alpha$, *if all ordinals less than* β *and different from* β *belong to* A, *then* β *also belongs to* A.

Then A = S.

Proof Suppose $A \neq S$, then S - A has a smallest element, say β. Since $I \in A$ we have $\beta \neq I$. But since all ordinals strictly less than β are in A, it follows from (ii) that β also belongs to A; a contradiction.

As in the case of the finite induction principle, the principle of transfinite induction can be used to prove that *a certain property* P *belongs to all elements of a well-ordered set* S *if it belongs to the smallest element of* S, *and if it belongs to an element* α *of* S *as soon as it belongs to all preceding elements of* α.

The Cesare Burali-Forti (1861-1931) Paradox

Let U be the "set" of all ordinal numbers. Then U is a well-ordered set, and it has an ordinal number, say α. We have $\alpha \in U$. By Lemma 22, the initial segment $U\alpha$ has ordinal number α. Hence $U\alpha$ is similar to U. But this contradicts the fact that a well-ordered set is not similar to any of its initial segments.

The only way to avoid this paradox seems to be to declare that the words that seem to define U do not define a set. Another way of saying this is that there is no set consisting exactly of all the ordinal numbers.

PROBLEMS

1. Prove all statements given in section 1.2.
2. Show that $A \cap B = \emptyset$ if and only if $A - B = A$ and $B - A = B$.
3. Let A be the set of all rational numbers that can be written as ratios of integers whose denominator is 6 and let B be the set of all rational numbers that can be written as fractions whose denominator is 8. Determine the sets $A \cup B$ and $A \cap B$.
4. The *symmetric difference* Δ is defined with respect to a pair of sets A and B as follows: $A \Delta B = (A - B) \cup (B - A)$. Prove that (a) $A \Delta A = \emptyset$, (b) $A \Delta B = B \Delta A$, (c) $A \Delta (B \Delta C) = (A \Delta B) \Delta C$, (d) for any sets A and B there exists a unique set X such that $A \Delta X = B$.
5. Use the principle of finite induction to prove that for every positive integer n,
$$1 + 2 + \ldots + n = \frac{n(n + 1)}{2}.$$
6. Prove that if $A \cap B = A$ and $A \cup B = A$ then $A = B$.
7. Let $\langle X_i \rangle_{i \varepsilon I}$ be an arbitrary family of subsets of a set E. Prove that De Morgan's laws hold. That is

$$\complement\left(\bigcup_{i\varepsilon I} X_i\right) = \bigcap_{i\varepsilon I} \complement X_i$$

$$\complement\left(\bigcap_{i\varepsilon I} X_i\right) = \bigcup_{i\varepsilon I} \complement X_i.$$

8. Let $f: X \to Y$. Let $\langle X_i \rangle_{i\varepsilon I}$ be a family of subsets of X and let $\langle Y_j \rangle_{j\varepsilon J}$ be a family of subsets of Y. Prove that

   (a) $f\left[\bigcup_{i\varepsilon I} X_i\right] = \bigcup_{i\varepsilon I} f\left[X_i\right]$

   (b) $f\left[\bigcap_{i\varepsilon I} X_i\right] \subset \bigcap_{i\varepsilon I} f\left[X_i\right]$

   (c) Equality in (b) need not occur.

   (d) $f^{-1}\left[\bigcup_{j\varepsilon J} Y_j\right] = \bigcup_{j\varepsilon J} f^{-1}\left[Y_j\right]$

   (e) $f^{-1}\left[\bigcap_{j\varepsilon J} Y_j\right] = \bigcap_{j\varepsilon J} f^{-1}\left[Y_j\right]$

9. Let $\langle A_n \rangle_{n=1}^{\infty}$ be a sequence of sets. Set $B_n = \bigcup_{k=n}^{\infty} A_k$ and $C_n = \bigcap_{k=n}^{\infty} A_k$. Show that

$$\bigcup_{n=1}^{\infty} A_n = \bigcup_{n=1}^{\infty} B_n \quad \text{and} \quad \bigcap_{n=1}^{\infty} A_n = \bigcap_{n=1}^{\infty} C_n.$$

10. If $\langle A_n \rangle_{n=1}^{\infty}$ is a sequence of sets show that

$$\bigcap_{n=1}^{\infty} A_n = \complement\left(\bigcup_{n=1}^{\infty} \complement A_n\right).$$

11. Let $f: X \to Y$ be a mapping. Prove that f is onto iff for every nonempty subset S of Y, $f^{-1}[S] \neq \emptyset$.

12. Let $f: X \to Y$ be a one-to-one mapping, and let A and B be subsets of X. Prove that

$$f[A \cap B] = f[A] \cap f[B]$$

13. Let $f: X \to Y$ be a one-to-one mapping and let $\langle A_i \rangle_{i \varepsilon I}$ be a family of sets in X. Prove that

$$f\left[\bigcap_{i \varepsilon I} A_i\right] = \bigcap_{i \varepsilon I} f[A_i]$$

14. Let $f: X \to Y$ be a mapping from X onto an infinite set Y. Prove that X is an infinite set.

15. If A, B, and C are sets such that $(A \cup B) - C$ is an infinite set, then either $A - C$ or $B - C$ is an infinite set. If $(A \cap B) - C$ is an infinite set, then both $A - C$ and $B - C$ are infinite sets.

16. Which of the following relations are equivalence relations?

(a) Let T be the set of all triangles in the plane. For $x,y \varepsilon T$ define xRy iff x and y are similar triangles.

(b) Let N be the set of all positive integers. For $x,y \varepsilon N$ define xRy iff x and y are relatively prime (i.e., the only common divisor of x and y is 1).

(c) Let X and Y be sets and $f: X \to Y$ a mapping. Let F be the subset of $X \times X$ consisting of all pairs $\langle a,b \rangle$ such that $f(a) = f(b)$.

17. Let F be a binary relation defined on the set N of positive integers by

$$xFy \text{ if and only if } x + y = \text{odd}.$$

Prove that

(a) F is not the graph of a function on N into N.

(b) F is not an equivalence relation.

(c) F is not a partial ordering.

18. Let F be an equivalence relation on a set A containing more than one point. Prove that F is not a linear ordering.

19. Let N be the set of positive integers, and put $A = N \times N$. On A define the relation F as follows:

$$\langle a,b \rangle \, F \, \langle c,d \rangle \text{ iff } a + d = b + c.$$

(a) Prove that F is an equivalence relation on A.

(b) Define the addition between two elements of A by $\langle a,b \rangle + \langle c,d \rangle = \langle a + c, b + d \rangle$. Then prove that $\langle a,b \rangle \, F \, \langle a',b' \rangle$ and $\langle c,d \rangle \, F \, \langle c',d' \rangle$ imply $\langle \langle a,b \rangle + \langle c,d \rangle \rangle \, F \, \langle \langle a',b' \rangle + \langle c',d' \rangle \rangle$. Thus we have in effect defined addition between equivalence classes of pairs of positive integers.

20. Let F be a partial ordering on a set A and let G be a binary relation on A defined by

$$xGy \text{ if and only if } yFx.$$

Prove that G is a partial ordering on A.

21. Let F be a binary relation on the set of all positive integers N. Suppose that F has the

following properties.

(a) If n is even and m is odd, then nFm.

(b) If n and m are both even or both odd, then nFm if and only if there exists a non-negative integer k such that $n = m + k$. Is F a partial ordering on N?

22. Prove that the polynomials $p(x)$, ordered by setting $p(x) \prec q(x)$ if and only if there is an x such that for every $\xi > x$, $p(\xi) < q(\xi)$, are linearly ordered.

23. Rephrase the order relation of Problem 22 in terms of the coefficients of the polynomials.

24. Why does the procedure of Problem 22 not provide a linear order relation for the continuous functions?

25. Prove Proposition 2.

26. Prove that if X and Y are finite sets, then $X \times Y$ is a finite set.

27. Show that if X and Y are sets, and $X \times Y$ is a finite set, then X and Y are not necessarily both finite.

28. Prove that a finite set is not equivalent to any of its proper subsets.

29. Let $f: X \to Y$ be a mapping of X onto Y. Is it true that Y uncountable implies X uncountable?

30. Prove without using Bernstein's theorem that an open interval and a closed interval of real numbers are equivalent.

31. Prove that the definitions of the sum and product of two cardinal numbers a and b are independent of the choices of the sets A and B (see section 1.8).

32. Show that if A is countable, then for each integer n the set of all ordered n-tuples of elements of A is a countable set.

33. The roots of any equation of the form

$$a_0 x^n + a_1 x^{n-1} + \ldots + a_{n-1} x + a_n = 0$$

where $a_0, a_1, \ldots, a_n$ are integers, are called *algebraic numbers*. Show that the set of all algebraic numbers is countable. (Hint: For every positive integer p, there is only a finite number of equations such that

$$n + |a_0| + |a_1| + \ldots + |a_n| = p.)$$

34. Prove that

(a) $c + c = c$

(b) $c + c + \ldots + c + \ldots = c$

35. Prove that $\aleph_0 \cdot \aleph_0 = \aleph_0$

36. Prove that $c \cdot c = c$

Here is an outline of the proof. The cardinal number of the set of points in the square $0 < x \leq 1$, $0 < y \leq 1$ is $c \cdot c$. We only need to show that there is a one-to-one correspondence between this square and the interval $(0,1]$. Write every $\langle x,y \rangle$ as a pair of non-terminating decimals $\langle 0.a_1 a_2, \ldots, a_n, \ldots, 0.b_1 b_2, \ldots, b_n, \ldots \rangle$. This may be done in one and only one way. Mate with this point the number $0.a_1 b_1 a_2 b_2, \ldots, a_n b_n, \ldots$. This is a one-to-one mapping between the square and a subset of the interval $(0,1]$. Bernstein's theorem may be used to complete the proof.

37. Show that every infinite set can be expressed as a denumerable union of disjoint sets each of which is infinite.

38. We associate with every set A the set 2^A of all functions on A with values 0 or 1. If a is the cardinal number of A, then the cardinal number of 2^A is denoted by 2^a.

(a) Prove that $\overline{\overline{\mathcal{P}(A)}} = 2^a$.

(Hint: Establish a one-to-one correspondence between 2^A and $\mathcal{P}(A)$ by mating with each function in 2^A the subset of A on which the function is 1.)

(b) Prove that $a \prec 2^a$.

(c) Prove that $2^{\aleph_0} = c$.

39. Does there exist an infinite σ-algebra of sets which has only countably many members?

40. Prove Proposition 18.

41. (a) Show that every nonempty subset of a well-ordered set is well-ordered under the induced ordering.

(b) Show that every ordered set, that is similar to a well-ordered set, is itself well-ordered.

42. Show that if W and W' are similar well-ordered sets, there is one and only one function f on W to W' which is one-to-one and order-preserving.

43. Prove that the condition of Problem 42 is satisfied by some orders which are not well-ordered sets.

44. If $\prec$ is a partial ordering on a set X with the property that every nonempty subset of X has a smallest element, then X is a well-ordered set.

CHAPTER 2

The Real Numbers

2.1 FUNDAMENTAL STRUCTURES IN MATHEMATICS

One of the essential achievements of the mathematicians of this century has been the recognition and formulation of the structures of mathematics. The most important of these structures can be found in the most classical mathematical systems. The set of real numbers, for example, possesses an algebraic structure, an order structure, a topological structure, a vector space structure, etc. The structure of a system is delineated when the axioms of the system have been set down. For a deeper discussion on the concept of structure and the role that it plays in mathematics the reader is referred to the works of N. Bourbaki [6].

A certain number of structures, being encountered in almost all questions of analysis, could be qualified as fundamental. For example, equivalence relation, group, ring, field, topological space. We shall consider some of them in this chapter and elsewhere in this book.

2.2 GROUPS--RINGS--FIELDS

Definition *A group is a pair,* $(G, \cdot)$ *such that* G *is a nonempty set and* $\cdot$ *, called the group operation, is a mapping from the Cartesian product* $G \times G$ *into* G*, such that the following axioms are satisfied.*

(a) The operation, $\cdot$ *, is associative, that is,* $x\cdot(y\cdot z) = (x\cdot y)\cdot z$ *for all elements* x, y *and* z *of* G.

(b) There is an element e*, called the* identity, *such that* $e\cdot x = x\cdot e = x$ *for each* x *in* G.

(c) For each x *in* G *there is an element* x^{-1}*, called the* inverse *of* x*, such that* $x\cdot x^{-1} = x^{-1}\cdot x = e$.

If the group operation is denoted by $+$, then the inverse of x is written $-x$. Also notice that the image of $\langle x,y\rangle$ under the mapping, $\cdot$, is written $x\cdot y$ instead of the usual notation $\cdot(\langle x,y\rangle)$. If no confusion is possible we shall write xy for $x\cdot y$. It is an easy matter to prove that the element e as described in (b) is unique, and that every x in G has a unique inverse.

A group $(G, \cdot)$ is said to be *abelian* or *commutative* if and only if $x\cdot y = y\cdot x$ for all elements x and y in G. If a group is finite, its *order* is the number of its elements.

Examples of Groups

1) The set $G = \{0,1\}$ with the operation $+$, defined by: $0+0 = 0$; $0+1 = 1+0 = 1$; $1+1 = 0$.

2) Let S be a set, and G the collection of all one-to-one mappings of S onto itself. We define the operation, $\cdot$, in G as follows. For any

two elements x and y in G we put $x \cdot y = xy = z$ if and only if $z(a) = x(y(a))$ for every element a of S. We now prove that $(G, \cdot)$ is a group. In fact, $\cdot$ is associative. For let x, y and z be any three elements in G, and a any element of S. We have

$$\{(xy)z\}(a) = (xy)(z(a)) = x(y(z(a))$$

$$\{x(yz)\}(a) = x((yz(a)) = x(y(z(a))).$$

Hence $(xy)z = x(yz)$. The identity element e of G is the identity mapping of S onto S. That is for every $a \in S$, $e(a) = a$. Finally if $x \in G$ then x^{-1} is defined to be the *inverse mapping* of x. This definition is consistent since x is one-to-one and onto.

Definition *A ring is a triple $(R,+,\cdot)$ such that R is a nonempty set and $+, \cdot$ are mappings (operations) on the Cartesian product $R \times R$ into R, such that the following conditions are satisfied.*

(a) $(R,+)$ is an abelian group.

(b) For every x,y,z in R, $x \cdot (y \cdot z) = (x \cdot y) \cdot z$ and

(c) $x \cdot (y+z) = x \cdot y + x \cdot z$, $(y+z) \cdot x = y \cdot x + z \cdot x$.

Examples of Rings

1) The set of all integers with the usual operations of addition and multiplication.

2) Let S be any nonempty set and $\mathcal{P}(S)$ the collection of all subsets of S. For $A \in \mathcal{P}(S)$, $B \in \mathcal{P}(S)$ define

$$A + B = (A \cup B) - (A \cap B) \quad \text{and} \quad A \cdot B = A \cap B$$

then $(\mathcal{P}(S),+,\cdot)$ is a ring.

In addition the ring $(\mathcal{P}(S),+,\cdot)$ satisfies the condition

$$A \cdot A = A.$$

Rings which satisfy this latter condition are called *Boolean rings*.

Definition *A* field *is a ring* $(F,+,\cdot)$ *such that* F *has at least two elements and* $(F - \{\theta\}, \cdot)$, *where* θ *is the identity element with respect to* $+$, *is a commutative group.*

Examples of Fields

1) Let F be a set containing only two elements a and b. Define the operations $+$ and $\cdot$ as follows:

$$a+a = a, \quad a+b = b, \quad b+a = b, \quad b+b = a$$
$$a \cdot a = a, \quad a \cdot b = a, \quad b \cdot a = a, \quad b \cdot b = b$$

then $(F,+,\cdot)$ is a field.

2) We shall see in this chapter that the set of real numbers and the set of rational numbers, with the usual operations of addition and multiplication, are fields.

Definition *An ordered field* $(F,+,\cdot)$ *is a field where an ordering* $\leq$ *is defined such that the following axioms are satisfied.*

(a) $x \leq y$ *and* $y \leq z$ *imply* $x \leq z$.

(b) "$x \leq y$ *and* $y \leq x$" *is equivalent to* $x = y$.

(c) for any two elements x,y *in* F, *either* $x \leq y$ *or* $y \leq x$.

(d) $x \leq y$ *implies* $x + z \leq y + z$.

(e) $\theta \leq x$ *and* $\theta \leq y$ *imply* $\theta \leq x \cdot y$.

We agree to write $y \geq x$ for every $x \leq y$. Also the relation "$x \leq y$ and $x \neq y$" is written $x < y$ or $y > x$.

2.3 ISOMORPHISM OF MATHEMATICAL SYSTEMS

An *ordered subfield* of an ordered field is a subset, which with the same operations and order relation is an ordered field.

Definition *Let* (G_1, o_1) *and* (G_2, o_2) *be two groups where* o_1 *and* o_2 *denote the group operations. Then* (G_1, o_1) *is isomorphic to* (G_2, o_2) *(in symbols* $(G_1, o_1) \sim (G_2, o_2)$*) iff there exists a one-to-one correspondence* ϕ *between* G_1 *and* G_2 *such that for every* $x, y \in G_1$, *we have* $\phi(x) o_2 \phi(y) = \phi(x\ o_1\ y)$.

It would be desirable to have a similar notion of isomorphism for any other mathematical systems. Now it is possible that a mathematical system may involve more than a given set and operation, so that the problem of including "all possible" mathematical systems in a general definition of isomorhpism would be meaningless. This is the reason for the repeated occurrence of definitions bearing this name.

Motivation for each definition of isomorphism lies in demanding that the sets involved be equivalent and that the tables representing the relations and functions involved be related in such a manner that the systems in question are "essentially" indistinguishable.

Definition *Two ordered fields* F_1 *and* F_2 *are said to be* isomorphic *iff there is a one-to-one corres-*

pondence ϕ *between the sets* F_1 *and* F_2 *such that*

(a) For every $x \in F_1$, $y \in F_1$, $\phi(x + y) = \phi(x) + \phi(y)$ *and* $\phi(xy) = \phi(x)\phi(y)$.

(b) For every $x \in F_1$, $y \in F_1$, $x \leq y$ *implies* $\phi(x) \leq \phi(y)$.

Let F be an ordered field, and S a subset of F. We say that b is an *upper bound* of S iff for each $x \in S$ we have $x \leq b$. An element a of F is called a *least upper bound*, or a *supremum* of S, iff a is an upper bound of S and iff $a \leq b$ for each upper bound b of S. We denote by sup S or $\sup_{x \in S} x$ the least upper bound of S.

The concept of *greatest lower bound* or *infimum* of S (denoted inf S or $\inf_{x \in S} x$) is similarly defined.

It is easy to see that *the supremum and infimum of a set are uniquely determined whenever they exist.*

The real number system $\mathbb{R}$ *(called also the real line) is defined to be an ordered field such that every nonempty subset* S *which has an upper bound has a least upper bound in* $\mathbb{R}$.

In the real number system we denote by 0 (called *zero*) and 1 the identity elements for the operations + and $\cdot$, respectively. If $x \neq 0$ then its inverse with respect to the multiplication is denoted by x^{-1} or $1/x$.

For the sake of convenience of the reader we restate below the axioms defining the real number system $\mathbb{R}$. For all $x, y,$ and z in $\mathbb{R}$ we have:

Axiom 1 $x + y = y + x$.

Axiom 2 $(x + y) + z = x + (y + z)$.

Axiom 3 There is $0 \in \mathbb{R}$ such that $x + 0 = x$.

Axiom 4 For each $x \in \mathbb{R}$, there is a $-x \in \mathbb{R}$ such that $x + (-x) = 0$.

Axiom 5 $xy = yx$.

Axiom 6 $(xy)z = x(yz)$.

Axiom 7 There is $1 \in \mathbb{R}$ such that $1 \neq 0$ and $x \cdot 1 = x$ for all $x \in \mathbb{R}$.

Axiom 8 For each x in $\mathbb{R}$, different from 0, there is an $x^{-1} \in \mathbb{R}$ such that $xx^{-1} = 1$.

Axiom 9 $x(y + z) = xy + xz$.

Axiom 10 $x \leq y$ and $y \leq z$ imply $x \leq z$.

Axiom 11 "$x \leq y$ and $y \leq x$" is equivalent to $x = y$.

Axiom 12 For any two elements x, y of $\mathbb{R}$, either $x \leq y$ or $y \leq x$.

Axiom 13 $x \leq y$ implies $x + z \leq y + z$.

Axiom 14 $0 \leq x$ and $0 \leq y$ imply $0 \leq xy$.

Axiom 15 (Completeness Axiom) Every nonempty set of $\mathbb{R}$ which has an upper bound has a least upper bound.

From these axioms we can derive all the usual laws of arithmetic and the rules for operating with inequalities.

One can prove that there exist systems satisfying the above fifteen axioms for real numbers. It is even possible to construct such a system (from the system of rationals) using the methods given by R. Dedekind ("Dedekind cuts") or G. Cantor ("Cantor fundamental sequences"). Although these constructions have great logical and historical interest they will not be presented here. We shall simply assume that such a system exists. Also it is possible to prove that the collection of axioms defining the real number system is *categorical*. By this we mean that if $\mathbb{R}'$ is any other system satisfying the same axioms as $\mathbb{R}$, then $\mathbb{R}$ and $\mathbb{R}'$ are isomorphic with respect to the operations and relations given in the axioms. In other words *the system* $\mathbb{R}$ *of all real numbers is in a sense unique.*

The reader who is familiar with the set N of natural numbers and the set Q of rational numbers, might wonder how, in which sense, these two sets are considered as subsets of the abstract set $\mathbb{R}$ of real numbers defined by means of the above fifteen axioms. The answer to this question lies in the following proposition. We prove that the real number system contains sets isomorphic to the sets N and Q.

1. Proposition *Every ordered field contains sets isomorphic to the natural numbers and the rational numbers.*

Proof We only give an outline of the proof. Certain routine portions of the proof are left to the reader. Let F be an ordered field. We denote by θ and e the identities of F with respect to the addition and multiplication. Define a function $f: N \to F$ by $f(1) = e$ and $f(n + 1) = f(n) + e$. The mapping f is one-to-one. In fact let $p,q \in N$ and $p < q$. Then there is an $n \in N$ such that $q = p + n$. We prove that $f(p) < f(q)$, by induction on n. For $n = 1$ we have $q = p + 1$ and $f(p) < f(p) + e = f(q)$. Suppose now that for $n = k$ we have $f(p) < f(p + k)$. Then

$f(p + k) < f(p + k) + e = f(p + k + 1)$, which implies $f(p) < f(p + k + 1)$. This proves that for every n, we have $f(p) < f(p + n)$, i.e., $f(p) < f(q)$.

Also by induction, one can prove that $f(p + q) = f(p) + f(q)$ and $f(pq) = f(p)\cdot f(q)$. It follows that f is a one-to-one correspondence between N and the range $f[N] \subset F$, and that f preserves sums, products, and the relation $<$. In other words N and $f[N]$ are isomorphic. Since F has at least two elements, there is an $x \in F$ such that $x \neq \theta$. But $xe = x$ and $x\theta = \theta$ so that $e \neq \theta$. Now note that for every $x \neq \theta$, $xx > \theta$. In particular $ee > \theta$. But $ee = e$, so that $e > \theta$. For every positive integer n, we denote by ne the sum $e + e \ldots + e$ obtained by adding e to itself n times.

For all integers $n, m, m \neq 0$, ne/me will designate the solution of the equation $(me)x = ne$. One easily proves that

(a) For any integers k, n, m with $k \neq 0$, $m \neq 0$ we have $kne/kme = ne/me$.

(b) If m_1, m_2, n_1, n_2 are integers, with $m_1 \neq 0$ and $m_2 \neq 0$, then

$$\frac{n_1e}{m_1e} + \frac{n_2e}{m_2e} = \frac{(n_1m_2 + m_1n_2)e}{m_1m_2e} \quad \text{and} \quad \frac{n_1e}{m_1e}\,\frac{n_2e}{m_2e} = \frac{n_1n_2e}{m_1m_2e}\,.$$

(c) If m_1, m_2, n_1, n_2 are integers, with $m_1 \neq 0$, $m_2 \neq 0$, then $\dfrac{n_1e}{m_1e} > \dfrac{n_2e}{m_2e}$ if and only if $\dfrac{n_1}{m_1} > \dfrac{n_2}{m_2}$.

Let ϕ be the mapping from the set of rational numbers Q into the reals, defined by $\phi(\frac{m}{n}) = \frac{me}{ne}$. It follows from (a), (b), (c) that ϕ is an isomorphism between Q and the subfield $\phi[Q]$ of F. The proposition is proved.

It follows from Proposition 1 that the real number system being an ordered field contains sets isomorphic to N and Q and it is in this sense that N and Q are considered as subsets of the set of real numbers. Another direct consequence of Proposition 1 is that *if an ordered field F has no ordered proper subfield then it is isomorphic with the field of rational numbers.*

2. Proposition (Axiom of Archimedes) *For any pair x,y of real numbers such that $0 < x$, $0 < y$ there is a positive integer n such that $y < nx$.*
Proof Suppose the proposition is not true and let N be the set of all positive integers. Then there are $x > 0$, $y > 0$ such that for every positive integer n we have $nx \leq y$. This means that y is an upper bound of the set $\{nx: n \in N\}$. By Axiom 15, let b be the least upper bound of $\{nx: n \in N\}$. Then for every $n \in N$ we have $(n+1)x \leq b$ or $nx \leq b - x$, which means that $b - x$ is an upper bound of $\{nx: n \in N\}$. Hence $b \leq b - x$ or $x \leq 0$, which contradicts the assumption $0 < x$.

3. Proposition *Let x and y be any two real numbers with $x < y$. Then there is a rational number r such that $x < r < y$.*
Proof First, suppose that $0 \leq x$, and put $y_1 = (x + y)/2$. We have $x < y_1 < y$. By Proposition 2, there is a positive integer q such that

$1/(y_1 - x) < q$. Then $1/q < y_1 - x$. Also, by Proposition 2, since $0 < y_1$, $0 < 1/q$, the set $\{n \in N: y_1 < n\cdot(1/q)\}$ is a nonempty subset of N, therefore it has a smallest element, say p. Hence $y_1 < p/q$ and $(p - 1)/q \leq y_1 < y$. Also $x = y_1 - (y_1 - x) < y_1 - 1/q < p/q - (1/q) = (p - 1)/q$. Hence for $r = (p - 1)/q$ we have $x < r < y$.

Next suppose $x < 0$. We have $0 < -x$. By Proposition 2, applied to $-x$ and 1, there is a positive integer n such that $-x < n$. Then $0 < n + x$, and there is a rational number r_1 such that $n + x < r_1 < n + y$. Hence for the rational number $r = r_1 - n$ we have $x < r < y$. The theorem is proved.

2.4 THE EXTENDED REAL NUMBER SYSTEM

Definition *The extended real number system* $\overline{\mathbb{R}}$, *consists of the real number system to which two symbols* $+\infty$ *and* $-\infty$ *(read:* plus infinity *and* minus infinity, *respectively) have been adjoined, satisfying the following properties:*

$-\infty < x < +\infty$	*if* $x \in \mathbb{R}$
$x + (+\infty) = (+\infty) + x$	
$\quad = x - (-\infty) = +\infty$	*if* $x \in \overline{\mathbb{R}}$ *and* $x \neq -\infty$
$x + (-\infty) = (-\infty) + x$	
$\quad = x - (+\infty) = -\infty$	*if* $x \in \overline{\mathbb{R}}$ *and* $x \neq +\infty$
$x\cdot(+\infty) = (+\infty)\cdot x = +\infty$	*if* $x \in \overline{\mathbb{R}}$ *and* $0 < x$

$$x\cdot(-\infty) = (-\infty)\cdot x = -\infty \qquad \text{if } x \in \overline{\mathbb{R}} \text{ and } 0 < x$$
$$x\cdot(+\infty) = (+\infty)\cdot x = -\infty \qquad \text{if } x \in \overline{\mathbb{R}} \text{ and } x < 0$$
$$x\cdot(-\infty) = (-\infty)\cdot x = +\infty \qquad \text{if } x \in \overline{\mathbb{R}} \text{ and } x < 0$$
$$\frac{x}{+\infty} = \frac{x}{-\infty} = 0 \qquad \text{if } x \in \mathbb{R}$$

$$-(+\infty) = -\infty; \quad -(-\infty) = +\infty; \quad |+\infty| = |-\infty| = +\infty$$
$$(+\infty) + (-\infty) = (-\infty) + (+\infty) = 0$$
$$0\cdot(\pm\infty) = (\pm\infty)\cdot 0 = 0.$$

The reader must realize that the symbols $+\infty$ and $-\infty$ are not real numbers. It is impossible to extend the operations of addition and multiplication to $+\infty$ and $-\infty$ and still maintain the validity of defining the real number system axioms as well as some of the properties of order that are involved. For example, we can realize that addition in $\overline{\mathbb{R}}$ is not always associative. Some authors refrain from defining $(+\infty) + (-\infty)$ or $0(\pm\infty)$, but these definitions are very useful in the theory of integration.

If S is a nonempty set of real numbers which has not an upper bound then, we define the $\sup S$ to be $+\infty$. Similarly the infimum of a nonempty set which has not a lower bound is defined to be $-\infty$. Now if $S = \emptyset$, and B any element in $\overline{\mathbb{R}}$, the statement "$x \in \emptyset \to x \leq B$" is true. This means that all elements of $\overline{\mathbb{R}}$ are upper bounds of $\emptyset$. In particular $-\infty$ is an upper bound of $\emptyset$, and since $-\infty$ is the least member of $\overline{\mathbb{R}}$ we define the $\sup \emptyset = -\infty$.

Similarly, we define $\inf \emptyset = +\infty$. Observe that the inequality $-\infty = \sup \emptyset < \inf \emptyset = +\infty$ is peculiar to the empty set.

Thus *in the extended real number system, every*

set (empty or not) *has a least upper bound and a greatest lower bound.* This is the main reason for introducing the symbols $+\infty$ and $-\infty$.

2.5 TOPOLOGY OF THE REAL LINE

For any two elements a,b in $\mathbb{R}$ with $a < b$ the set $\{x \in \mathbb{R}: a < x < b\}$ is called an open interval and is written (a,b). Similarly the set $\{x \in \mathbb{R}: a \leq x \leq b\}$ is called a closed interval and is written $[a,b]$. For $a = b$, the notation $[a,a]$ means the set $\{a\}$. Also, for $a < b$, the sets $\{x \in \mathbb{R}: a \leq x < b\}$ and $\{x \in \mathbb{R}: a < x \leq b\}$ called *half open* intervals are written $[a,b)$ and $(a,b]$ respectively. We also consider the infinite open intervals $(a,+\infty) = \{x \in \mathbb{R}: x > a\}$, $(-\infty,b) = \{x \in \mathbb{R}: x < b\}$. The closed infinite intervals $[a,+\infty)$, $(-\infty,b]$ are defined in a similar way. We often write $(-\infty,+\infty) = \mathbb{R}$.

Definition *A subset* O *of* $\mathbb{R}$ *is said to be open if it is empty or if for every* $x \in O$, *there exists an open interval containing* x *and contained in* O.

The following propositions derive directly from this definition.

O_1: *Every union* (finite or infinite) *of open sets is open.*

O_2: *Every finite intersection of open sets is open.*

O_3: *The set* $\mathbb{R}$ *and the empty set* $\emptyset$ *are open sets.*

To prove O_1 let $\{A_\alpha\}$ be any collection of open sets, and put $A = \bigcup A_\alpha$. Let $x \in A$. Then there is an α such that $x \in A_\alpha$. Since A_α is open there is an open interval say I_x such that $x \in I_x \subset A_\alpha$. Hence $x \in I_x \subset A$ which proves that A is open. To prove O_2 let $A_1, A_2, \ldots, A_n$ be a finite collection of open sets and put $B = A_1 \cap A_2 \cap .. \cap A_n$. If B is empty then, B by definition is open. So suppose $B \neq \emptyset$ and let $x \in B$. Then $x \in A_i$ for $i = 1,2,\ldots,n$. Since the A_i are open there must exist open intervals $I_i (i = 1,2,\ldots,n)$ such that $x \in I_i \subset A_i$ for $i = 1,2,\ldots,n$. Put $I = \bigcap_{i=1}^{n} I_i$. Then I is an open interval containing x and contained in B. Hence B is open. Finally O_3 is obvious.

It is easy to see that every open interval is an open set, while a closed interval $[a,b]$ is not an open set.

The intersection of the collection $\{(-1/n, 1/n)\}$ $(n = 1,2,\ldots)$ of open sets is the set $\{0\}$ consisting of the single point 0, which is not open. This example proves that *the intersection of an arbitrary collection of open sets is not always an open set.*

The following proposition shows the structure of open sets. First we prove a lemma.

4. Lemma *Let $\mathcal{C}$ be a collection of disjoint nonempty open sets. Then $\mathcal{C}$ is a countable collection.*

Proof Put $B = \bigcup_{A \in \mathcal{C}} A$, and let S be the set of all rational numbers contained in A. The set B as an open set contains open intervals. Since by the corollary of the axiom of Archimedes every open interval contains a rational number, it follows that $S \neq \emptyset$. The sets of the collection $\mathcal{C}$ being disjoint, every $r \in S$ is contained in one and only one set of the collection $\mathcal{C}$, say A_r. Since every A in $\mathcal{C}$ contains a rational number, $r \to A_r$ is a mapping of S onto $\mathcal{C}$. Thus $\mathcal{C}$ is countable as the image of the countable set S.

5. Proposition *A nonempty set A of real numbers is open if and only if A is the union of a unique nonempty countable collection of disjoint open intervals (finite or infinite).*

Proof First suppose that A is the union of a countable collection of disjoint open intervals. Then, by O_1, A is open. Next let A be a nonempty open set, and $x \in A$. Then there exist y and z in $\mathbb{R}$, such that $(x,y) \subset A$ and $(z,x) \subset A$. Put $a = \inf\{z \in R: (z,x) \subset A\}$ and $b = \sup\{y \in R: (x,y) \subset A\}$. Then we prove that the open interval $I_x = (a,b)$, which obviously contains x, is contained in A. In fact let $w \in I_x$. Then either $x \leq w < b$ or $a < w \leq x$. Suppose $x \leq w < b$. Then there is a number $y > w$ such that $(x,y) \subset A$. This implies $w \in A$. If $a < w \leq x$ then we prove in the same way that there is a $z < w$ such that $(z,w) \subset A$, and so $w \in A$. We also observe that a and b do not belong to A. For if $b \in A$ then for some $\varepsilon > 0$, we have $(b - \varepsilon, b + \varepsilon) \subset A$, which implies that

$(x, b + \varepsilon) \subset A$, which contradicts the definition of b. Likewise $a \notin A$. Consider now the collection $\mathcal{C}$ of all intervals I_x obtained as above. Since each element of A is contained in some interval of the collection $\mathcal{C}$ and each interval is contained in A, we have that $A = \bigcup_{x \varepsilon A} I_x$. Finally, we prove that if (a,b), (c,d) are any two distinct intervals in $\mathcal{C}$ then $(a,b) \cap (c,d) = \emptyset$. For suppose, on the contrary, that these intervals have a nonempty intersection, and let $x \varepsilon (a,b) \cap (c,d)$. Then at least one of the endpoints of one of these open intervals would belong to the other open interval, which would imply that one of the end points would belong to A. But this would contradict the fact proven above, that no end point of an interval I_x belongs to A. Thus $\mathcal{C}$ is a collection of disjoint open sets, and by Lemma 4 it must be countable.

We finally prove that the collection $\mathcal{C}$ is unique. Suppose that $A = \bigcup \mathcal{J}$ where $\mathcal{J}$ is a collection of pairwise disjoint open intervals. Let $(a',b') \varepsilon \mathcal{J}$. We first prove that the points a' and b' are not in A. For suppose $a' \varepsilon A$. Then there is an open interval (c,d) in $\mathcal{J}$ such that $a' \varepsilon (c,d)$. It follows that $(a',b') \neq (c,d)$ and so $(a',b') \cap (c,d) \neq 0$. This contradiction shows that $a' \notin A$. Similarly $b' \notin A$. Next let $x \varepsilon (a',b')$. Then $(a',x] \subset A$ and $[x,b') \subset A$ so that $(a',b') \subset I_x \subset A$. This shows that $\mathcal{J} \subset \mathcal{C}$. Let $I_y \varepsilon \mathcal{C}$ and such that $I_y \notin \mathcal{J}$. Then y would be in A while $y \notin \bigcup \mathcal{A} = A$, a contradiction. The theorem is proved.

Definition *A subset* F *of* $\mathbb{R}$ *is said to be* **closed** *iff its complement* $\complement F$ *is open.*

Every closed interval $[a,b]$ is a closed set, for the complement of $[a,b]$ is the union of the two open intervals $(-\infty,a)$ and $(b,+\infty)$ and is therefore an open set. The set of positive integers is another example of closed set.

It is possible for a set to be neither open nor closed. Example, $[0,1)$. The following propositions are direct consequences of the propositions O_1, O_2, O_3 and the definition of closed set. The proofs are left to the reader.

F_1: *The intersection of any collection* (finite or infinite) *of closed sets is a closed set.*

F_2: *The union of any finite collection of closed sets is a closed set.*

F_3: *The set* $\mathbb{R}$ *and the empty set* $\emptyset$ *are closed sets.*

Definition *Let* $E \subset \mathbb{R}$ *and* $x \in \mathbb{R}$.

(a) We call x *a* point of closure *of* E, *iff every open interval containing* x *contains a point of* E.

(b) We call x *an* accumulation point *of* E, *iff every open interval containing* x *contains at least one point of* E *other than* x *(*x *itself need not belong to* E*). The set of all accumulation points of* E *is denoted by* E'. E' *is called the* derived set *of* E.

(c) We call x *an* isolated point *of* E *iff*

there is an open interval I_x *containing* x*, such that* $I_x \cap E = \{x\}$.

Observe that every point of a set E is a point of closure of E; an isolated point of E is a point of E which is not an accumulation point of E.

Example

Let $E = [1,2) \cup \{3\}$. Then every point x, such that $1 \leq x < 2$ is an accumulation point of E and it belongs to E. The point 2 is an accumulation point of E but it does not belong to E. Finally the point 3 is an isolated point of E.

The set of points of closure of a set E is denoted by $\overline{E}$ and it is called the *closure* of E.

6. Proposition *Properties of the closure:*

(a) $\overline{\emptyset} = \emptyset$

(b) $A \subset \overline{A}$

(c) $A \subset B$ *implies* $\overline{A} \subset \overline{B}$

(d) $\overline{\overline{A}} = \overline{A}$

(e) $\overline{A \cup B} = \overline{A} \cup \overline{B}$

(Note: *It is clear that here* $\overline{\overline{A}}$ *denotes the closure of* $\overline{A}$ *and not the cardinal number of A.)*

Proof Properties (a), (b), and (c) are obvious. To prove (d) let $x \in \overline{\overline{A}}$, and I_x any open interval containing x. Then I_x contains a point $y \in \overline{A}$. Let I_y be an open interval containing y, such that $I_y \subset I_x$. Then there is a point z of A contained in I_y. Hence $z \in I_x$, which means that $x \in \overline{A}$. Hence $\overline{\overline{A}} \subset \overline{A}$. By (b) we have $\overline{A} \subset \overline{\overline{A}}$.

Therefore $\overline{\overline{A}} = \overline{A}$.

To prove (e), we note that since $A \subset A \cup B$ and $B \subset A \cup B$ we have by (c), $\overline{A} \subset \overline{A \cup B}$ and $\overline{B} \subset \overline{A \cup B}$. Hence $\overline{A} \cup \overline{B} \subset \overline{A \cup B}$. To finish, we prove that $\overline{A \cup B} \subset \overline{A} \cup \overline{B}$. Indeed, suppose this is not true. Then there is an $x \in \overline{A \cup B}$ such that $x \notin \overline{A}$ and $x \notin \overline{B}$. This implies that there are open intervals I'_x and I''_x containing x such that $I'_x \cap A = \emptyset$ and $I''_x \cap B = \emptyset$. Hence the open interval $I_x = I'_x \cap I''_x$ which contains x, does not contain any point of $A \cup B$. But this contradicts the fact that x is a point of closure of $A \cup B$. Hence $\overline{A \cup B} \subset \overline{A} \cup \overline{B}$ and (e) is proved.

7. Proposition *A set* F *is closed if and only if* $F = \overline{F}$.

Proof First suppose F closed. Then we only need to prove $\overline{F} \subset F$, or equivalently $\complement F \subset \complement \overline{F}$. Let $x \in \complement F$. Then $\complement F$, as the complement of the closed set F, is open, so that there is an open interval I_x containing x, such that $I_x \subset \complement F$. This means that I_x contains no points of F, which implies that $x \notin \overline{F}$, or equivalently $x \in \complement \overline{F}$.

Next suppose $F = \overline{F}$. We prove that $\complement F$ is open, by showing that if $x \in \complement F$ then, there is an open interval I_x containing x such that $I_x \subset \complement F$. For if such an interval did not exist then, for every open interval I_x containing x we would have $I_x \cap F \neq \emptyset$, which would imply $x \in \overline{F}$, or, since $F = \overline{F}$, we would have $x \in F$ which would contradict the assumption $x \in \complement F$. The proposition is proved.

A collection $\mathcal{C}$ of open sets is said to be an *open covering* of a set E iff $E \subset \bigcup\{O: O \in \mathcal{C}\}$.

If the collection $\mathcal{C}$ is finite then $\mathcal{C}$ is called a *finite open covering* of E.

The following theorem states a very important property of closed and bounded intervals.

8. Theorem (Heine-Borel-Lebesgue) *Every open covering of a bounded and closed interval* $[a,b]$ *has a finite subcovering. That is, if* $\mathcal{C}$ *is any collection of open sets such that* $[a,b] \subset \bigcup\{O: O \in \mathcal{C}\}$, *then there is a finite number of sets* $O_1, O_2, \ldots, O_n$ *in* $\mathcal{C}$ *such that* $[a,b] \subset O_1 \cup O_2 \cup \ldots \cup O_n$.

Proof The theorem is obvious if $a = b$. So suppose $a < b$. Let S be the set of points x of $[a,b]$ such that the interval $[a,x]$ can be covered by a finite number of sets in the collection $\mathcal{C}$. The set S is not empty since it contains the point $x = a$, and it is obviously bounded. Therefore it has a least upper bound, say m. Since b is an upper bound of S we have that $m \in [a,b]$. Then there is an open set $O \in \mathcal{C}$ containing m. Since O is open there is an open interval $(m-\varepsilon, m+\varepsilon)$ $(\varepsilon > 0)$ contained in O. Since $m = \sup S$, there must exist an $x \in S$ with $m - \varepsilon < x$. From the definition of the set S it follows that there is a finite number $O_1, O_2, \ldots, O_k$ of sets in $\mathcal{C}$ such that $[a,x] \subset O_1 \cup O_2 \cup \ldots \cup O_k$.

Consequently, the finite collection $\{O_1,\ldots,O_k,O\}$ covers $[a,m+\varepsilon]$. Now no point of $[a,b]$ can belong to the interval $(m,m+\varepsilon)$, for in this case $[a,m]$ would be covered by a finite subcollection of sets in $\mathcal{C}$, so that m would not be the least upper bound of S. It follows that $m = b$, which means that $[a,b]$ can be covered by a finite collection of sets in $\mathcal{C}$, namely, $\{O_1,O_2,\ldots,O_k,O\}$. The theorem is proved.

9. Corollary *Let F be a closed and bounded set of real numbers. Then every open covering of F has a finite subcovering.*

Proof Let $\mathcal{C}$ be an open covering of F, and $[a,b]$ a bounded and closed interval containing F. Then since F is closed the collection $\mathcal{C}$ and the complement $\complement F$ of F form an open covering of $[a,b]$, say $\mathcal{C}'$. By Theorem 8, there is a finite collection of sets in $\mathcal{C}'$, say $\{O_1',O_2',\ldots,O_n'\}$ which cover $[a,b]$. If $\{O_1',O_2',\ldots,O_n'\}$ does not contain $\complement F$, it is a finite subcollection of $\mathcal{C}$, and the corollary holds in this case. If one of the sets O_i' $(i = 1,2,\ldots,n)$ say O_n', happens to be equal to $\complement F$ (i.e., $O_n' = \complement F$), then the subcollection $\{O_1',O_2',\ldots,O_{n-1}'\}$ is a subcollection of $\mathcal{C}$ and obviously covers F, since no point of F belongs to $\complement F$. The corollary is proved.

10. Corollary *Let $\mathcal{C}$ be a collection of closed and bounded sets of real numbers, such that the intersection of every finite subcollection of $\mathcal{C}$ is nonempty. Then $\bigcap_{F\in\mathcal{C}} F \neq \emptyset$.*

Proof Suppose on the contrary $\bigcap_{F \in \mathcal{C}} F = \emptyset$. By taking the complements of both sides we have $\bigcup_{F \in \mathcal{C}} \complement F = \mathbb{R}$. Let F_o be a fixed member of $\mathcal{C}$. Then $\{\complement F\}_{F \in \mathcal{C}}$ is an open covering of F_o (since it is an open covering of $\mathbb{R}$) and by Corollary 9 there must be a finite number of sets in $\{\complement F\}_{F \in \mathcal{C}}$, say $\complement F_1, \complement F_2, \ldots, \complement F_n$, such that $F_o \subset \bigcup_{i=1}^{n} \complement F_i$. This implies that $\bigcap_{i=1}^{n} F_i \subset \complement F_o$. It follows that $F_1 \cap F_2 \cap \ldots \cap F_n \cap F_o = \emptyset$, which contradicts our assumption that the intersection of every finite subcollection of $\mathcal{C}$ is nonempty.

11. Proposition (Converse of Corollary 9) *Let F be a set of real numbers such that every open covering $\mathcal{C}$ of F contains a finite subcovering. Then F is bounded and closed.*

Proof Let $\mathcal{C}$ be the collection of all open intervals $I_n = (-n,n)$ $(n = 1,2,\ldots)$. Then $\mathcal{C}$ is an open covering of F, therefore there is a finite subcovering of F. Let N be the largest subscript of any I_n in this finite subcollection. Clearly $F \subset I_N$, hence F is bounded.

To prove that F is closed we prove that $\complement F$ is open. Let $x \in \complement F$, and denote by G_n the complement of the closed interval $[x - (1/n), x + (1/n)]$. Then the collection $\{G_n\}$

$(n = 1,2,\ldots)$ is an open covering of F. It follows that a finite subcollection of $\{G_n\}$ covers F and, as in the previous paragraph, we have for some integer N that $F \subset G_N$. Thus the open interval $(x - (1/N), x + (1/N)) \subset \complement F$ which proves that $\complement F$ is open.

12. Theorem (Bolzano-Weierstrass) *Every infinite subset* E *of a bounded and closed interval* $[a,b]$ *has at least one accumulation point in* $[a,b]$.

Proof Suppose the theorem is not true. Then no point of $[a,b]$ is an accumulation point of E. This implies that for every $x \in [a,b]$ there is an open interval I_x containing only one point of E, namely x itself. The I_x form an open covering of $[a,b]$. By Theorem 8 there is a finite number of I_x which cover $[a,b]$. But since each of the I_x contains only one point of E, this implies that E is a finite set, which contradicts the assumption that E is infinite. The theorem is proved.

It follows immediately from the above theorem that *a subset of* $[a,b]$ *which has no accumulation points in* $[a,b]$ *is necessarily a finite set.*

Notice that the assertion of Theorem 8 does not necessarily hold if $[a,b]$ is replaced by a nonclosed interval or by an unbounded interval. The reader is invited to give counterexamples for both cases.

A sequence $\langle x_n \rangle$ is said to be *increasing* (*decreasing*) iff $x_n \leq x_{n+1}$ $(x_n \geq x_{n+1})$ for

$n = 1,2,\ldots$. A sequence is called *monotonic* iff it is increasing or iff it is decreasing.

We observe that a monotonic sequence converges to a real number if and only if it is bounded (i.e., there is a positive number M such that $|x_n| < M$ for $n = 1,2,\ldots$). For if $\langle x_n \rangle$ is increasing then $\lim_{n\to\infty} x_n = \sup \{x_n : n = 1,2,\ldots\}$. If $\langle x_n \rangle$ is decreasing then $\lim_{n\to\infty} x_n = \inf \{x_n : n = 1,2,\ldots\}$. On the other hand, if $\langle x_n \rangle$ converges to a real number it is bounded (see Prob. 2.16).

In this book the single unaccompanied word "convergent" will mean "convergent to a real limit." The infinite cases will always be expressed in terms of "convergence in the extended real number system."

An infinite sequence $\langle x_n \rangle$ *of real numbers is said to converge to a real number* ℓ, *(or that* ℓ *is the limit of* $\langle x_n \rangle$*) iff for every* $\varepsilon > 0$, *we have* $x_n \in (\ell - \varepsilon, \ell + \varepsilon)$ *except for at most a finite number of values of* n. If the sequence converges to ℓ, we write $\lim_{n\to\infty} x_n = \ell$. Unless otherwise specified all sequences in this book are infinite sequences.

The limit ℓ, *if it exists, is unique*. For suppose that $\ell_1 \neq \ell$. Let $\varepsilon = |\ell_1 - \ell|/2$. Since for every n, except at most finitely many, one has $x_n \in (\ell - \varepsilon, \ell + \varepsilon)$, then $x_n \in (\ell_1 - \varepsilon, \ell_1 + \varepsilon)$ for only finite number of the x_n. Therefore ℓ_1 can not be a limit of $\langle x_n \rangle$.

A sequence $\langle x_n \rangle$ *is said to converge to* $+\infty$, *iff for every given number* k *there is an* N *such that* $n \geq N$ *implies* $x_n > k$. In this case we write $\lim_{n\to\infty} x_n = +\infty$. A similar definition can be given for the expression $\lim_{n\to\infty} x_n = -\infty$. An infinite sequence $\langle x_n \rangle$ of real numbers is said to be a *Cauchy sequence* iff *for every* $\varepsilon > 0$ *there is an* N (depending on ε) *such that the two inequalities* $n \geq N$ *and* $m \geq N$ *imply* $|x_n - x_m| < \varepsilon$.

The proof of the following propositions can be found in any rigorous, advanced calculus book. "*An infinite sequence* $\langle x_n \rangle$ *of real numbers converges to a real number, if and only if* $\langle x_n \rangle$ *is a Cauchy sequence*" (see [1], 2d ed., p. 66).

A real number ℓ *is said to be a cluster point of the infinite sequence* $\langle x_n \rangle$ *iff every open interval containing* ℓ, *contains infinitely many terms of the sequence* $\langle x_n \rangle$. *Thus* ℓ is a cluster point of $\langle x_n \rangle$ if and only if, for any $\varepsilon > 0$ and N, there is an $n \geq N$ such that $|x_n - \ell| < \varepsilon$.

The point $+\infty$ *is a cluster point of* $\langle x_n \rangle$ *iff given* k *and* N, *there is an* $n \geq N$ *such that* $x_n \geq k$. An analogous definition applies to $-\infty$. We notice that if an infinite sequence converges to an extended real number ℓ, then ℓ is a cluster point of the sequence. On the other hand an infinite sequence may have several cluster points. For example, the sequence $\langle x_n \rangle$, where $x_n = n$ if n is even and $x_n = 1$, if n is odd, has two cluster points, namely, 1 and $+\infty$.

Roughly speaking the largest and the smallest cluster points of an infinite sequence are called respectively the limit superior and the limit inferior of the sequence. More precisely we have the following definition.

Definition *A real number* $L(\ell)$ *is the* limit superior (limit inferior) *of the infinite sequence* $\langle x_n \rangle$, *iff for evrey* $\varepsilon > 0$ *there is an infinite number of* x_n *in the open interval* $(L - \varepsilon, L + \varepsilon)$ $((\ell - \varepsilon, \ell + \varepsilon))$ *and only a finite number of* x_n *are greater (less) than* $L + \varepsilon$ $(\ell - \varepsilon)$.

We extend the notion of limit superior (limit inferior) to include $+\infty$ $(-\infty)$ as follows: The extended real number $+\infty$ $(-\infty)$ is the limit superior (limit inferior) of the infinite sequence $\langle x_n \rangle$ iff given k and N there is an $n \geq N$ such that $x_n > k$ $(x_n < k)$.

We denote by $\overline{\lim}_{n\to\infty} x_n$ and $\underline{\lim}_{n\to\infty} x_n$ the limit superior and the limit inferior of the infinite sequence $\langle x_n \rangle$, respectively.

The following proposition asserts the existence of the limit superior and the limit inferior of any sequence of real numbers.

13. Proposition *If* $\langle x_n \rangle$ *is an infinite sequence of real numbers and* $t_n = \sup_{k \geq n} x_k$, *then*

$$\overline{\lim}_{n\to\infty} x_n = \lim_{n\to\infty} t_n = \inf_n (\sup_{k \geq n} x_k).$$

Proof If $\langle x_n \rangle$ is unbounded above then $t_n = +\infty$ and

$\overline{\lim}_{n\to\infty} x_n = +\infty$ and the proposition follows.

Next suppose $\langle x_n \rangle$ is bounded above. It follows that $\langle t_n \rangle$ is a bounded above and nonincreasing sequence of real numbers. Hence $\lim_{n\to\infty} t_n$ exists and equals the $\inf_n (\sup_{k\geq n} x_k)$. Put $L = \lim_{n\to\infty} t_n$. Then, if $L \in \mathbb{R}$, for every $\varepsilon > 0$ and every positive integer n, we have $t_n > L - \varepsilon$. Since $t_n = \sup_{k\geq n} x_k$, there is an $m \geq n$ such that $L - \varepsilon < x_m < L + \varepsilon$, which means that there are infinitely many x_n in the open interval $(L - \varepsilon, L + \varepsilon)$.

Since $\langle t_n \rangle$ is nonincreasing, there is an n_o such that $t_{n_o} < L + \varepsilon$. Hence $x_{n_o}, x_{n_o+1}, \ldots$ are all less than $L + \varepsilon$, which implies that only a finite number of x_n is greater than $L + \varepsilon$. This proves that L is the limit superior of $\langle x_n \rangle$. It can be easily seen that the theorem also holds if $L = -\infty$. The theorem is proved.

Given a sequence $\langle x_n \rangle$ we associate with it the sequence $\langle s_n \rangle$ (where $s_n = x_1 + x_2 + \ldots + x_n$) or $\sum_{n=1}^{\infty} x_n$. This last symbol is called an infinite series and the numbers s_n are called the partial sums of the series. If $\langle s_n \rangle$ converges to a real number s we say that the series *converges* to s and write $\sum_{n=1}^{\infty} x_n = s$. The number s is called the *sum* of the series. If $\langle s_n \rangle$ does not converge to a real number we say that the series *diverges*. We assume that the reader is familiar with the elementary theory of infinite series.

2.6 CONTINUOUS FUNCTIONS

A real-valued function f defined on a set E of real numbers is said to be *continuous* at a point x of E iff for every open interval V containing the number $f(x)$ there is an open interval I containing x, such that $f[I \cap E] \subset V$ (i.e., for every $y \in I \cap E$ we have $f(y) \in V$).

This is equivalent to saying that f is continuous at $x \in E$ if and only if given $\varepsilon > 0$ there is a $\delta > 0$ such that $y \in E$, $|y - x| < \delta$ imply $|f(x) - f(y)| < \varepsilon$.

It follows from the above definition that *a real-valued function defined on a set* E *of real numbers is not continuous at a point* x *of* E *if and only if there is an* $\varepsilon > 0$ *such that for every* $\delta > 0$ *there is a* $y_0 \in E$ *with* $|y_0 - x| < \delta$, *such that* $|f(x) - f(y_0)| \geq \varepsilon$.

The function f *is said to be continuous on the whole set* E *iff it is continuous at every point of* E.

14. Theorem *Let* f *be a real-valued function defined on the whole real line. Then* f *is continuous if and only if for each open set* O *of real numbers,* $f^{-1}[O]$ *is also an open set.*

Proof Suppose f is continuous on $\mathbb{R}$, and let O be an open set of real numbers. Let $x \in f^{-1}[O]$. Then $y = f(x) \in O$. Since O is open there is an open interval V containing y such that $V \subset O$. But f is continuous, so there is an open interval I containing x such that $f[I] \subset V$, which means that $I \subset f^{-1}[O]$. Therefore $f^{-1}[O]$ is open. Suppose now that for every open set O, $f^{-1}[O]$ is open and let V be an open interval containing f(x). Then $f^{-1}[V]$ is an open set containing x. So there is an open interval I containing x and contained in $f^{-1}[V]$. We have $f[I] \subset V$. Hence f is continuous at x. The theorem is proved.

15. Theorem *A function* $f: \mathbb{R} \to \mathbb{R}$ *is continuous at a point* x_o *if and only if, for every infinite sequence* $\langle x_n \rangle$ *which converges to* x_o, *the infinite sequence* $\langle f(x_n) \rangle$ *converges to* $f(x_o)$.

Proof Suppose that f is continuous at x_o and let $\langle x_n \rangle$ be an infinite sequence such that $\lim_{n\to\infty} x_n = x_o$. Let V be any open interval containing $f(x_o)$. Let I be an open interval containing x_o, such that $f[I] \subset V$. Since $\langle x_n \rangle$ converges to x_o, there is an N such that $n \geq N$ implies $x_n \in I$. Hence for $n \geq N$ we have $f(x_n) \in V$. Therefore $\lim_{n\to\infty} f(x_n) = f(x_o)$.

Next suppose that for every infinite sequence $\langle x_n \rangle$ such that $\lim_{n\to\infty} x_n = x_o$, we have $\lim_{n\to\infty} f(x_n) = f(x_o)$. We prove that f is continuous at x_o. Suppose it is not. Then there is an open interval V_o containing $f(x_o)$ such that for every open interval I containing x_o, we have $f[I] \not\subset V_o$. This means that there is an x_1, such that $|x_1 - x_o| < 1$ and $f(x_1) \notin V_o$. Then there is an x_2 such that $|x_2 - x_o| < \min\{|x_1 - x_o|, 1/2\}$ and $f(x_2) \notin V_o$. Continuing in this way, we get an infinite sequence $\langle x_n \rangle$ such that $\lim_{n\to\infty} x_n = x_o$ and $f(x_n) \notin V_o$ for every positive integer n, which means that the sequence $\langle f(x_n) \rangle$ cannot converge to $f(x_o)$. But this contradicts our assumption that whenever $x_n \to x_o$ then $f(x_n) \to f(x_o)$. The theorem is proved.

16. Theorem *Let* $f: S \to \mathbb{R}$ *be a continuous function on a closed and bounded set* S. *Then the set* $f[S]$ *is bounded and closed.*

Proof By Proposition 11 it suffices to prove that every open covering of $f[S]$ has a finite subcovering.

Let $\mathcal{C}$ be any open covering of $f[S]$, and let $x \in S$. Then there is at least an open set in $\mathcal{C}$, say O_x, such that $f(x) \in O_x$. Since O_x is open there is an open interval V_x such that $f(x) \in V_x \subset O_x$. By continuity there is an open interval I_x such that $x \in I_x$ and $f[I_x \cap S] \subset V_x$. The collection of all I_x is an open covering of S. Since S is bounded and closed, a finite number of I_x, say $I_{x_1}, I_{x_2}, \ldots, I_{x_n}$ cover S, and so $S \subset I_{x_1} \cup I_{x_2} \cup \ldots \cup I_{x_n}$.

We have for $k = 1,2,\dots,n$

$$f[I_{x_k} \cap S] \subset V_{x_k} \subset O_{x_k}$$

where $O_{x_k} \in \mathcal{G}$. Hence

$$f[S] \subset O_{x_1} \cup O_{x_2} \cup \dots \cup O_{x_n}.$$

The theorem is proved.

Definition *A real-valued function* f *defined on a set* S *of real numbers is said to be* uniformly continuous *on* S *if and only if for every* $\varepsilon > 0$ *there is a* $\delta > 0$ *such that* $x \in S$, $y \in S$, $|x - y| < \delta$ *imply* $|f(x) - f(y)| < \varepsilon$.

It follows from this definition that f *is not uniformly continuous on* S *if and only if there is an* $\varepsilon_0 > 0$ *such that for every* $\delta > 0$ *there are* $x_0 \in S$, $y_0 \in S$, *with* $|x_0 - y_0| < \delta$ *and* $|f(x_0) - f(y_0)| \geq \varepsilon_0$.

It is clear that if f is uniformly continuous on S then f is also continuous on S. The converse of this statement is false in general. To see this let $f\colon \mathbb{R} \to \mathbb{R}$ with $f(x) = x^2$, $x \in \mathbb{R}$. Obviously f is continuous on $\mathbb{R}$. Let $\varepsilon_0 = 2$. For $\delta > 0$, let $0 < \delta' < \delta$, $x_0 = \delta' + (1/\delta')$, $y_0 = (1/\delta')$. Then $|x_0 - y_0| = \delta' < \delta$ and $|f(x_0) - f(y_0)| = |(\delta' + (1/\delta'))^2 - (1/\delta')^2| > \varepsilon_0$.

17. Theorem *If* $f\colon F \to \mathbb{R}$ *is continuous, and* F *is a bounded and closed set of real numbers, then* f *is uniformly continuous on* F.

Proof Let $\varepsilon > 0$ be given. Then for every $x \in F$ there is a $\delta_x > 0$ such that $y \in F$, $|x - y| < \delta_x$ imply $|f(x) - f(y)| < (1/2)\varepsilon$. We now observe that the family of the open intervals $\{(x - (1/2)\delta_x, x + (1/2)\delta_x)\}$ form an open covering of the bounded and closed set F, therefore, by Heine-Borel-Lebesgue theorem there is a finite subcovering of F, say $\{(x_i - (1/2)\delta_{x_i}, x_i + (1/2)\delta_{x_i})\}_{i=1}^{n}$. Let $\delta = \min\{(1/2)\delta_{x_1}, (1/2)\delta_{x_2}, \ldots, (1/2)\delta_{x_n}\}$ and let $x \in F$, $y \in F$ with $|x - y| < \delta$. Since y is a point of F, it belongs to some open interval $(x_i - (1/2)\delta_i, x_i + (1/2)\delta_i)$ of the finite subcovering of F, so that $|y - x_i| < (1/2)\delta_i < \delta_i$. We observe now that $|x - x_i| < \delta_i$. For $|x - x_i| \le |x - y| + |y - x_i| < \delta + (1/2)\delta_i \le \delta_i$. But

$$|x - x_i| < \delta_i \quad \text{implies} \quad |f(x) - f(x_i)| < \tfrac{1}{2}\varepsilon$$

and

$$|y - x_i| < \delta_i \quad \text{implies} \quad |f(y) - f(x_i)| < \tfrac{1}{2}\varepsilon.$$

Hence

$$|f(x) - f(y)| < |f(x) - f(x_i)| + |f(x_i) - f(y)| < \varepsilon$$

which proves that f is uniformly continuous on F.

Let $\langle f_n \rangle$ be an infinite sequence of real-valued functions, all defined on a set E of real numbers. The sequence $\langle f_n \rangle$ is said to *converge pointwise* on E to a real-valued function f (defined on E) if and only if for every $x \varepsilon E$ we have $\lim_{n\to\infty} f_n(x) = f(x)$.

A family $\mathcal{F}$ of functions defined on a set E is said to be *uniformly bounded* iff there exists a number M such that $|f(x)| < M$ for all x in E and all f in $\mathcal{F}$.

The sequence $\langle f_n \rangle$ is said to *converge uniformly* to f on E if and only if given $\varepsilon > 0$ there is an N such that $x \varepsilon E$, $n \geq N$ imply $|f_n(x) - f(x)| < \varepsilon$. It follows from the last definition that $\langle f_n \rangle$ *fails to converge uniformly to* f *on* E *if and only if there is an* $\varepsilon_o > 0$ *such that for every* N, *there are* $n_o \geq N$, $x_o \varepsilon E$ *such that* $|f_{n_o}(x_o) - f(x_o)| \geq \varepsilon_o$.

We observe that if $\langle f_n \rangle$ converges uniformly to F on E, then f_n converges also pointwise to F on E. Again the converse of this statement is false. For let $E = \mathbb{R}$ and $f_n(x) = 1/(1 + nx^2)$. We easily see that the sequence $\langle f_n \rangle$ converges pointwise to the function f which is 0 for $x \neq 0$ and 1 for $x = 0$. However, the convergence is not uniform. For let $\varepsilon_o = 1/4$ and N be arbitrary. If $x_o = 1/\sqrt{N}$ and $n_o = 2N$ then

$$|f_{n_o}(x_o) - f(x_o)| = \frac{1}{3} \geq \varepsilon_o = \frac{1}{4} .$$

Hence $\langle f_n \rangle$ does not converge uniformly on $\mathbb{R}$.

18. Theorem *If* $\langle f_n \rangle$ *is a sequence of real-valued functions defined and continuous on a set* E *of real numbers and uniformly convergent on* E *to a function* f*, then* f *is continuous on* E.

Proof Let $a \in E$ and $\varepsilon > 0$. Since $\langle f_n \rangle$ converges uniformly on E, there is an N such that for every $n \geq N$, $|f_n(x) - f(x)| < \varepsilon/3$ for every $x \in E$. Let $n_0 > N$. Since f_{n_0} is continuous at $x = a$ there is a $\delta > 0$ such that if $|x - a| < \delta$ and $x \in E$ then $|f_{n_0}(x) - f_{n_0}(a)| < \varepsilon/3$. Now for $x \in E$ and $|x - a| < \delta$ we have

$$
\begin{aligned}
|f(x) - f(a)| &\leq |f(x) - f_{n_0}(x)| + |f_{n_0}(x) - f_{n_0}(a)| \\
&\quad + |f_{n_0}(a) - f(a)| \\
&< \frac{\varepsilon}{3} + \frac{\varepsilon}{3} + \frac{\varepsilon}{3} = \varepsilon.
\end{aligned}
$$

This shows that f is continuous at $x = a$.

(*Note:* Uniform convergence of $\langle f_n \rangle$ is sufficient but not necessary to transmit continuity from the individual terms f_n to the limit function f. To see this let $f_n(x) = nx(1 - x)^n$, $E = [0,1]$, and $f(x) = 0$ for $x \in [0,1]$. The reader is invited to verify that $\langle f_n \rangle$ converges to the continuous function f, but that the convergence is not uniform. Another counterexample is the sequence $\langle f_n \rangle$, where $f_n(x) = x/n$ for $x \in \mathbb{R}$.)

2.7 SEMICONTINUOUS FUNCTIONS

Let f be a real-valued function defined on the whole real line. We defined f to be continuous at a point $x = a$ iff for every $\varepsilon > 0$ there is a $\delta > 0$ such that $|x - a| < \delta$ implies $|f(x) - f(a)| < \varepsilon$. If the last inequality is written $f(a) - \varepsilon < f(x) < f(a) + \varepsilon$ then, the above definition of continuity is equivalent to the following two conditions:

(i) For any number $A < f(a)$, there is an open interval V containing x_0 such that for every $x \in V$ we have $A < f(x)$.

(ii) For any number $B > f(a)$, there is an open interval V containing x_0 such that for every $x \in V$ we have $f(x) < B$.

If we keep only one of the above conditions then we have the notion of semicontinuity. More precisely we have the following definition.

Definition *A function* $f\colon \mathbb{R} \to \mathbb{R}$ *is said to be* lower (upper) semicontinuous *at the point* $a \in \mathbb{R}$ *iff for every* $A < f(a)$ $(f(a) < A)$ *there is an open interval* I *containing* a, *such that for every* $x \in I$ *we have* $A < f(x)$ $(f(x) < A)$. *When this condition holds at every point of* $\mathbb{R}$ *then* f *is said to be lower (upper) semicontinuous on* $\mathbb{R}$.

We shall also consider functions which assume the values $\pm\infty$.

If $f(a) = -\infty$ then f, by definition, is lower semicontinuous at the point a.

If $f(a) = -\infty$ then f is said to be upper semicontinuous at the point a iff for every $A > 0$ there is an open interval containing a in which $f(x) < -A$.

If $f(a) = +\infty$, then f, by definition, is upper semicontinuous at the point a.

If $f(a) = +\infty$, then f is said to be lower semicontinuous at the point a iff for every $A > 0$ there is an open interval containing a, in which $f(x) < A$.

The following propositions are immediate consequences of this definition.

(a) *A function* $f: \mathbb{R} \to \mathbb{R}$ *is continuous at a point* $a \in \mathbb{R}$ *if and only if* f *is both upper semicontinuous and lower semicontinuous at* a.

(b) *A function* $f: \mathbb{R} \to \mathbb{R}$ *is continuous if and only if it is both upper semicontinuous and lower semicontinuous.*

(c) *The lower semicontinuity of* f *at a point* a *is equivalent to the upper semicontinuity of* $-f$ *at the same point.*

Examples

1) Let $f(0) = 0$ and $f(x) = 1/x^2$ for $x \neq 0$. Then f is lower semicontinuous at the point $x = 0$.

2) Characteristic functions of open sets are lower semicontinuous functions; characteristic functions of closed sets are upper semicontinuous.

3) Let $f(x) = 0$ if x = rational and $f(x) = 1$ if x = irrational. Then f is lower semicontinuous at every rational number, and upper semicontinuous at every irrational number.

Definition *Let* f *be a real (or extended-real)-valued function* f *defined for all* x *in an open interval containing* y. *Then the* limit superior *and the* limit inferior *of* f, *when* $x \to y$ *are defined respectively as follows.*

$$\overline{\lim}_{x \to y} f(x) = \inf_{\delta > 0} \{\sup_{0<|x-y|<\delta} f(x)\}$$

$$\underline{\lim}_{x \to y} f(x) = \sup_{\delta > 0} \{\inf_{0<|x-y|<\delta} f(x)\}$$

It can be easily seen that a real-valued function f *is lower (upper) semicontinuous at the point* y *iff*

$$f(y) \leq \underline{\lim}_{x \to y} f(x) \quad (f(y) \geq \overline{\lim}_{x \to y} f(x)).$$

2.8 THE CANTOR TERNARY SET

Let F_0 be the interval $[0,1]$. Remove the open interval $((1/3,(2/3))$ from F_0 and denote the remaining closed set by F_1. Then remove the open intervals $((1/3^2),(2/3^2))$ and $((7/3^2),(8/3^2))$ from F_1 and denote the remaining closed set, consisting of the points of the intervals

$$\left[0,\frac{1}{3^2}\right], \left[\frac{2}{3^2},\frac{3}{3^2}\right], \left[\frac{6}{3^2},\frac{7}{3^2}\right], \left[\frac{8}{3^2},1\right]$$

by F_2. From each of these four intervals remove the middle interval of length $(1/3)^3$, and so forth. Continuing in this way we obtain a sequence of closed and bounded sets F_n such that $F_n \supset F_{n+1}$

for all n. The family $\langle F_n \rangle$ satisfies the hypothesis of Corollary 10, therefore $F = \bigcap_{n=1}^{\infty} F_n \neq \emptyset$.

The set F is called the *Cantor ternary set.*

The set F can be characterized in the following manner. We write each of the numbers x, $0 \leq x \leq 1$, in the triadic system

$$x = \sum_{n=1}^{\infty} \frac{a_n}{3^n}$$

where the numbers a_n can only assume the values 0, 1 and 2. It is easily checked that F contains those and only those numbers x, $0 \leq x \leq 1$ which can be written in at least one way in the triadic system such that the number 1 does not appear in the sequence $\langle a_n \rangle$.

We now prove that *the cardinal number of* F *is* c. To see this consider the following mapping (see Prob. 48) of F onto [0,1]. To each $x = \sum_{n=1}^{\infty} (a_n/3^n)$ in F, assign the number $\sum_{n=1}^{\infty} (b_n/2^n)$ in the interval [0,1], where $b_n = 0$ if $a_n = 0$ and $b_n = 1$ if $a_n = 2$. Clearly this mapping is onto [0,1]. This proves that the cardinal number of [0,1] is not less than the cardinal number of [0,1], i.e., c. On the other hand since F is a subset of [0,1], its cardinal number is less than or equal to c. It follows by Bernstein's theorem (Chap. 1) that $\overline{\overline{F}} = c$.

Let F' be the set of accumulation points of F. Since F is closed we have $F' \subset F$. If $x \in F$ let I be an open interval containing x. Let I_n be that interval of F_n which contains x. Choose n large enough so that $I_n \subset I$. Let x_n be an end point of I_n such that $x_n \neq x$. It follows from the construction of F that $x_n \in F$. Hence x is an accumulation point of F. Hence $F = F'$.

A set of real numbers which equals the set of its accumulation points is called a *perfect set* (see Prob. 31).

Thus F *is a perfect set.*

2.9 BOREL SETS

Let $\mathcal{O}$ be the collection of all open sets of the real line $\mathbb{R}$. Then by Chapter 1, Proposition 15 there is a smallest σ-algebra $\mathcal{B}$ of subsets of $\mathbb{R}$ which contains $\mathcal{O}$. The elements of $\mathcal{B}$ are called *Borel sets*. Borel sets are sometimes defined as the members of the smallest σ-algebra containing all closed sets or as the members of the smallest σ-algebra containing all bounded and closed sets. The reader can easily verify that all these definitions are equivalent.

A set which can be expressed as a countable (finite or infinite) intersection of open sets is called of type G_δ or, simply, a G_δ set.

A set which can be written as a countable (finite or infinite) union of closed sets is called of type F_σ or, simply, an F_σ set.

Clearly sets of type G_δ or F_σ are Borel sets. Every closed set is an F_σ set, as is every countable set. A countable union of sets of type F_σ is again an F_σ set. An open finite interval (a,b) is an F_σ set because it can be expressed as the countable union of the closed intervals $[a + 1/n, b - 1/n]$, $n = 1,2,\ldots.$ Since an open set can be expressed as the countable union of open intervals it follows that an open set is an F_σ set. Also one easily sees that the complement of an F_σ set is a G_δ set and vice versa.

Starting from the classes F_σ and G_δ we can construct more complicated classes of Borel sets as follows. A set is of type $F_{\sigma\delta}$ iff it can be expressed as the countable intersection of sets of type F_σ. A set is of type $F_{\sigma\delta\sigma}$ iff it can be expressed as a countable union of sets of type $F_{\sigma\delta}$, and so on. Similarly a set is of type $G_{\delta\sigma}$ iff it can be written as the countable union of sets of type G_δ. A set is of type $G_{\delta\sigma\delta}$ iff it is the countable intersection of sets of type $G_{\delta\sigma}$, and so fotrh.

Thus, the classes in the two sequences

$$F_\sigma, F_{\sigma\delta}, F_{\sigma\delta\sigma}, \ldots.$$

$$G_\delta, G_{\delta\sigma}, G_{\delta\sigma\delta}, \ldots.$$

are all classes of Borel sets. However, these two classes do not exhaust all Borel sets. There are

Borel sets which do not belong to either of these classes (see [18]).

PROBLEMS

1. Show that

   (a) Every ring A has a unique element θ such that $x + \theta = \theta + x = x$ for every $x \in A$.

   (b) For every $x \in A$, $\theta \cdot x = x \cdot \theta = \theta$.

2. Show that every field F has a unique element e such that for every $x \in F$, $xe = ex = x$.

3. Prove that in an ordered field, for every $x \neq \theta$ we have $xx > \theta$.

4. Prove that for any positive integer n there is a ring which has exactly n elements.

5. Prove that there is a field which has exactly five elements but there is no field which has exactly ten elements.

6. Let X and Y be two nonempty sets of real numbers bounded above, with $a = \sup X$ and $b = \sup Y$. Let Z be the set $Z = \{x + y: x \in X, y \in Y\}$. Prove that $a + b = \sup Z$.

7. Let X be a nonempty set of real numbers bounded above and let $a = \sup X$. Prove that for every given $\varepsilon > 0$ there exists an $x \in X$ such that $a - \varepsilon < x \leq a$. State and prove a similar result if X is bounded below.

8. Let X and Y be two sets of real numbers bounded above. Put $a = \sup X$ and $b = \sup Y$. Let Z denote the set $Z = \{xy: x \in X, y \in Y\}$. Prove that although $\sup Z$ is not in general

equal to ab, equality holds in case X and Y consist of nonnegative numbers.

9. Let us denote by $x \vee y$ and $x \wedge y$ the larger and the smaller of the real numbers x and y, respectively. Prove that

(a) $(x \wedge y) \wedge z = x \wedge (y \wedge z)$.

(b) $x \wedge y + x \vee y = x + y$.

(c) $(-x) \wedge (-y) = -(x \vee y)$.

(d) $x \vee y + z = (x + z) \vee (y + z)$

(e) $z(x \vee y) = (zx) \vee (zy)$ if $z \geq 0$.

(f) $|x| = x \vee (-x)$.

(g) $x \vee y = \frac{1}{2}(x + y + |x - y|)$.

(a') $(x \vee y) \vee z = x \vee (y \vee z)$.

(b') $x \vee y - x \wedge y = |x - y|$.

(c') $(-x) \vee (-y) = -(x \wedge y)$.

(d') $x \wedge y + z = (x + z) \wedge (y + z)$.

(e') $z(x \wedge y) = (zx) \wedge (zy)$ if $z \geq 0$.

(f') $-|x| = x \wedge (-x)$.

(g') $x \wedge y = \frac{1}{2}(x + y - |x - y|)$.

10. Give a few examples of sets of real numbers that are neither open nor closed.

11. Prove that

(a) The set of irrational numbers is neither open nor closed.

(b) For every real number a the set of numbers $\{x: x \geq a\}$ is closed.

12. (a) Let E be a set of real numbers and denote by E' the set of all accumulation points of E (E' is called the *derived set* of E). Prove that E' is a closed set, and that $(E')' \subset E'$.

(b) Put $E^{(2)} = (E')'$, and $E^{(n)} = (E^{(n-1)})'$

for $n = 3,4,\ldots$. Let n_o be an arbitrary but fixed positive integer greater than 1. Give an example of a set E such that $E^{(n_o)}$ is a proper subset of $E^{(n_o-1)}$.

(c) Generalize (b): Give an example of a set E such that $E^{(n)}$ is a proper subset of $E^{(n-1)}$ for all $n \geq 2$.

13. Show that if A and B are sets of real numbers such that $A \subset B$ and $B - A$ is finite, then $A' = B'$.
14. Show that every monotone infinite bounded sequence of real numbers converges to a real number.
15. Show that every monotone infinite sequence of real numbers is convergent in the extended real number system.
16. (a) Show that if an infinite sequence $\langle x_n \rangle$ converges to a real number ℓ, then $\langle x_n \rangle$ is a Cauchy sequence.

    (b) Show that every Cauchy sequence is bounded.

    (c) Show that if a subsequence of a Cauchy sequence converges to a real number ℓ, then the original sequence converges to ℓ.
17. Use Problem 16 to prove that a sequence of real numbers is convergent if and only if it is a Cauchy sequence.
18. Prove that if $\langle x_n \rangle$ is convergent, then $\langle |x_n| \rangle$ is also convergent. Is the converse true?

19. Prove that a sequence $\langle x_n \rangle$ is convergent in the extended real number system if and only if there is exactly one member of the extended real number system which is a cluster point of $\langle x_n \rangle$. Prove that this statement is not always true if the word "extended" is omitted.

20. Prove that a sequence $\langle x_n \rangle$ converges to a member ℓ of the extended real number system if and only if

$$\ell = \underline{\lim}_{n\to\infty} x_n = \overline{\lim}_{n\to\infty} x_n .$$

21. Let $\langle x_n \rangle$ and $\langle y_n \rangle$ be sequences of real numbers such that neither of the expressions

$$\overline{\lim}_{n\to\infty} x_n + \underline{\lim}_{n\to\infty} y_n,$$

$$\overline{\lim}_{n\to\infty} x_n + \overline{\lim}_{n\to\infty} y_n$$

is of the form $+\infty - \infty$. Then prove that

$$\overline{\lim}_{n\to\infty} x_n + \underline{\lim}_{n\to\infty} y_n \le \overline{\lim}_{n\to\infty} (x_n + y_n)$$

$$\le \overline{\lim}_{n\to\infty} x_n + \overline{\lim}_{n\to\infty} y_n .$$

(Hint: First prove $\overline{\lim}_{n\to\infty} (x_n + y_n) \le \overline{\lim}_{n\to\infty} x_n + \overline{\lim}_{n\to\infty} y_n$ then apply this inequality to the sequences $\langle x_n + y_n \rangle$ and $\langle -y_n \rangle$.)

22. Find $\overline{\lim}_{n\to\infty} x_n$ and $\underline{\lim}_{n\to\infty} x_n$ if x_n is given by

(a) $\cos n$,

(b) $(-1)^n n/(1+n)^n$,

(c) $(n/3) - [n/3]$, [$[x]$ means the greatest integer $\leq x$].

23. Let $\langle x_n \rangle$ and $\langle y_n \rangle$ be sequences of nonnegative real numbers such that $(\overline{\lim}_{n\to\infty} x_n)\cdot(\overline{\lim}_{n\to\infty} y_n)$ is not of the form $0\cdot\infty$. Prove that

$$\overline{\lim}_{n\to\infty}(x_n y_n) \leq (\overline{\lim}_{n\to\infty} x_n)\cdot(\overline{\lim}_{n\to\infty} y_n).$$

24. Prove that a series $\sum_{n=1}^{\infty} x_n$ converges if and only if for every $\varepsilon > 0$ there is an integer N such that $|x_n + x_{n+1} + \ldots + x_m| \leq \varepsilon$ if $m \geq n \geq N$.

25. (a) Prove that if $\sum_{n=1}^{\infty} |x_n|$ converges, then $\sum_{n=1}^{\infty} x_n$ converges.

(b) Give a counterexample to disprove the converse of (a).

26. Let E be a set of positive real numbers. For every finite subset F or E define $S(F)$ to be the sum of all the elements of F. Let us denote by $\sum x$ the least upper bound of the set of all numbers $S(F)$ where F runs over the set of all finite subsets of E.

(a) Prove that if $\sum x$ is finite, then E is a countable set.

(b) Prove that if E is countable and if $\langle x_n \rangle$ is a sequence whose range is E, then $\sum x = \sum_{n=1}^{\infty} x_n$.

27. (a) Give an example of a set S of real numbers which is closed but not bounded and find an open covering of S which has no finite subcovering.

    (b) Find an open covering of the set $E = \{x: 0 < x \leq 1\}$ which has no finite subcovering.

28. Let E be a set of real numbers with the property that for every $x \in E$ there is an open interval I_x containing x, and such that $I_x \cap E$ is a countable set. Prove that E is countable.

29. Let E be a set of real numbers. A point $a \in E$ is said to be a *condensation point* of E iff every open interval containing a, contains an uncountable subset of E. Prove that if E is not countable it contains at least one condensation point.

30. Let E be an uncountable set of real numbers and let S be the set of all condensation points of E. Prove that

    (a) $E - S$ is countable.

    (b) $E \cap S$ is not countable.

    (c) S is closed.

    (d) S contains no isolated points.

31. A set E of real numbers is said to be *perfect* iff $E = E'$. Prove that every uncountable closed set can be expressed as the union

of a perfect set and a countable set.

32. Let $\mathcal{O}$ denote the set of all open sets of the real line $\mathbb{R}$.

(a) Show that the cardinal number of $\mathcal{O}$ is at least c.

(Hint: Consider the open intervals of length 1.)

(b) Show that the cardinal number of $\mathcal{O}$ is at most c.

(Hint: There are $\aleph_0$ open intervals with rational endpoints. Call them $I_1, I_2, \ldots$. If O is any open set, then every $a \in O$ is in some of the intervals I_n, say $I_{n(a)}$. The intervals $\langle I_{n(a)} \rangle_{a \in O}$ form a subsequence $\langle I_n \rangle$. Hence the set of open sets can be put in one-to-one correspondence with a set of subsequences of the set of positive integers.)

(c) Show that the cardinal number of $\mathcal{O}$ is c.

33. Let $f\colon [a,b] \to \mathbb{R}$ be continuous and let $f(x) = 0$ when x is irrational. Prove that $f(x) = 0$ for every $x \in [a,b]$.

34. Let $f\colon [a,b] \to [c,d]$ be a strictly increasing function from $[a,b]$ onto $[c,d]$, with $[a,b]$ and $[c,d]$ closed and bounded intervals of real numbers. Prove that f is continuous. Prove that the inverse function f^{-1} exists and is also continuous on $[c,d]$.

35. Let E be a bounded and closed set of real numbers and let $f\colon E \to \mathbb{R}$ be continuous. Prove that f is bounded on E, and assumes

its maximum and minimum on E; that is there are points x_1 and x_2 in E such that $f(x_1) \leq f(x) \leq f(x_2)$ for all x in E. (Hint: Use Theorem 16.)

36. Let f and g be real-valued continuous functions on the whole real line. Prove that
    (a) $|f|$ is continuous.
    (b) $f + g$ and fg are continuous.
    (c) If $f \vee g$ is the function defined by $(f \vee g)(x) = f(x) \vee g(x)$ (see Prob. 9) then $f \vee g$ is continuous.
    (d) Define $f \wedge g$ in a similar manner and prove that it is continuous.
37. Let $f: [0,1] \to [0,1]$ be continuous. Prove that there exists a point $x \in [0,1]$ such that $f(x) = x$.
    (Hint: Consider the function $f(x) - x$.)
38. Prove that a real-valued function which is uniformly continuous on a bounded set of real numbers is bounded.
39. Let S be a set of real numbers which is not both closed and bounded.
    (a) Give an example of a function which is continuous on S but not bounded.
    (b) Give an example of a function which is continuous and bounded on S but which has no maximum on S.
    (c) Give an example of a function which is continuous on S but which is not uniformly continuous.
40. Does there exist a continuous real-valued function f on [0,1] such that f(x) is

rational for x irrational, and $f(x)$ is irrational for x rational?

41. Prove the assertions stated in Examples 1, 2, and 3 (see p. 88).

42. Show that a function $f: (a,b) \to \mathbb{R}$ is lower semicontinuous on (a,b) if and only if the set $\{x: f(x) > \lambda\}$ is open for each real number λ.

43. Let $S = \{x_1, x_2, \ldots\}$ be any denumerable (infinite countable) set of real numbers. Let $f(x_n) = 1/n$ $(n = 1,2,\ldots)$, and $f(x) = 0$ for $x \notin S$. Prove that f is upper semicontinuous everywhere but is discontinuous on S.

44. Show that if f and g are lower semicontinuous functions on the real line, so are $f \vee g$ and $f + g$.

45. Let $f_n: \mathbb{R} \to \mathbb{R}$ be lower semicontinuous for $n = 1,2,\ldots.$ Then the function f defined by

$$f(x) = \sup_n f_n(x)$$

is also lower semicontinuous.

46. Prove that for every lower semicontinuous function from a bounded and closed interval $[a,b]$ into the extended real number system there exists at least one point x_0 in $[a,b]$ such that

$$f(x_0) = \inf \{f(x): x \in [a,b]\}.$$

47. Prove that a set E is perfect (see Prob. 31) if and only if E is the complement of a

countable (finite or infinite) number of disjoint open intervals, no two of which have an endpoint in common.

48. (a) Prove that the Cantor set F has the following property: For any open interval I, there is an open subinterval J of I such that $J \cap F = \emptyset$. Sets with this property are called *nowhere dense*.

(b) Prove that every nowhere dense nonempty perfect set E of real numbers has cardinal number c (see Prob. 47).

(Hint: Establish a one-to-one correspondence between E and the Cantor set.)

49. Let $f: F \to [0,1]$ be a function defined on the Cantor set F in the following way. For $x \in F$ write the unique ternary expansion of x without using the digit 1:

$$x = 0.(2c_1)(2c_2)\ldots.(2c_n)\ldots,$$

with $c_n = 0$ or 1. Set

$$f(x) = 0.c_1\, c_2 \ldots c_n \ldots.$$

(a) Evaluate $f(1/3)$, $f(2/3)$ and $f(3/4)$.

(b) Show that the mapping f is onto $[0,1]$.

(c) Show that if (α_n, β_n) is one of the intervals removed in the nth stage of the construction of the set F, then

$$f(\alpha_n) = f(\beta_n).$$

(d) Show that if $x_1 \in F$ and $x_2 \in F$ with

$x_1 < x_2$, then $f(x_1) \leq f(x_2)$.

(e) *The Cantor function* $k: [0,1] \to [0,1]$ is defined by

$$k(x) = f(x) \qquad \text{if } x \in F$$

$$k(x) = f(\alpha_n) = f(\beta_n) \quad \text{if } \alpha_n < x < \beta_n.$$

Show that the Cantor function is continuous.

50. Show that

(a) The union of a set of intervals, closed, open, or semiopen, is a Borel set.

(b) Every countable set is an F_σ set.

(c) A countable intersection of sets of type G_δ is again a G_δ set.

(d) A countable union of sets of type F_σ is again an F_σ set.

51. Let $\mathcal{C}$ be a collection of sets of real numbers such that the cardinal number of $\mathcal{C}$ is $\leq c$. Prove that the cardinal number of the collection of the sets in the smallest σ-algebra which contains $\mathcal{C}$, is also $\leq c$.

52. Show that there is no sequence $\langle f_n \rangle_{n=1}^{\infty}$ of continuous functions on $\mathbb{R}$ such that

$$\lim_{n\to\infty} f_n(x) = 0 \quad \text{if } x \text{ is rational}$$

$$\lim_{n\to\infty} f_n(x) = 1 \quad \text{if } x \text{ is irrational.}$$

CHAPTER 3

Lebesgue Integral

3.1 INTRODUCTION

The generalization of the Riemann integral due to Henri Lebesgue is based on the notion of the "measure" of a subset of the real line. The task in the case of the Lebesgue integral is to give a useful generalization of the "length of an interval" but extended to more complicated sets of real numbers. Let us be more explicit. The Riemann integral of a bounded function f over a closed and bounded interval $[a,b]$ is the limit of sums (called *Riemann sums*) of the form

$$S = \sum_{i=1}^{n} f(t_i)\ell(I_i)$$

where $I_1, I_2, \ldots, I_n$ are disjoint intervals whose union is $[a,b]$, $\ell(I_i)$ denotes the length of I_i and t_i is an arbitrary point in I_i, for $i = 1, 2, \ldots, n$. In other words we require that the sums S approximate the Riemann integral of f no matter how we choose the point t_i in I_i, which means that varying t_i in I_i must have no significant influence on the value of the sum S. But this is possible only if variation of the points t_i produces in general little change in the value of $f(t_i)$. So we observe that one can expect the sums S to approximate the integral, if a sufficient proximity of different values of t implies arbitrary proximity of the values of the function f. If the function f is continuous or if f is not "too discontinuous" these conditions are satisfied. For discontinuous functions this need not be the case at all. We shall see that a function is Riemann integrable if and only if it is continuous on [a,b] except on a set of "measure zero" (see Theor. 36). Hence, roughly speaking, the definition of the Riemann integral seems to be justifiable for continuous functions but that is rather artificial for all other functions.

Considerations of this general nature, made it clear to the mathematicians of the end of the nineteenth century that another type of integral, more general and more useful for dealing with limit processes, was needed. The successful attempt in this direction was made by Lebesgue. Lebesgue's idea was to combine the points t into sets I_i, not because these points happened to be close on the x-axis, but rather by an insistance on closeness of the corresponding values of the function. So instead of subdividing the interval [a,b] of the x-axis into subintervals he subdivided a closed interval [A,B] of the y-axis containing the range of f:

$$A = y_0 < y_1 < \dots < y_n = B.$$

This time the sets I_i $(i = 1,2,\dots,n)$ are defined by

$$I_i = \{t: y_i \le f(t) < y_{i+1}\}.$$

Clearly to all points t in I_i correspond values of f lying close together, although unlike the Riemann process the points t within a single I_i can be very far from one another. It now becomes clear that in order to obtain sums similar to the Riemann sums S, we need an extension of "length" which can be applied to sets of a sufficient generality to include the sets I_i. So a number mI_i was associated with each set I_i, called the measure of I_i, and sums of the form

$$\sum_{i=1}^{n} y_i' mI_i,$$

with $y_{i-1} \le y_i' < y_i$, replaced the sums S of the Riemann theory.

Later in this chapter we will follow a slightly different but equivalent approach, to the one given above, in defining the Lebesgue integral.

The concept of measure of a set also has applications to many problems in the theory of functions, in probability, topology, functional analysis, and other areas of mathematics. The Lebesgue integral permits the solution of numerous problems that the

Riemann integral is unable to handle. In particular the Lebesgue integral is crucial in the theory of trigonometric series, the theory of function spaces, as well as in other branches of mathematics.

3.2 LEBESGUE MEASURE

Definition *A* set function *is a function defined on a collection of sets, and having values either in the set of real or complex numbers, or in the extended real number system. In the last case its range contains at most one of the improper values* $+\infty$ *and* $-\infty$.

A positive set function is a real-valued or extended real-valued set function which has no negative values.

The function ℓ which associates to every interval I its length $\ell(I)$ is an example of a set function with domain the collection of all intervals of the real line. If I is an infinite interval then $\ell(I) = +\infty$. We now want to extend the notion of length to a collection of sets larger than the collection of intervals, and we want this new collection to be the largest possible. Therefore it is natural to ask if there is an extended real-valued set function, m, with the following properties:

1) $m(E)$ (or in short mE) is defined for every set E of real numbers, and it is a member of the nonnegative extended real number system.

2) If E is an interval then, $mE = \ell(E)$.

3) If $\langle E_n \rangle$ is a sequence of disjoint subsets

of real numbers then, $m\left(\bigcup_{n=1}^{\infty} E_n\right) = \sum_{n=1}^{\infty} mE_n$.

4) If we denote by $E + x = \{y + x : y \in E\}$ the translation of a set E by x, then $m(E + x) = mE$.

In section 3.3 we shall prove that the answer to this question is in the negative. Consequently at least one of the above four properties must be weakened. Without going into further considerations of the various existing approaches to weakening these properties we just state that it is generally regarded as most useful to retain properties 2), 3), and 4), and weaken property 1). In the following sections we shall construct a set function m defined on a certain collection $\mathcal{M}$ of subsets of the real line and satisfying properties 2), 3), and 4). It turns out that the collection $\mathcal{M}$ is a σ-algebra of sets of real numbers. With this in mind we give the following definitions.

Definition *A* measure *is an extended real-valued set function* μ *on a* σ*-algebra* $\mathcal{M}$ *of sets, such that:*

(a) $\mu(E) \geq 0$ *if* $E \in \mathcal{M}$

(b) $\mu(\emptyset) = 0$

(c) If $\langle E_n \rangle$ *is a sequence of pairwise disjoint members of* $\mathcal{M}$, *then* $\mu\left(\bigcup_{n=1}^{\infty} E_n\right) = \sum_{n=1}^{\infty} \mu E_n$.

The set function m that we are about to construct (called Lebesgue measure) is a translation-invariant measure with the property that $m(I) = \ell(I)$, for each interval I. The members of the collection $\mathcal{M}$ are called *Lebesgue measurable sets* or simply *measurable sets*. We shall see in section 3.3 that $\mathcal{M} \neq \mathcal{P}(\mathbb{R})$. In other words there are sets of real numbers which are not measurable.

Let O be a nonempty open set. Then by Proposition 5 of Chapter 2, we have $O = \bigcup_{n=1}^{\infty} I_n$ where $\langle I_n \rangle$ is a countable collection of disjoint open intervals (finite or infinite). Observe that this collection is unique. Let us denote by $|O|$ the finite or infinite sum of the series $\sum_{n=1}^{\infty} \ell(I_n)$. Then $|O|$ is well defined since it is the sum of a series with nonnegative terms. We set $|\emptyset| = 0$.

Remark *Clearly if* O_1 *and* O_2 *are open sets then* $|O_1 \cup O_2| \leq |O_1| + |O_2|$. *If in addition* $O_1 \cap O_2 = \emptyset$, *then* $|O_1 \cup O_2| = |O_1| + |O_2|$ *(see Prob. 1).*

Definition *If* E *is a subset of real numbers, the* outer measure *of* E *is defined as*

$$m^*E = \inf_{O \supset E}\{|O|\}$$

where O *varies over all open sets containing* E.

Notice that if E is an open set then $m^*E = |E|$, for the infimum is obtained for $O = E$. It follows directly from this definition that the outer measure of the empty set and the outer measure of a set consisting of one single point are both zero. Also if $A \subset B$ then $m^*A \leq m^*B$, for the collection of open sets which cover B, also cover A.

1. Proposition *If* $\langle E_n \rangle$ *is a countable collection of sets of real numbers then*

$$m^*\left(\bigcup_{n=1}^{\infty} E_n\right) \leq \sum_{n=1}^{\infty} m^*E_n.$$

Proof We may assume that for each n, the outer measure m^*E_n is finite, for otherwise there is nothing to prove. Let $\varepsilon > 0$ be given. Then for each n, there is an open set O_n containing E_n, such that $m^*O_n = |O_n| < m^*E_n + \varepsilon/2^n$. Set $O = \bigcup_{n=1}^{\infty} O_n$. Then $\bigcup_{n=1}^{\infty} E_n \subset O$, so that $m^*\left(\bigcup_{n=1}^{\infty} E_n\right) \leq m^*O$. Let I_j $(j = 1,2,\ldots)$ be the components of the open set O. Then each I_j is the union of some components of the open sets O_n. Since the I_j's are disjoint, we have $|O| \leq \sum_{n=1}^{\infty} |O_n|$. It follows that

$$m^*\left(\bigcup_{n=1}^{\infty} E_n\right) \leq m^*O = |O| \leq \sum_{n=1}^{\infty} |O_n|$$

$$< \sum_{n=1}^{\infty} m^*E_n + \varepsilon.$$

Since ε was arbitrary the theorem follows.

Remark *The outer measure of any countable set of real numbers is zero.* This follows from the fact (above) that the outer measure of a set consisting of a single point is zero.

Definition (Carathéodory) *A set* E *of real numbers is said to be* measurable *iff for each set* A, $m^*A \geq m^*(A \cap E) + m^*(A - E)$.

This definition is equivalent to saying that E is measurable iff for each set A, $m^*A =$

$m^*(A \cap E) + m^*(A - E)$. This follows from the fact that the inequality

$$m^*A \le m^*(A \cap E) + m^*(A - E)$$

is always true.

An immediate consequence of the above definition is that if $m^*E = 0$ then E is measurable. For if A is any set we have $A - E \subset A$ which implies $m^*(A - E) \le m^*A$. Also $A \cap E \subset E$ implies $m^*(A \cap E) = 0$. Hence $m^*A \ge m^*(A \cap E) + m^*(A - E)$ and therefore E is measurable. In particular the empty set is measurable.

2. Lemma *The interval* $I = (a,\infty)$ *is measurable.*
Proof Let A be any set such that $m^*A < +\infty$. Then for any $\varepsilon > 0$, there is an open set O containing A, with $|O| < m^*A + \varepsilon$. Put $O_1 = O \cap (a,\infty)$ and $O_2 = O \cap (-\infty,a)$. The sets O_1 and O_2 are disjoint open sets. We have $A \cap I \subset O_1$ and $A - I \subset O_2 \cup \{a\}$ so that

$$m^*(A \cap I) \le |O_1|$$

and

$$m^*(A - I) \le |O_2| + m^*\{a\} = |O_2|$$

Hence

$$m^*(A \cap I) + m^*(A - I) \le |O_1| + |O_2| = |O| < m^*A + \varepsilon$$

Since ε was arbitrary we have

$$m^*A \ge m^*(A \cap I) + m^*(A - I)$$

for every A such that $m^*A < +\infty$. Furthermore the above inequality also holds if $m^*A = +\infty$. Hence I is measurable.

3. Proposition *The class $\mathcal{M}$ of measurable sets is a σ-algebra.*

Proof We must prove: (i) $E \in \mathcal{M} \Rightarrow \complement E \in \mathcal{M}$; (ii) If $\langle E_n \rangle$ is a sequence of disjoint sets in $\mathcal{M}$ then $\bigcup_{n=1}^{\infty} E_n \in \mathcal{M}$. The proof of (i) follows from the fact that the definition of measurability is symmetric in E and $\complement E$. That is if in the inequality $m^*A \geq m^*(A \cap E) + m^*(A - E)$ we replace E by $\complement E$ then the inequality remains unchanged. To prove (ii) we first prove that the union of two measurable sets E_1, E_2 is measurable. Let A be an arbitrary set. Since $E_1 \in \mathcal{M}$ we have $m^*A \geq m^*(A \cap E_1) + m^*(A - E_1)$. Since $E_2 \in \mathcal{M}$ we also have $m^*(A - E_1) \geq m^*[(A - E_1) \cap E_2] + m^*[(A - E_1) - E_2] = m^*[(A - E_1) \cap E_2] + m^*[A - (E_1 \cup E_2)]$. But $[(A - E_1) \cap E_2] \cup (A \cap E_1) = A \cap (E_1 \cup E_2)$ so that

$$m^*[A \cap (E_1 \cup E_2)] \leq m^*(A \cap E_1) + m^*[(A - E_1) \cap E_2].$$

Thus

$$\begin{aligned} m^*A &\geq m^*(A \cap E_1) + m^*[(A - E_1) \cap E_2] \\ &\qquad + m^*[A - (E_1 \cup E_2)] \\ &\geq m^*[A \cap (E_1 \cup E_2)] + m^*[A - (E_1 \cup E_2)] \end{aligned}$$

which proves that $E_1 \cup E_2 \in \mathcal{M}$. Using the principle of mathematical induction we conclude that *any finite union of measurable sets is measurable.* Next we notice that if $A, B \in \mathcal{M}$ then $A - B \in \mathcal{M}$. For $A - B$ is the complement of the set $\complement A \cup B$, which is the union of two measurable sets, therefore it is measurable. Define now the sequence $\langle B_n \rangle$ by

$B_1 = E_1$, $B_n = E_n - \bigcup_{k=1}^{n-1} E_k$ for $n \geq 2$. Then each B_n is measurable. Using Proposition 16 of Chapter 1 we have that for each n, $\bigcup_{k=1}^{n} E_k = \bigcup_{k=1}^{n} B_k$, $\bigcup_{n=1}^{\infty} E_n = \bigcup_{n=1}^{\infty} B_n$ and that the B_n's are pairwise disjoint. We prove that if A is any arbitrary set then

$$m^*\left(A \cap \bigcup_{k=1}^{n} B_k\right) = \sum_{k=1}^{n} m^*(A \cap B_k)$$

for every integer n. The proof proceeds by induction. For $n = 1$ this equality is trivially true. Assume then it is true for n. Since $\bigcup_{k=1}^{n} B_k$ is measurable we have

$$m^*\left(A \cap \bigcup_{k=1}^{n+1} B_k\right) \geq m^*\left[A \cap \left(\bigcup_{k=1}^{n+1} B_k\right) \cap \left(\bigcup_{k=1}^{n} B_k\right)\right] + m^*\left[A \cap \left(\bigcup_{k=1}^{n+1} B_k\right) - \bigcup_{k=1}^{n} B_k\right]$$

$$= m^*\left(A \cap \bigcup_{k=1}^{n} B_k\right) + m^*(A \cap B_{n+1}) = \sum_{k=1}^{n+1} m^*(A \cap B_k).$$

Now since $\bigcup_{k=1}^{n} E_k$ is measurable, for each n we have

$$m^*A \geq m^*\left(A \cap \bigcup_{k=1}^{n} E_k\right) + m^*\left(A - \bigcup_{k=1}^{n} E_k\right)$$

$$\geq m^*\left(A \cap \bigcup_{k=1}^{n} B_k\right) + m^*\left(A - \bigcup_{n=1}^{\infty} E_n\right)$$

$$= \sum_{k=1}^{n} m^*(A \cap B_k) + m^*\left(A - \bigcup_{n=1}^{\infty} E_n\right)$$

Since the left-hand side of the inequality is independent of n, we have, by letting $n \to \infty$,

$$m^*A \geq \sum_{n=1}^{\infty} m^*(A \cap B_n) + m^*\left(A - \bigcup_{n=1}^{\infty} E_n\right)$$

$$\geq m^*\left(A \cap \bigcup_{n=1}^{\infty} B_n\right) + m^*\left(A - \bigcup_{n=1}^{\infty} E_n\right)$$

$$= m^*\left(A \cap \bigcup_{n=1}^{\infty} E_n\right) + m^*\left(A - \bigcup_{n=1}^{\infty} E_n\right)$$

This proves (ii).

4. Proposition *Every Borel set is measurable.*
Proof By Lemma 2 for each a, the interval (a,∞) is measurable, that is $(a,\infty) \in \mathcal{M}$. Since $\mathcal{M}$ is a σ-algebra the complement of (a,∞), that is $(-\infty,a]$, is in $\mathcal{M}$. Also since $(-\infty,a) = \bigcup_{n=1}^{\infty} (-\infty, a - (1/n)]$ we have $(-\infty,a) \in \mathcal{M}$. For each

open interval (a,b) we have $(a,b) = (-\infty,b) \cap (a,\infty)$, hence $(a,b) \in \mathcal{M}$. Also, since every open set is the union of a countable number of open intervals it is measurable. Thus $\mathcal{M}$ contains all open sets. Therefore $\mathcal{M}$ contains the smallest σ-algebra which contains all open sets, that is $\mathcal{M}$ contains all Borel sets. Finally, if F is a closed set then $\complement F$ is open, so $\complement F \in \mathcal{M}$, which implies $F \in \mathcal{M}$. (*Note*: We have seen [Chap. 2, Prob. 32] that the cardinal number of the set of all open sets of the real line $\mathbb{R}$ is c. It follows [see Chap. 2, Prob. 51] that the collection of all Borel sets in $\mathbb{R}$ has cardinal number c, the same as that of the Cantor set F. Now let $\mathcal{F}$ be the class of all subsets of F. We know that all members of $\mathcal{F}$ are Lebesgue measurable sets with measure 0, because F has measure 0 [Prob. 12]. But the cardinal number of $\mathcal{F}$ is 2^c which is greater than the cardinal number of the collection of Borel sets. Hence some members of $\mathcal{F}$ are not Borel sets. In other words *there are Lebesgue measurable sets which are not Borel sets*. For an example of a measurable non-Borel set see Gelbaum and Olmsted [10].)

Definition *The function* m *defined for every* $E \in \mathcal{M}$ *by* $mE = m^*E$ *is called the* Lebesgue measure. *Thus the Lebesgue measure of a measurable set* E *is the outer measure of* E.

5. Proposition *Let* $\langle E_n \rangle$ *be a sequence of pairwise disjoint measurable sets. Then*

$$m\left(\bigcup_{n=1}^{\infty} E_n\right) = \sum_{n=1}^{\infty} mE_n$$

Proof We have $\bigcup_{k=1}^{n} E_k \subset \bigcup_{n=1}^{\infty} E_n$ so that $m^*\left(\bigcup_{k=1}^{n} E_k\right) = \sum_{k=1}^{n} m^*E_k \leq m^*\left(\bigcup_{n=1}^{\infty} E_n\right)$ or $\sum_{k=1}^{n} mE_k \leq m\left(\bigcup_{n=1}^{\infty} E_n\right)$. Letting $n \to \infty$ in the left-hand side of this inequality we get $\sum_{n=1}^{\infty} mE_n \leq m\left(\bigcup_{n=1}^{\infty} E_n\right)$. Since $m\left(\bigcup_{n=1}^{\infty} E_n\right) = m^*\left(\bigcup_{n=1}^{\infty} E_n\right) \leq \sum_{n=1}^{\infty} mE_n$, we have $m\left(\bigcup_{n=1}^{\infty} E_n\right) = \sum_{n=1}^{\infty} mE_n$.

6. Proposition *Let $\langle E_n \rangle$ be an infinite sequence of measurable sets such that $E_{n+1} \subset E_n$ for each n. Let mE_1 be finite. Then $m\left(\bigcap_{n=1}^{\infty} E_n\right) = \lim_{n\to\infty} mE_n$.*

Proof Put $\bigcap_{n=1}^{\infty} E_n = F$. Then F is measurable. Also for each n, the set $E_n - E_{n+1} = E_n \cap \complement E_{n+1}$ is measurable. We easily see that the sets $E_n - E_{n+1}$ are pairwise disjoint and that

$$E_1 - F = \bigcup_{n=1}^{\infty} (E_n - E_{n+1}).$$

Thus

$$E_1 = F \cup (E_1 - F); \quad F \cap (E_1 - F) = \emptyset$$

$$E_n = E_{n+1} \cup (E_n - E_{n+1});$$

$$E_{n+1} \cap (E_n - E_{n+1}) = \emptyset \quad (n = 1,2,\ldots)$$

$$mE_n \leq mE_1 < +\infty \qquad n = (1,2,\ldots)$$

We have $mE_1 = mF + m(E_1 - F)$; $mE_n = mE_{n+1} + m(E_n - E_{n+1})$. Hence $mE_1 - mF = \sum_{n=1}^{\infty} (mE_n - mE_{n+1})$. The sum of the telescoping series being $mE_1 - \lim_{n\to\infty} mE_n$, it follows that $m \bigcap_{n=1}^{\infty} E_n = \lim_{n\to\infty} mE_n$. The proposition is proved.

7. Proposition *The Lebesgue measure* m *is a translation invariant measure with the property that* $m(I) = \ell(I)$ *for each interval* I.

Proof It is left to the reader.

8. Proposition *A set* E *is measurable if and only if for any given* $\varepsilon > 0$ *there is an open set* $O \supset E$ *such that* $m^*(O - E) < \varepsilon$.

Proof Let E be a measurable set. First assume $mE < +\infty$. Let $\varepsilon > 0$ be given. Then by the definition of outer measure there is an open set $O \supset E$ such that $m^*E \leq m^*O < m^*E + \varepsilon$ or $0 \leq m^*(O) - m^*E < \varepsilon$. Since $O = (O - E) \cup E$ and $m^*E < +\infty$ we have $m^*O - m^*E = m^*(O - E)$. Hence $m^*(O - E) < \varepsilon$. Next consider the case $mE = +\infty$, and put $E_n = E \cap (-n,n)$. Then E_n is measurable for each n, and by the above result there is an open set $O_n \supset E_n$ such that $m^*(O_n - E_n) < \varepsilon/2^n$. Since $O_n - E \subset O_n - E_n$ we also have $m^*(O_n - E) < \varepsilon/2^n$.

Set $\bigcup_{n=1}^{\infty} O_n = O$. Clearly O is open, $O \supset E$ and $\bigcup_{n=1}^{\infty} (O_n - E) = O - E$. It follows that $m^*(O - E) \leq \sum_{n=1}^{\infty} m^*(O_n - E) < \sum_{n=1}^{\infty} \varepsilon/2^n = \varepsilon$.

To prove the second half of the theorem assume that given $\varepsilon > 0$ there is an open set $O \supset E$ such that $m^*(O - E) < \varepsilon$. Thus for any given positive integer n, there is an open set $O_n \supset E$ with $m^*(O_n - E) < 1/n$. Set $G = \bigcap O_n$. Then G is measurable and $G \supset E$. Since $G \subset O_n$ we have $G - E \subset O_n - E$. Thus $m^*(G - E) \leq m^*(O_n - E)$ and $0 \leq m^*(G - E) < 1/n$ for all n. It follows that $m^*(G - E) = 0$. Let A now be any arbitrary set. We have

$$A \cap G \supset A \cap E \quad \text{and} \quad m^*(A \cap G) \geq m^*(A \cap E).$$

Since G is measurable we have $m^*A \geq m^*(A \cap G) + m^*(A \cap \complement G)$. But $A \cap \complement E = A \cap [\complement G \cup (G - E)] = (A \cap \complement G) \cup (A \cap (G - E))$. Hence $m^*(A \cap \complement E) \leq m^*(A \cap \complement G) + m^*(A \cap (G - E))$. Since $A \cap (G - E) \subset G - E$ we have $m^*[A \cap (G - E)] = 0$. Thus $m^*A \geq m^*(A \cap G) + m^*(A \cap \complement G) \geq m^*(A \cap E) + m^*(A \cap \complement E)$. Hence E is measurable.

Remark *In proving Proposition 8 we also proved that if E is measurable, then there is a set G of type G_δ such that $G \supset E$ and $m^*(G - E) = 0$. The converse of this fact is also true (see Prob. 7).*

9. Proposition *A set* E *is measurable if and only if for any given* $\varepsilon > 0$ *there is a closed set* $F \subset E$ *such that* $m^*(E - F) < \varepsilon$.

The proof is similar to the proof of Proposition 8 and is left to the reader.

3.3 EXISTENCE OF A NONMEASURABLE SET

Let E be any subset of the interval $[0,1)$, and let $a \in [0,1)$. We denote by E_a the set of all y such that $y = x + a$ if $x \in E$ and $x + a < 1$, or $y = x + a - 1$ if $x \in E$ and $x + a \geq 1$. That is, $E_a = \{y: y = x + a, x \in E, x + a < 1\} \cup \{y: y = x + a - 1, x \in E, x + a \geq 1\} = E_a^1 \cup E_a^2$. E_a is called the translation modulo 1 by a of E, and is also denoted by $E_a = E + a[\text{mod. } 1]$. Clearly $E_a \subset [0,1)$ and $E_a^1 \cap E_a^2 = \emptyset$. Next we prove that if E is measurable then E_a is also measurable. In fact, since E_a^1 is the translation of the set $E \cap [0,1 - a)$ by a, and E_a^2 is the translation of $E \cap [1 - a,1)$ by $a - 1$, both E_a^1, E_a^2 must be measurable. Also since the Lebesgue measure is translation invariant we have

$$mE_a^1 = m\{E \cap [0,1 - a)\}, \quad mE_a^2 = m\{E \cap [1 - a,1)\}$$

so that $mE_a = mE_a^1 + mE_a^2 = mE$. To proceed with the proof of the existence of a nonmeasurable set, let S be the set of all rational numbers in $[0,1)$. For each $x \in [0,1)$ consider the set S_x, and let $\mathcal{C}$ be the class of all sets S_x. Then two sets S_{x_1} and S_{x_2} are either identical or disjoint. In fact assume that $S_{x_1} \cap S_{x_2} \neq \emptyset$. Let $y \in S_{x_1} \cap S_{x_2}$ and $z \in S_{x_1}$. Then the numbers $z - x_1$, $x_1 - y$, $y - x_2$ being rational, their sum $z - x_2$ must also be rational, so that $z \in S_{x_2}$, which implies $S_{x_1} \subset S_{x_2}$. Similarly we have $S_{x_2} \subset S_{x_1}$, so that $S_{x_1} = S_{x_2}$. By the axiom of choice there is a set K consisting of one point from each of the sets of $\mathcal{C}$. Let $\langle r_n \rangle_{n=1}^{\infty}$ be an enumeration of all rational numbers in $(0,1)$, and for each n consider the set K_{r_n}, the translation modulo 1 by r_n of the set K. The sets K_{r_n} are disjoint. For let $z \in K_{r_{n_1}} \cap K_{r_{n_2}}$. Then $z - r_{n_1}$ and $z - r_{n_2}$ either belong to K or differ by 1 from points in K, so we have two points in K whose difference is rational. But this is impossible because these two points must belong to the same S_x, and K contains only one point from each set in $\mathcal{C}$. Thus the K_{r_n} are disjoint. Now for each n we have $K_{r_n} \subset [0,1)$, so that $\bigcup_{n=0}^{\infty} K_{r_n} \subset [0,1)$, with $K_{r_0} = K$. On the other hand if x is any arbitrary point in $[0,1)$ then for some rational number r_n, the number $x - r_n$ is (or differs by 1 from) the point of S_x that belongs to K. Thus $x \in K_{r_n}$. It follows that $\bigcup_{n=0}^{\infty} K_{r_n} = [0,1)$. The nonmeasurable set that we are looking for is K. For if K were measurable then all the sets K_{r_n} would be measurable and have the same measure as K. But in this case we would have

$$m[0,1) = \sum_{n=0}^{\infty} mK_{r_n}$$

and the right-hand side would be either zero or infinity, depending on whether mK is zero or positive. But this is impossible because $m[0,1) = 1$. Thus K is not a measurable set.

3.4 MEASURABLE FUNCTIONS

Definition *Let f be an extended real-valued function whose domain E is a set of real numbers. Then f is said to be* measurable *iff its domain is measurable and for each real number a, the set $\{x \in E: f(x) > a\}$ is measurable.*

Let E be a set. Then the characteristic function of E is defined by

$$\chi_E(x) = \begin{cases} 1 & \text{for } x \in E \\ 0 & \text{for } x \notin E \end{cases}$$

Clearly χ_E is a measurable function if and only if E is a measurable set.

The following proposition provides equivalent definitions of a measurable function.

10. Proposition *Let f be an extended real-valued function defined on a measurable set E. Then each of the following four statements implies the other three.*

1) $\{x \in E: f(x) > a\}$ is measurable for every real number a.

2) $\{x \in E: f(x) \geq a\}$ is measurable for every real number a.

3) $\{x \in E: f(x) < a\}$ is measurable for every real number a.

4) $\{x \in E: f(x) \leq a\}$ is measurable for every real number a.

Proof The relations

$$\{x \in E: f(x) \geq a\} = \bigcap_{n=1}^{\infty} \{x \in E: f(x) > a - \frac{1}{n}\}$$

and

$$\{x \in E: f(x) < a\} = E - \{x \in E: f(x) \geq a\}$$

show respectively that 1) implies 2) and 2) implies 3). Similarly the relations

$$\{x \in E: f(x) \leq a\} = \bigcap_{n=1}^{\infty} \{x \in E: f(x) < a + \frac{1}{n}\}$$

and

$$\{x \in E: f(x) > a\} = E - \{x \in E: f(x) \leq a\}$$

show respectively that 3) implies 4) and 4) implies 1). Hence any one of the statements 1), 2), 3), and 4) may be taken to define the measurability of f.

11. Corollary *Let* f *be a measurable function whose domain is* E. *Then for each* α *in the extended real number system the set* $\{x \in E: f(x) = \alpha\}$ *is measurable.*

Proof First assume that α is real. Then $\{x \in E: f(x) = \alpha\} = \{x \in E: f(x) \geq \alpha\} \cap \{x \in E: f(x) \leq \alpha\}$. Since f is measurable it follows from Proposition 10 that both sets on the right-hand side are measurable so that their intersection is measurable. If $\alpha = +\infty$, then the corollary follows from the relation

$$\{x \in E: f(x) = +\infty\} = \bigcap_{n=1}^{\infty} \{x \in E: f(x) \geq n\}$$

since for each n, $\{x \in E: f(x) \geq n\}$ is measurable.

Warning *The converse of Corollary 11 is not true. To see this the reader is invited to construct a function* f *such that* $\{x \in E: f(x) > 0\}$ *is a non-measurable set and such that* f *assumes each value once at the most.*

Remark *It should be noted that every continuous real-valued function with measurable domain* E *is measurable. For the set* $\{x: f(x) > a\}$ *as the inverse image of the open set* $(a,+\infty)$ *is open, so that its intersection with* E *is measurable.*

A real-valued function f whose domain is the closed interval $[a,b]$ is said to be a *step-function* iff the interval $[a,b]$ can be subdivided by points

$c_0 = a < c_1 < c_2 < \ldots < c_n = b$ into a finite number of subintervals such that in the open intervals (c_k, c_{k+1}) $(k = 0,1,2,\ldots,n-1)$ f is constant. It is easily seen that a step-function is measurable.

12. Theorem *Let f and g be measurable functions with the same domain, and let c be a real number. Then the functions f + c, cf, f + g, f - g, and fg are all measurable.*

Proof Let a be a real number. Then

$$\{x: f(x) + c > a\} = \{x: f(x) > a - c\}.$$

Since f is measurable, $\{x: f(x) > a - c\}$ must be measurable, so that $\{x: f(x) + c > a\}$ is measurable Thus f + c is measurable. The measurability of cf is shown in a similar manner. We now prove that f + g is measurable. Let Q be the set of all rational numbers. Then

$$\{x: f(x) + g(x) < a\} = \{x: f(x) < a - g(x)\}$$

$$= \bigcup_{r \varepsilon Q} [\{x: f(x) < r\} \cap \{x: g(x) < a - r\}].$$

For if $x \varepsilon \{x: f(x) + g(x) < a\}$ then $f(x) < a - g(x)$ and by the Archimedean axiom there is an $r \varepsilon Q$ such that $f(x) < r < a - g(x)$, so that $x \varepsilon \{x: f(x) < r\} \cap \{x: g(x) < a - r\}$, thus x belongs to the second member of the last equality. Also if x belongs to the second member it satisfies the inequality $f(x) + g(x) < a$ so that $x \varepsilon \{x: f(x) + g(x) < a\}$. Since f and g are measurable, $\{x: f(x) < r\}$, $\{x: g(x) < a - r\}$ as well as their intersection are measurable. Finally since Q is countable, $\{x: f(x) + g(x) < a\}$, as the countable union of measurable sets, is measurable. To prove that $f - g$ is measurable we simply notice that f and $(-1)g$ are measurable so that their sum $f - g$ must also be measurable. To prove that fg is measurable we first prove that f^2 is measurable. We have

$$\{x: f^2(x) > a\} = \{x: f(x) > \sqrt{a}\} \cup \{x: f(x) < -\sqrt{a}\}$$

if $a \geq 0$, and

$$\{x: f^2(x) > a\} = E = \text{the domain of } f$$

if $a < 0$. In both cases we see that $\{x: f^2(x) > a\}$ is measurable. The measurability of fg follows from the identity:

$$f(x)g(x) = \frac{1}{4}\{[f(x) + g(x)]^2 - [f(x) - g(x)]^2\}.$$

Observe that the conventions $(+\infty) + (-\infty) = 0$ and $0 \cdot (\pm\infty) = 0$ of Chapter 2 are used here; it may happen that for some x we have $f(x) = +\infty$, $g(x) = -\infty$ or $f(x) = 0$ and $g(x) = \pm\infty$.

13. Theorem *Let* $\langle f_n \rangle_{n=1}^{\infty}$ *be a sequence of measurable functions, defined on some common domain* E. *Then the functions* $\sup_n f_n$ *and* $\inf_n f_n$ *are both measurable.*

Proof Put $H(x) = \sup_n f_n$. Then if a is real the measurability of $H(x)$ follows from the equality

$$\{x: H(x) \leq a\} = \bigcup_{n=1}^{\infty} \{x: f_n(x) \leq a\}.$$

By Theorem 12, applied for $c = -1$, we have that $-f_n$ is measurable for each n. The result for $\inf_n f_n$ follows from the fact that $\inf_n f_n = -\sup(-f_n)$.

14. Corollary *If* $\langle f_n \rangle_{n=1}^{\infty}$ *is a sequence of measurable functions defined on some common domain, then*

$$\overline{\lim}_{n\to\infty} f_n, \qquad \underline{\lim}_{n\to\infty} f_n$$

are both measurable.

Proof It follows from Theorem 13 and the fact that

$$\overline{\lim}_{n\to\infty} f_n = \inf_n(\sup_{k\geq n} f_k);$$

$$\underline{\lim}_{n\to\infty} f_n = \sup_n(\inf_{k\geq n} f_k).$$

We shall see that in the theory of Lebesgue integration it frequently happens that a statement about a function, which is true except on a set of measure zero, implies the same consequences as if it were true everywhere. This is the reason that we are introducing below the term *almost everywhere*.

If a certain property holds for all the points of a set E except for the points of a subset E_1 of the set E such that $mE_1 = 0$, then we say that the property holds *almost everywhere* (a.e.) on the set E, or for almost all points of E. Notice that E may be the empty set. That is if the property holds everywhere it holds almost everywhere.

Definition *Two functions f and g defined on the same set E are said to be* equivalent *iff they are equal almost everywhere on E (i.e., $m\{x: f(x) \neq g(x)\} = 0$).*

3.5 APPROXIMATION THEOREMS

The representation of a function exactly or approximately by means of simpler functions is very useful and arises naturally. Examples of such representations are the decomposition of a polynomial into binomials of first degree, the expansion of a function into a power series, and many others. Such representations throw more light onto the structural properties of the function. For example, the definition of a measurable function given above is quite formal. It does not give much information about the nature of such a function. However, after learning Luzin's theorem (Theor. 22) the reader will realize that even though a measurable function can be everywhere discontinuous, still in a certain sense every measurable function is very "close" to a continuous function.

Definition *A function f defined on the whole real line is said to be* simple *iff it is measurable and its range is a finite set of real numbers.*

Among the simplest examples of simple functions are constant functions and characteristic functions of measurable sets. A simple function is clearly bounded.

Let f be a real-valued function defined on the whole real line. Then f is simple if and only if there exists a finite number of measurable sets $A_1, A_2, \ldots, A_n$ and real numbers $a_1, a_2, \ldots, a_n$ such that for each real number x we have

$$f(x) = \sum_{k=1}^{n} a_k \chi_{A_k}(x).$$

To see this suppose first that f is a simple function and let $c_1, c_2, \ldots, c_n$ be the distinct non-zero values of f. Define the sets $A_k = \{x\colon f(x) = c_k\}$ for $k = 1, 2, \ldots, n$. Then the A_k's are measurable and clearly $f(x) = \sum_{k=1}^{n} a_k \chi_{A_k}(x)$. Conversely if $A_1, A_2, \ldots, A_n$ are measurable sets and $a_1, a_2, \ldots, a_n$ are real numbers, the range of the function cannot contain more than 2^n elements, so it is finite. Also since the characteristic function of a measurable set is measurable, the linear combination $\Sigma a_k \chi_{A_k}$ of measurable functions must be measurable. It follows from the above discussion that the representation of a simple function under the form $f = \sum_{k=1}^{n} a_k\ A_k$ is not unique. However, if the real numbers $a_1, a_2, \ldots, a_n$ are distinct and nonzero and $A_k = \{x: f(x) = a_k\}$ $(k = 1,2,\ldots,n)$ then the representation $\sum_{k=1}^{n} a_k \chi_{A_k}$ with these properties, is unique and is sometimes called the *canonical representation*. The A_k's in this case are pairwise disjoint nonempty measurable sets.

It is an easy matter to prove that, if f and g are any simple functions and c any real number then, $|f|$, cf, $f + g$ are also simple functions.

15. Theorem *Let f be a nonnegative measurable function defined on a measurable set E. Then there exists a sequence $\langle f_n \rangle$ of nonnegative simple functions with $f_{n+1} \geq f_n$ and such that for every $x \in E$*

$$\lim_{n\to\infty} f_n(x) = f(x).$$

Proof For each positive integer n and each $x \in E$ define

$$f_n(x) = \begin{cases} \dfrac{i-1}{2^n} & \text{if } \dfrac{i-1}{2^n} \leq f(x) < \dfrac{i}{2^n} \\ n & \text{if } f(x) \geq n \end{cases} ; \; i = 1,2,\ldots,n2^n.$$

Then f_n is a nonnegative simple function. Moreover the sequence $\langle f_n \rangle$ is nondecreasing. For let $x \in E$. Then, either

$$\frac{i-1}{2^n} \leq f(x) < \frac{i}{2^n}$$

for some i, or $f(x) \geq n$. If the double inequality is the case we have $f_n(x) = (i-1)/2^n$. Since $(2i-2)/2^{n+1} \leq f(x) < 2i/2^{n+1}$ we have

$$f_{n+1}(x) = \frac{2i-2}{2^{n+1}} = f_n(x), \text{ if } \frac{2i-2}{2^{n+1}} \leq f(x) < \frac{2i-1}{2^{n+1}},$$

or

$$f_{n+1}(x) = \frac{2i-1}{2^{n+1}} > f_n(x), \text{ if } \frac{2i-1}{2^{n+1}} \leq f(x) < \frac{2i}{2^{n+1}}$$

so that $f_n(x) \leq f_{n+1}(x)$. If $f(x) \geq n$ is the case then either $n \leq f(x) < n+1$ or $n+1 \leq f(x)$. Hence in either case we have $f_n(x) = n \leq f_{n+1}(x)$.

We now prove that

$$\lim_{n\to\infty} f_n(x) = f(x).$$

If $f(x) = +\infty$ then $f_n(x) = n$ for each n. If $f(x) < +\infty$ then for $n > f(x)$ we have

$$0 \leq f(x) - f_n(x) < \frac{1}{2^n}$$

so in either case

$$\lim_{n\to\infty} f_n(x) = f(x).$$

The principle expressed by the following theorem is that a pointwise convergent sequence of measurable functions can be converted to a uniformly convergent sequence if a set of arbitrarily small measure is removed from the domain of convergence.

16. Theorem (Egoroff) *Let E be a measurable set of finite measure, and $\langle f_n \rangle$ be a sequence of*

a.e. real-valued measurable functions on E. Assume that $\langle f_n \rangle$ converges a.e. on E to a real-valued function f. Then for every $\varepsilon > 0$, there exists a measurable subset F of E such that $mF < \varepsilon$ and such that $\langle f_n \rangle$ converges to f uniformly on $E - F$.

Proof Since $\langle f_n \rangle$ converges to f a.e. on E, there is a subset S of E such that $mS = 0$, and $\langle f_n \rangle$ converges to f everywhere on $A = E - S$. For n and k positive integers set $A_{n,k} = \{x \in A: |f_n(x) - f(x)| \geq 1/k\}$, and $E_N^k = \bigcup_{n=N}^{\infty} A_{n,k}$. Clearly $E_{N+1}^k \subset E_N^k$. We prove that $\bigcap_N E_N^k = \emptyset$. Let $x \in A$. Then since $f_n(x) \to f(x)$ the difference $|f_n(x) - f(x)|$ is smaller than $1/k$ if n is taken large enough. This means that there is an N such that $x \notin A_{n,k}$ if $n \geq N$. This implies that $x \notin E_N^k$, so that $\bigcap_N E_N^k = \emptyset$. By Proposition 6 it follows that $\lim_{N\to\infty} mE_N^k = 0$. Let η be an arbitrary positive real number. Then there is an N depending on k and η such that $mE_N^k < \eta/2^k$. Set $F_1 = \bigcup_k E_N^k$. Then $mF_1 \leq \sum_{k=1}^{\infty} mE_N^k < \eta$. We prove that $f_n \to f$ uniformly on $A - F_1$. For all $x \in A - F_1$ we have $x \notin F_1$ or equivalently $x \notin E_N^k$ for all k. This means that for the ε given in the statement of the theorem, there is a k such that $1/k < \varepsilon$, and there is an N corresponding to this k such that for $n \geq N$ and all $x \in A - F_1$ we have $|f_n(x) - f(x)| < 1/k < \varepsilon$. This proves that $f_n \to f$ uniformly on $A - F_1$ or equivalently $f_n \to f$ uniformly on $E - F$ where $F = F_1 \cup S$ and $mF \leq mF_1 + mS = mF_1 < \eta$.

17. Theorem *Let f be an almost everywhere real-valued measurable function defined on a set E of finite measure. Then for any given $\varepsilon > 0$ there exists a measurable bounded function g defined on E such that $m\{x \in E: f(x) = g(x)\} < \varepsilon$.*

Proof Let n be a positive integer and set

$$E_n = \{x \in E: f(x) > n\};$$

$$A = \{x \in E: |f(x)| = +\infty\}.$$

Since f is assumed to be real-valued almost everywhere we have that $mA = 0$. Also one easily sees that $\bigcap_{n=1}^{\infty} E_n = A$. On the other hand the sequence $\langle E_n \rangle$ is decreasing, i.e., for each n we have $E_{n+1} \subset E_n$, and $mE_1 < +\infty$. It follows from Proposition 6, that $\lim_{n\to\infty} mE_n = mA = 0$. Hence there is a positive integer n_o such that $mE_{n_o} < \varepsilon$. Let g be the function defined on E such that $g(x) = f(x)$ if $x \in E - E_{n_o}$ and $g(x) = 0$ if $x \in E_{n_o}$. Then g is measurable and for every $x \in E$ we have $|g(x)| \leq n_o$. Furthermore, $\{x \in E: f(x) \neq g(x)\} \subset E_{n_o}$, so that $m\{x: f(x) \neq g(x)\} < \varepsilon$. The theorem is proved.

18. Proposition *Let F be a closed set of real numbers contained in the closed interval $[a,b]$, and let f be a real-valued function defined and continuous on F. Then there is a continuous function g defined on $[a,b]$ such that $f(x) = g(x)$ for each $x \in F$ and $\max|g(x)| = \max|f(x)|$.*

Proof We only give an outline of the proof, leaving the remaining portions of it to the reader. Since F is closed $[a,b] - F$ is open and it consists of a countable number of pairwise disjoint open intervals the end points of which belong to F. Define the function g to be linear in each of these open intervals, and to be equal to f on F. Then since g is obviously continuous on $[a,b] - F$ it remains to prove that g is continuous on F. One can prove the continuity at $x \in F$ by proving that the right and left limits of g at x exist and are equal. Moreover g being continuous on the closed and bounded interval $[a,b]$ it assumes its maximum at some point x_0 of $[a,b]$. The point x_0 cannot be in $[a,b] - F$, because g is linear on the intervals which compose this set, thus $x_0 \in F$. Therefore $\max|g(x)| = \max|f(x)|$. The theorem is proved.

19. Theorem *Let f be a measurable function defined on a bounded and closed interval $E = [a,b]$. Assume that f is real-valued almost everywhere on $[a,b]$. Then for any given $\varepsilon > 0$ there is a continuous function g on $[a,b]$ such that*

$$m\{x \in E: |f(x) - g(x)| \geq \varepsilon\} < \varepsilon.$$

If $|f(x)| \leq M$ *we can choose* g *so that* $|g(x)| \leq M$.

Proof Suppose first that $|f(x)| \leq M$ for each $x \in E$. Let k be a positive integer such that $M/k < \varepsilon$. Consider the sets

$$E_i = \{x \in E: \frac{i-1}{k} M \leq f(x) < \frac{i}{k} M\}$$

$$(i = 1-k, 2-k, \ldots, k-1)$$

$$E_k = \{x \in E: \frac{k-1}{k} M \leq f(x) \leq M\}$$

Clearly these sets are measurable and $E = \bigcup_{i=1-k}^{k} E_i$. By Proposition 9, for each i, there is a closed set $F_i \subset E_i$ such that $mF_i > mE_i - (\varepsilon/2k)$. Set $F = \bigcup_{i=1-k}^{k} F_i$. We have $E - F = \bigcup (E_i - F_i)$ so that $mE - mF < \varepsilon$. Define the function h on the set F by $h(x) = (i/k)M$ for $x \in F_i$, $(i = 1-m, \ldots, m)$. The fact that h is constant on each closed set F_i implies that h is continuous on F (Prove this!) We also have that for each $x \in F$, $|f(x) - h(x)| < \varepsilon$ and $|h(x)| \leq M$. Then by Proposition 18, there is a continuous function g on E, which coincides with h on F, and such that $|g(x)| \leq M$ for all $x \in E$. Since $\{x \in E: |f(x) - g(x)| \geq \varepsilon\} \subset E - F$ we conclude that g is the function that satisfies the requirement of the theorem in the case of f bounded.

Assume now that f is not bounded. Then by Theorem 17 there is a bounded function φ such that $m\{x \in E: f(x) \neq \varphi(x)\} < \varepsilon/2$. Since φ is bounded, by the above result, there is a continuous function g such that $m\{x \in E: |g(x) - \varphi(x)| \geq \varepsilon\} < \varepsilon/2$. But $\{x \in E: |f(x) - g(x)| \geq \varepsilon\} \subset \{x \in E: f(x) \neq \varphi(x)\} \cup \{x \in E: |g(x) - \varphi(x)| \geq \varepsilon\}$ so that g is the required function. The theorem is proved.

Definition *Let* $\langle f_n \rangle$ *be a sequence of measurable functions defined on a measurable set* E. *Assume that each* f_n *is almost everywhere real-valued on* E. *Let* f *be a measurable and almost everywhere real-valued function on* E. *Then* f_n *is said to* converge in measure *to* f *iff*

$$\lim_{n\to\infty} m\{x \in E: |f_n(x) - f(x)| \geq \varepsilon\} = 0$$

for all positive numbers ε.

It is an easy matter to prove that the sequence $\langle f_n \rangle$ converges to f in measure if and only if for every $\varepsilon > 0$ there is an N such that for all $n \geq N$ we have

$$m\{x \in E: |f_n(x) - f(x)| \geq \varepsilon\} < \varepsilon.$$

It can be proven (see Prob. 22) that if the sequence $\langle f_n \rangle$ converges pointwise almost everywhere to f on a set of finite measure, then $\langle f_n \rangle$ converges to f in measure. However, as the following example shows, the converse of the above statement is not true.

Example

Let $I = [0,1)$. For each positive integer k define

the functions $f_1^{(k)}, f_2^{(k)}, \ldots, f_k^{(k)}$ as follows

$$f_i^{(k)}(x) = \begin{cases} 1 & \text{if } x \in \left[\frac{i-1}{k}, \frac{i}{k}\right) \\ 0 & \text{if } x \notin \left[\frac{i-1}{k}, \frac{i}{k}\right) \end{cases} \qquad (i = 1,2,\ldots,k).$$

If k varies over all positive integers, we get a sequence $\langle h_n \rangle$ of functions such that:

$$h_1 = f_1^{(1)},\ h_2 = f_1^{(2)},\ h_3 = f_2^{(2)},\ h_4 = f_1^{(3)},\ \ldots,$$

and so on.

We prove that $\langle h_n \rangle$ converges to the function zero in measure. In fact if $h_n = f_i^{(k)}$ we have $\{x: |h_n(x)| \geq \varepsilon\} = [(i-1)/k, (2/k))$ for all ε such that $0 < \varepsilon \leq 1$. Then $\lim_{n\to\infty} m\{x: |h_n(x)| \geq \varepsilon\} = 0$. But on the other hand for no x in $[0,1)$, $\lim_{n\to\infty} h_n(x) = 0$. For if $x \in [0,1)$ we can find an i for every k such that $x \in [(i-1)/k, i/k)$ so that $f_i^{(k)}(x) = 1$. This means that $\langle h_n \rangle$ does not converge pointwise to zero. The theorem is proved.

Although convergence in measure does not imply pointwise convergence, we do have the following theorem.

20. Theorem (F. Riesz) *Let $\langle f_n \rangle$ be a sequence of measurable functions which converges in measure to a function f. Then there is a subsequence $\langle f_{n_k} \rangle$ which converges to f almost everywhere.*

Proof Let $\langle \varepsilon_k \rangle$ be a decreasing sequence of positive real numbers such that $\sum \varepsilon_k < +\infty$. Since $\langle f_n \rangle$ converges in measure to f, for a given positive integer k there is an integer n_k such that for all $n \geq n_k$, $m\{x: |f_{n_k}(x) - f(x)| \geq \varepsilon_k\} < \varepsilon_k$. Set $S = \bigcap_{p=1}^{\infty}\left[\bigcup_{k=p}^{\infty}\{x: |f_{n_k}(x) - f(x)| \geq \varepsilon_k\}\right]$ and let $x \notin S$. Then there is a p such that $x \notin \bigcup_{k=p}^{\infty}\{x: |f_{n_k}(x) - f(x)| \geq \varepsilon_k\}$. This implies that for $k \geq p$ we have $|f_{n_k}(x) - f(x)| < \varepsilon_k$ which means that $f_{n_k}(x) \to f(x)$. Hence for any $x \notin S$, $\lim_{k\to\infty} f_{n_k}(x) = f(x)$. But

$$mS \leq m\left[\bigcup_{k=p}^{\infty}\{x: |f_{n_k}(x) - f(x)| \geq \varepsilon_k\}\right] < \sum_{k=p}^{\infty}\varepsilon_k$$

for all p, and since the series $\sum \varepsilon_k$ is convergent the remainder $\sum_{k=p}^{\infty}\varepsilon_k$ can be made arbitrarily small. Hence $mS = 0$. Thus $\langle f_{n_k} \rangle$ converges to f except on a set S of measure zero, which means that $f_{n_k} \to f$ pointwise a.e. The theorem is proved.

21. Proposition *Let f be a measurable function defined and real-valued a.e. on the closed and bounded interval $[a,b]$. Then there exists a sequence*

of continuous functions converging to f almost everywhere.

Proof First we prove that there exists a sequence of continuous functions $\langle f_n \rangle$ converging in measure to f. Let $\langle \varepsilon_n \rangle$ be a decreasing sequence of positive real numbers such that $\lim_{n\to\infty} \varepsilon_n = 0$. By Theorem 19, for each n there is a continuous function f_n such that $m\{x: |f(x) - f_n(x)| \geq \varepsilon_n\} < \varepsilon_n$. If $\varepsilon > 0$ is given then there is an N such that for $n \geq N$ we have $\varepsilon_n < \varepsilon$ and

$$m\{x: |f(x) - f_n(x)| \geq \varepsilon\}$$

$$\leq m\{x: |f(x) - f_n(x)| \geq \varepsilon_n\} < \varepsilon_n < \varepsilon.$$

This proves that $\langle f_n \rangle$ converges to f in measure. Now by Theorem 20 a subsequence of $\langle f_n \rangle$ converges to f a.e. The theorem is proved.

22. Theorem (N. Luzin) *Let f be a measurable function defined and almost everywhere real-valued on the closed and bounded interval $[a,b]$. Then for every $\varepsilon > 0$ there exists a continuous function g such that $m\{x: f(x) \neq g(x)\} < \varepsilon$. If $|f(x)| \leq M$ for $x \in [a,b]$ then g can be defined so that $|g(x)| \leq M$ for $x \in [a,b]$.*

Proof By Proposition 21, there is a sequence $\langle f_n \rangle$ of continuous functions such that $\lim_{n\to\infty} f_n(x) = f(x)$ a.e. on $[a,b]$. By Theorem 16, there is a measurable subset F of $[a,b]$ such that $mF < \varepsilon/2$

and such that $\langle f_n \rangle$ converges to f uniformly on $[a,b] - F$. Since the convergence is uniform, we have by Theorem 18 of Chapter 2, that the restriction of f to the set $[a,b] - F$ is continuous. Let E be a closed subset of $[a,b] - F$ such that $mE > m([a,b] - F) - (\varepsilon/2)$. Again the restriction of f to E is continuous. By Proposition 18 we get a continuous function g defined on $[a,b]$ which coincides with f on E. Hence

$$\{x: f(x) \neq g(x)\} \subset [a,b] - E$$

and

$$m\{x: f(x) \neq g(x)\} < \varepsilon.$$

Thus g is the function sought by the theorem. If $|f(x)| \leq M$ then, again by Proposition 18, we have $|g(x)| \leq M$. The theorem is proved.

3.6 THE LEBESGUE INTEGRAL OF A BOUNDED FUNCTION OVER A SET OF FINITE MEASURE

Let us first recall the definition of the definite Riemann integral of a bounded real-valued function on the bounded and closed interval $[a,b]$. For each subdivision $a = t_0 < t_1 < \ldots < t_n = b$ of $[a,b]$ define the upper and lower Darboux sums S and s, respectively, as follows:

$$S = \sum_{i=1}^{n} (t_i - t_{i-1}) M_i$$

and

$$s = \sum_{i=1}^{n} (t_i - t_{i-1})m_i,$$

where

$$M_i = \sup f(x) \; ; \; t_{i-1} < x \le t_i$$

$$m_i = \inf f(x) \; ; \; t_{i-1} < x \le t_i \; .$$

The upper and lower Riemann integrals of f are defined by

$$R\overline{\int_a^b} f(x)dx = \inf S \; ; \; R\underline{\int_a^b} f(x)dx = \sup s.$$

If $\inf S = \sup s$, then we say that f is *Riemann integrable* and the common value of the upper and lower integrals is denoted by $R\int_a^b f(x)dx$, and is called the Riemann integral of f. It is a real number. The letter R before the symbol $\int$ is used here to indicate that we are dealing with the Riemann integral.

According to the above definition of the integral the upper sum S is the integral of the step function ψ defined on $[a,b]$ with subdivision points $t_1, t_2, \ldots, t_n$, and equal to M_i in the interval $(t_{i-1}, t_i]$ $(i = 1,2,\ldots,n)$. Similarly the lower sum s is the integral of the step function

ϕ defined on the same interval with the same subdivision points, and which is equal to m_i in the interval $(t_{i-1}, t_i]$. In other words we have

$$S = \int_a^b \psi(x)dx \quad ; \quad s = \int_a^b \phi(x)dx.$$

Notice that $f(x) \leq \psi(x)$ and $\phi(x) \leq f(x)$. Hence the Riemann integral of f can be defined by

$$\int_a^b f(x)dx = \inf \int_a^b \psi(x)dx = \sup \int_a^b \phi(x)dx$$

with the infimum taken over all step functions ψ such that $\psi(x) \geq f(x)$, and the supremum over all step functions ϕ such that $\phi(x) \leq f(x)$.

The above definition of the Riemann integral suggests a definition of the Lebesgue integral along similar lines.

23. Lemma *Let f be a simple function and $f = \sum_{k=1}^{n} a_k \chi_{A_k}$, $f = \sum_{i=1}^{m} b_i \chi_{B_i}$ be two different representations of f. Assume that $a_k \neq 0$, $b_i \neq 0$, $mA_k < +\infty$, $mB_i < +\infty$ for $k = 1,2,\ldots,n$ and $i = 1, 2,\ldots,m$. Then*

$$\sum_{k=1}^{n} a_k mA_k = \sum_{i=1}^{m} b_i mB_i.$$

Proof It is left to the reader (see Prob. 26).

Definition *Let* $f = \sum_{k=1}^{n} a_k \chi_{A_k}$ *be a simple function with* $a_k \neq 0$ $(k = 1,2,\ldots,n)$. *If* f *vanishes outside a set of finite measure the integral of* f *is defined by*

$$\int f(x)dx = \sum_{k=1}^{n} a_k mA_k.$$

Also if E *is any measurable set we define* $\int_E f(x)dx = \int f(x)\cdot\chi_E(x)dx.$

Note that by Lemma 23 the above definition of the integral is independent of the representation of f. Also, the fact that f vanishes outside of a set of finite measure ensures that all mA_k are finite. We sometimes simplify by writing $\int f$ instead of $\int f(x)dx$. The integral of the zero function is defined to be zero. By the above definition, if E is a measurable set of finite measure then $\int \chi_E = mE$.

24. Theorem *Let* f *and* g *be simple functions which vanish outside a set of finite measure, and let* a *and* b *be real numbers. Then*

(a) $\int (af + bg) = a \int f + b \int g$

(b) *If* f *is nonnegative a.e., then* $\int f \geq 0$

(c) *If* $f \geq 0$ *a.e. and* $\int f = 0$, *then* $f = 0$ *a.e.*

(d) *If* $f \geq g$ *a.e., then* $\int f \geq \int g$

Proof Suppose $f = \sum_{k=1}^{n} a_k \chi_{A_k}$ and $g = \sum_{i=1}^{m} b_i \chi_{B_i}$, where the a_k and b_i are real and nonzero. Since f and g vanish outside of a set of finite measure, mA_k, mB_i are finite. We have

$$f + g = \sum_{k=1}^{n} a_k \chi_{A_k} + \sum_{i=1}^{m} b_i \chi_{B_i}$$

and by the definition of the integral

$$\int f + g = \sum_{k=1}^{n} a_k mA_k + \sum_{i=1}^{m} b_i mB_i = \int f + \int g.$$

For $a \neq 0$ we have $af = \sum_{k=1}^{n} aa_k \chi_{A_k}$ so that

$$\int af = \sum_{k=1}^{n} aa_k mA_k = a \sum_{k=1}^{n} a_k mA_k = a \int f.$$

This proves (a).

To prove (b) we write $f = \sum_{j=1}^{N} c_j \chi_{E_j}$ in its canonical form. Thus $c_1, c_2, \ldots, c_N$ are distinct nonzero real numbers and $E_1, E_2, \ldots, E_N$ disjoint nonempty measurable sets, where E_j $(j = 1,2,\ldots,N)$ is the set on which f takes the value c_j. Since $f \geq 0$ a.e. we have $mE_j = 0$ whenever $c_j < 0$. Hence for all j we have $c_j \cdot mE_j \geq 0$ which implies $\int f = \sum_{j=1}^{N} c_j mE_j \geq 0$. If in addition $\int f = 0$, then

$mE_j = 0$ for all j, which implies that $f = 0$ a.e. This proves (b) and (c). Finally to prove (d) we apply (a) and (b) to $f - g$. The theorem is proved.

The definition of the Lebesgue integral of a bounded real-valued measurable function f defined on a set E of finite measure is quite analogous to the definition of the Riemann integral. In the Riemann case we consider integrals of the form $\int_a^b \psi(x)dx$ and $\int_a^b \phi(x)dx$ with ψ and ϕ step functions such that $\psi \geq f$, $\phi \leq f$ and then, we consider the $\inf \int \psi$ and $\sup \int \phi$. In the case of the Lebesgue integral we replace ψ and ϕ by simple functions and proceed in the same way as in the Riemann case. Obviously the class of simple functions contains the class of step functions. However, in the Lebesgue case things work nicely, for the numbers $\inf \int \psi$ and $\sup \int \phi$ are always equal, provided f is taken to be measurable and bounded. More precisely we have the following theorem.

25. Theorem *If f is a real-valued bounded measurable function defined on a set E of finite measure then*

$$\inf \int_E \psi = \sup \int_E \phi$$

with the supremum taken over all simple functions ϕ such that $\phi \leq f$ and with the infimum taken over all simple functions ψ such that $\psi \geq f$.

Proof Since f is bounded, there is a positive real number M such that $|f(x)| \leq M$ for all x. Put

$b_k = -M + [(k - 1)M/n]$ with $k = 0,1,2,\ldots,2n$, and set

$$E_k = \{x\colon b_k < f(x) \leq b_{k+1}\}$$

Clearly since f is measurable the E_k's are measurable pairwise disjoint sets with $\bigcup_{k=0}^{2n} E_k = E$. Hence $\sum_{k=0}^{2n} mE_k = mE$.

Consider the simple functions ψ_n and ϕ_n defined as follows:

$$\psi_n(x) = \sum_{k=0}^{2n} b_{k+1}\chi_{E_k}$$

$$\phi_n(x) = \sum_{k=0}^{2n} b_k\chi_{E_k}.$$

We have for every x, $\phi_n(x) \leq f(x) \leq \psi_n(x)$. By part (d) of Theorem 24 we get

$$\int_E \phi_n \leq \int_E \psi_n.$$

Hence

$$\inf \int_E \psi \leq \int_E \psi_n \quad \text{and} \quad \int_E \phi_n \leq \sup \int_E \phi$$

with the infimum (respectively supremum) taken over

all simple functions greater than or equal (respectively less than or equal) to f. Moreover observe that for any arbitrary simple functions ϕ and ψ such that $\phi \leq f \leq \psi$ we have $\int_E \phi \leq \int_E \psi$. This implies $\sup \int_E \phi \leq \inf \int_E \psi$. It follows that

$$0 \leq \inf \int_E \psi - \sup \int_E \phi \leq \int_E \psi_n - \int_E \phi_n.$$

But

$$\int_E \psi_n = \sum_k b_{k+1} mE_k \quad ; \quad \int_E \phi_n = \sum_k b_k mE_k$$

and

$$\int_E \psi_n - \int_E \phi_n = \int_E (\psi_n - \phi_n)$$

$$= \sum_{k=0}^{2n} (b_{k+1} - b_k) mE_k = \frac{M}{n} \sum_{k=0}^{2n} mE_k = \frac{M}{n} mE$$

Since $mE < +\infty$ we get by letting $n \to \infty$, $\inf \int \psi = \sup \int \phi$. The theorem is proved.

Remark *It is possible to prove (see Prob. 27) that if f is a bounded real-valued function defined on a set E of finite measure and if $\inf \int \psi = \sup \int \phi$ with ψ and ϕ as above, then f is necessarily measurable.*

Definition *The Lebesgue integral of a real-valued bounded measurable function f, whose domain is a set E of finite measure is defined by*

$$\int_E f(x)dx = \sup \int_E \phi(x)dx$$

with the supremum taken over all simple functions ϕ*, such that* $\phi \leq f$*. If* E *is the closed interval* $[a,b]$ *it is customary to write* $\int_a^b f(x)dx$ *instead of* $\int_{[a,b]} f(x)dx$*. We also sometimes write* $\int_E f$ *for* $\int_E f(x)dx$*.*

Let us now compare the Riemann and the Lebesgue integrals of a bounded function f. The Riemann *lower integral* is taken to be $\underline{\int}_a^b f(x)dx$ $= \sup \int_a^b \phi(x)dx$ with ϕ running over all step functions such that $\phi \leq f$. Since the class of simple functions contains all step functions we have that

$$\text{(1)} \qquad \underline{\int}_a^b \phi(x)dx \leq \sup \int_a^b \phi(x)dx$$

where the supremum is taken over all simple functions ϕ with $\phi \leq f$.

In a similar manner we get

$$\text{(2)} \qquad \inf \int_a^b \psi(x)dx \leq \overline{\int}_a^b f(x)dx$$

where ψ runs over all simple functions ψ with $\psi \geq f$. Now if the Riemann integral exists then(1) and(2) provide the equality

$$\inf \int_a^b \psi = \sup \int_a^b \phi$$

which by the remark made at the end of Theorem 25 (see Prob. 27) implies that f is measurable, and that its Lebesgue integral exists and is equal to its Riemann integral. On the other hand, there are bounded measurable functions which are not integrable in the Riemann sense. For example, the Dirichlet function f defined on the interval $[0,1]$ such that $f(x) = 0$ for x irrational and $f(x) = 1$ for x rational is not integrable in the Riemann sense, because $\underline{\int_a^b} f = 0$ and $\overline{\int_a^b} f = 1$ so that $\underline{\int_a^b} f \neq \overline{\int_a^b} f$. However, f being a simple function it is integrable in the Lebesgue sense with $\int_a^b f = 0$.

It follows from the above discussion that the Lebesgue integral of a bounded function is indeed a generalization of the Riemann integral. Later in this chapter we shall have further opportunity to compare the two integrals and discuss various defects of the Riemann integral. These defects disappear when the integral is taken in the Lebesgue sense.

3.7 THE LEBESGUE INTEGRAL OF A NONNEGATIVE FUNCTION

Let $\phi = \sum_{k=1}^{n} a_k \chi_{A_k}$ be a simple function with

$a_k > 0$ $(k = 1,2,\ldots,n)$. Then the integral of ϕ is defined to be $\int \phi(x)dx = \sum_{k=1}^{n} a_k mA_k$. It is a member of the nonnegative extended real number system. If E is a measurable set, then we define $\int_E \phi = \int \phi\chi_E$.

Definition *Let f be a nonnegative measurable function defined on a measurable set E. We define*

$$\int_E f(x)dx = \sup \int_E \phi$$

where the supremum is taken over all simple functions ϕ, such that $0 \le \phi \le f$.

It should be noted that this definition agrees with the definition given earlier of the integral of a bounded measurable function (see Prob. 31).

The following propositions are immediate consequences of the definitions and they are left to the reader. We assume that the functions and sets occurring in them are measurable.

(a) If $0 \le f \le g$, then $\int_E f \le \int_E g$.

(b) If $A \subset B$ and $f \ge 0$, then $\int_A f \le \int_B f$.

(c) If $f \ge 0$ and c is a nonnegative real number, then $\int_E cf = c \int_E f$.

(d) If $f(x) = 0$ for all $x \in E$, then $\int_E f = 0$.

(e) If $mE = 0$, then $\int_E f = 0$.

(f) If $f \ge 0$ then $\int_E f = \int f\chi_E$.

26. Proposition *Let f be a nonnegative measurable function defined on a measurable set E. Let $\langle E_n \rangle$ be a sequence of pairwise disjoint measurable sets such that $\bigcup_{n=1}^{\infty} E_n = E$. Then*

$$\int_E f = \sum_{k=1}^{\infty} \int_{E_n} f.$$

Proof First we shall prove the theorem if f is the characteristic function of some measurable set A. Next we shall assume f to be a simple function and then we shall proceed to the general case. If $f = \chi_A$ we have, because of the countable additivity of the Lebesgue measure

$$\int_E f = m(E \cap A) = \sum_{n=1}^{\infty} m(E_n \cap A) = \sum_{n=1}^{\infty} \int_E f.$$

If f is simple, then it is of the form $\sum_k c_k \chi_{A_k}$ and the conclusion still holds. In the general case, for every simple function ϕ such that $0 \le \phi \le f$, we have

$$\int_E \phi = \sum_{n=1}^{\infty} \int_{E_n} \phi \le \sum_{k=1}^{\infty} \int_{E_n} f.$$

Therefore by the definition of the integral of a nonnegative function we get

$$\int_E f \le \sum_{k=1}^{\infty} \int_{E_n} f.$$

Now if for some n, $\int_{E_n} f = +\infty$, then the theorem follows since $\int_{E_n} f \le \int_E f$. So suppose $\int_{E_n} f < +\infty$ for all n. Then, again by the definition of the integral, given $\varepsilon > 0$ there is a simple function ϕ, such that $0 \le \phi \le f$ and such that

$$\int_{E_1} \phi \ge \int_{E_1} f - \varepsilon \text{ , } \int_{E_2} \phi \ge \int_{E_2} f - \varepsilon.$$

Hence

$$\int_{E_1 \cup E_2} f \ge \int_{E_1 \cup E_2} \phi = \int_{E_1} \phi + \int_{E_2} \phi \ge$$

$$\int_{E_1} f + \int_{E_2} f - 2\varepsilon.$$

Since $\varepsilon > 0$ was arbitrary we get

$$\int_{E_1} f + \int_{E_2} f \le \int_{E_1 \cup E_2} f.$$

This inequality can be extended so that

$$\int_{E_1} f + \int_{E_2} f + \dots + \int_{E_n} \le \int_{\bigcup_{i=1}^{n} E_i} f$$

for every positive integer n.

Since $\bigcup_{i=1}^{n} E_i \subset E$ it follows that

$$\int_{E_1} f + \int_{E_2} f + \ldots + \int_{E_n} f \leq \int_E f.$$

Letting $n \to \infty$ we get $\sum_{n=1}^{\infty} \int_{E_n} f \leq \int_E f$ which, together with the inequality $\sum_{n=1}^{\infty} \int_{E_n} f \geq \int_E f$, proves the theorem.

27. Corollary *Let $f \geq 0$ be a measurable function defined on a measurable set E. Then*

(a) If $E_1 \subset E$ with $mE_1 = 0$, then $\int_E f = \int_{E-E_1} f$.

(b) If $\int_E f = 0$, then $f = 0$ a.e.

Proof (a) $\int_E f = \int_{E_1 \cup (E-E_1)} f = \int_{E_1} f + \int_{E-E_1} f = \int_{E-E_1} f.$

Proof (b) Put $S = \{x \in E\colon f(x) > 0\}$. We want to prove $mS = 0$. Assume on the contrary that $mS > 0$. Consider, for each positive integer n the set $S_n = \{x \in S\colon f(x) > 1/n\}$. Clearly S_n is measurable and $S = \bigcup_{n=1}^{\infty} S_n$. For each n we have $(1/n)\cdot mS_n < \int_{S_n} f \leq \int_E f = 0$ or $mS_n = 0$. But this implies that $mS = 0$, which contradicts our assumption $mS > 0$. Hence $f = 0$ a.e.

Part *(a)* of the above corollary tells us that the removal of a subset of measure zero from the domain E of the function f does not affect the value of the integral $\int_E f$.

28. Theorem (Lebesgue Monotone Convergence Theorem) *Let* $\langle f_n \rangle$ *be a sequence of measurable functions defined on a measurable set* E, *and such that* $0 \le f_1(x) \le f_2(x) \le \dots \le f_n(x) \le \dots \quad (x \in E)$. *Then*

$$\lim_{n\to\infty} \int_E f_n = \int_E (\lim_{n\to\infty} f_n).$$

Proof For each $x \in E$ the sequence $\langle f_n(x) \rangle$ is nondecreasing so that the $\lim_{n\to\infty} f_n(x)$ exists. Put $f(x) = \lim_{n\to\infty} f_n(x)$. Then by Corollary 14 f is measurable. Also by the definition of the integral of a nonnegative function, the sequence $\langle \int_E f_n \rangle$ converges as $n \to \infty$, to some member of the extended real number system, say a. Since for every n we have $\int_E f_n \le \int_E f$ it follows that $a \le \int_E f$. Let λ be such that $0 < \lambda < 1$, and consider a simple function ϕ such that $0 \le \phi \le f$. For each positive integer n set $E_n = \{x\colon f_n(x) \ge \lambda\phi(x)\}$. Since the sequence $\langle f_n \rangle$ is nondecreasing we have for each n $E_n \subset E_{n+1}$. We prove that $E = \bigcup_{n=1}^{\infty} E_n$. We only need to prove $E \subset \bigcup E_n$. Let $x \in E$. Then if $f(x) = 0$ we also have $\phi(x) = 0$ so that $x \in E_n$ for all n, which implies $x \in \bigcup E_n$. If $f(x) > 0$, then $\lambda\phi(x) < f(x)$ since $0 < \lambda < 1$. Thus for some sufficiently large n_0, since $f_n(x) \to f(x)$, we have $f_{n_0}(x) > \lambda\phi(x)$. This proves that $x \in \bigcup_{n=1}^{\infty} E_n$. By Proposition 16 of Chapter 1 there is a sequence $\langle A_n \rangle$ of pairwise disjoint measurable sets such that for each n, $E_n = \bigcup_{i=1}^{n} E_i = \bigcup_{i=1}^{n} A_i$ and $E = \bigcup_{n=1}^{\infty} E_n = \bigcup_{n=1}^{\infty} A_n$. For any n we have

$$\int_E f_n \geq \int_{E_n} f_n \geq \lambda \int_{E_n} \phi = \lambda \int_{\bigcup_{i=1}^{n} E_i} \phi .$$

By Proposition 26, if we let $n \to \infty$, we get

$$a \geq \lambda \int_E \phi$$

and by letting $\lambda \to 1$ we have

$$a \geq \int_E \phi.$$

By the definition of $\int_E f$ we get $a \geq \int_E f$. This inequality combined with the previously obtained $\int_E f \leq a$, gives $\int_E f = a = \lim_{n\to\infty} \int_E f_n$. The theorem is proved.

29. Proposition *If* $\langle f_n \rangle$ *is a sequence of nonnegative measurable functions defined on a measurable set* E, *then*

$$\int_E \left(\sum_{n=1}^{\infty} f_n \right) = \sum_{n=1}^{\infty} \int_E f_n .$$

Proof We first prove that

$$\int_E f_1 + f_2 = \int_E f_1 + \int_E f_2 .$$

By Theorem 15 there are nondecreasing sequences $\langle \phi_n \rangle$ and $\langle \psi_n \rangle$ of nonnegative simple functions such that $\phi_n \to f_1$, $\psi_n \to f_2$. We have

$$\int_E \phi_n + \psi_n = \int_E \phi_n + \int_E \psi_n .$$

By the monotone convergence theorem we get, by letting $n \to \infty$,

$$\int_E f_1 + f_2 = \int_E f_1 + \int_E f_2 .$$

It follows that for any positive integer n we have

$$\int_E f_1 + f_2 + \dots + f_n$$

$$= \int_E f_1 + \int_E f_2 + \dots + \int_E f_n .$$

Next put $S_n = \sum_{i=1}^{n} f_i$. The proposition now follows if we apply Theorem 28 to the sequence $\langle S_n \rangle$.

30. Theorem (Fatou's Lemma) *Let* $\langle f_n \rangle$ *be a sequence of nonnegative measurable functions defined on a measurable set* E. *Then*

$$\int_E \left(\underline{\lim}_{n\to\infty} f_n\right) \leq \underline{\lim}_{n\to\infty} \int_E f_n \, .$$

Proof For each positive integer n define g_n, by $g_n(x) = \inf_{k \geq n} f_k(x)$. By Theorem 13 g_n is measurable. The sequence $\langle g_n \rangle$ is nondecreasing and clearly $g_n \leq f_n$. Recalling the definition of the limit inferior of a sequence, we see that $g_n \to \underline{\lim}_{n\to\infty} f_n(x)$. By applying Theorem 28 to the sequence $\langle g_n \rangle$ we have

$$\int_E g_n \to \int_E \left(\lim_{n\to\infty} g_n\right) = \int_E \underline{\lim}_{n\to\infty} f_n \, .$$

Since $g_n \leq f_n$, it follows from the definition of the integral of a nonnegative function that $\int_E g_n \leq \int_E f_n$, so that by letting $n \to \infty$, and taking the limit inferior of both members in the last inequality we get

$$\int_E \underline{\lim}_{n\to\infty} f_n \leq \underline{\lim}_{n\to\infty} \int_E f_n \, .$$

The theorem is proved.

Definition *A nonnegative measurable function* f *is said to be* integrable *over a measurable set* E *iff its integral over* E *is finite, i.e.,* $\int_E f < +\infty$.

(Note: Observe that every nonnegative measurable function has an integral. But the term *integrable* is reserved for those functions with finite integral.*)*

31. Proposition *If* f *is a nonnegative integrable function over a measurable set* S, *then* $m\{x: f(x) = +\infty\} = 0$.

Proof Put $A = \{x: f(x) = +\infty\}$. Since f is measurable, A is also measurable. Assume $mA > 0$. Then the definition of the integral of f implies $\int_A f = +\infty$. On the other hand by Proposition 26, we have, since $A \cup (S - A) = S$ and $A \cap (S - A) = \emptyset$

$$\int_S f = \int_A f + \int_{S-A} f = +\infty$$

which contradicts the assumption that f is integrable. The theorem is proved.

32. Proposition *Let* f *be a nonnegative measurable function integrable over a measurable set* E. *Then given* $\varepsilon > 0$ *there exists a* $\delta > 0$ *such that for every measurable set* $S \subset E$ *with* $mS < \delta$ *we have* $\int_S f < \varepsilon$.

Proof Consider the sequence $\langle f_n \rangle$ of functions defined as follows

$$f_n(x) = \begin{cases} f(x) & \text{if} \quad f(x) \leq n \\ n & \text{if} \quad f(x) > n \, . \end{cases}$$

Then each f_n is a nonnegative bounded measurable function, and for each $x \in E$ we have

$$\lim_{n\to\infty} f_n(x) = f(x) .$$

The monotone convergence theorem applied to $\langle f_n \rangle$ gives

$$\lim_{n\to\infty} \int_E f_n = \int_E f .$$

Hence given $\varepsilon > 0$ there is a positive integer N such that

$$\int_E f - \frac{\varepsilon}{2} < \int_E f_n .$$

Now if we choose $\delta < (\varepsilon/2N)$ we have for every measurable set S with $mS < \delta$

$$\int_S f = \int_S (f - f_N) + \int_S f_N$$

$$< \int_E (f - f_N) + NmS < \frac{\varepsilon}{2} + \frac{\varepsilon}{2} = \varepsilon .$$

The proposition is proved.

3.8 THE GENERAL LEBESGUE INTEGRAL

With each real-valued function f we associate the functions f^+ and f^- defined as follows

$$f^+(x) = \max\{f(x),0\} ,$$

$$f^-(x) = \max\{-f(x), 0\} .$$

If f is measurable then it is easily seen that f^+ and f^- are also measurable.

We also have

$$f = f^+ - f^- \quad \text{and} \quad |f| = f^+ + f^- .$$

Hence if f is measurable $|f|$ is also measurable.

Definition *A measurable function* f *is said to be* integrable *over the measurable set* E, *iff both* f^+ *and* f^- *are integrable over* E. *The integral of an integrable function* f *is defined by*

$$\int_E f = \int_E f^+ - \int_E f^- .$$

Remark *The reader can easily verify that if a function* f *is measurable and bounded on a measurable set* E *of finite measure then* f *is integrable and its integral equals the integral of* f *taken in the sense of section 3.6. Thus, it seems that we could define the integral of a nonnegative function first, and then define the integral of a bounded measurable function as the integral of an integrable function. As a matter of fact several authors do so. However, in spite of this duplication we find the approach followed here pedagogically more desirable.*

33. Theorem *Let* f *and* g *be integrable functions on the real line and let* c *be a real number. Then*

(a) $f + g$ *and* cf *are both integrable*

(*b*) $\int f + g = \int f + \int g$.

(*c*) $\int cf = c\int f$.

Proof (*a*) By Theorem 12, $f + g$ and cf are measurable. By the properties of the integral of a nonnegative function we have

$$\int |f + g| \leq \int (|f| + |g|) = \int |f| + \int |g| < +\infty$$

$$\int |cf| = |c|\int |f| < +\infty \ .$$

Proof (*b*) Put $h = f + g$. Then

$$h = h^+ - h^- = f^+ - f^- + g^+ - g^-$$

or

$$h^+ + f^- + g^- = f^+ + g^+ + h^- \ .$$

By Proposition 29 we get

$$\int h^+ + \int f^- + \int g^- = \int f^+ + \int g^+ + \int h^- \ .$$

Since all these integrals are finite we have

$$\int f + g = \int h = \int h^+ - \int h^-$$

$$= \int f^+ - \int f^- + \int g^+ - \int g^- = \int f + \int g \ .$$

Proof (*c*) If $c \geq 0$, then

$$\int cf = \int (cf)^+ - \int (cf)^-$$

$$= \int cf^+ - \int cf^- = c\int f^+ - c\int f^- = c\int f \ .$$

Next assume $c \leq 0$. We first observe that $(-f)^+ = f^-$ and $(-f)^- = f^+$. We have

$$\int -f = \int (-f)^+ - \int (-f)^-$$

$$= \int f^- - \int f^+ = -(\int f^+ - \int f^-) = -\int f .$$

Since $-c \geq 0$ we also have

$$\int cf = \int (-c)(-f) = (-c)\int (-f)$$

$$= (-c)(-\int f) = c\int f .$$

34. Lebesgue Dominated Convergence Theorem. *Let* $\langle f_n \rangle$ *be a sequence of measurable functions such that*

$$\lim_{n\to\infty} f_n(x) = f(x) \quad a.e.$$

Suppose there exists an integrable function g *such that* $|f_n(x)| \leq g(x)$ $(n = 1,2,\ldots,)$ *for all* x. *Then* f *is integrable and*

$$\lim_{n\to\infty} \int |f_n - f| = 0 \quad ; \quad \lim_{n\to\infty} \int f_n = \int f .$$

Proof Since $|f| \leq g$ and f is measurable as the limit of a sequence of measurable functions, we have that f is integrable. We also have

$$|f_n - f| \leq 2g .$$

By applying Fatou's lemma to the sequence

of nonnegative functions $\langle 2g - |f_n - f| \rangle$ we get

$$\int 2g \leq \underline{\lim}_{n\to\infty} \int (2g - |f_n - f|)$$

$$= \int 2g + \underline{\lim}_{n\to\infty} (-\int |f_n - f|)$$

$$= \int 2g - \overline{\lim}_{n\to\infty} \int |f_n - f| .$$

The function $2g$ being integrable the $\int 2g$ is finite so that

$$\overline{\lim}_{n\to\infty} \int |f_n - f| \leq 0$$

which implies

$$\lim_{n\to\infty} \int |f_n - f| = 0 .$$

Finally since

$$|\int f_n - \int f| \leq \int |f_n - f|$$

we get by letting $n \to \infty$

$$\lim_{n\to\infty} \int f_n = \int f .$$

35. Bounded Convergence Theorem *Let* $\langle f_n \rangle$ *be a sequence of measurable functions defined on a measurable set* E *of finite measure. Suppose there exists a real number* M *such that* $|f_n(x)| \leq M$ *for all* n *and all* $x \in E$. *Suppose also that*

$$\lim_{n\to\infty} f_n(x) = f(x)$$

for each $x \in E$. *Then*

$$\lim_{n\to\infty} \int_E f_n = \int_E f \,.$$

Proof Apply Theorem 34 for $g = M\cdot\chi_E$.

Before we close this section we prove a theorem, due to Lebesgue, which provides a necessary and sufficient condition for a bounded real-valued function defined on a bounded interval to be Riemann integrable.

36. Theorem *Let* $f\colon [a,b] \to \mathbb{R}$ *be a bounded function and let* S *be the set of points at which* f *is discontinuous. Then* f *is Riemann integrable if and only if* $mS = 0$.

Proof Let $P = \{x_0, x_1, x_2, \ldots, x_n\}$ be a partition of $[a,b]$ with $a = x_0 < x_1 < \ldots < x_n = b$. Put $\|P\| = \max|x_i - x_{i-1}|$ for $i = 1,2,\ldots,n$.

$$\left.\begin{aligned} m_i &= \inf f(x) \\ M_i &= \sup f(x) \end{aligned}\right\} \text{ for } x_{i-1} \le x \le x_i,\ i = 1,2,\ldots,n$$

$$E_i = \{x\colon x_{i-1} \le x < x_i\} \quad \text{if} \quad i = 1,2,\ldots,n-1$$

$$E_n = \{x\colon x_{n-1} \le x \le x_n\} \,.$$

With the partition P we associate the functions

$$g_P = \sum_i m_i\chi_{E_i} \,, \quad G_P = \sum_i M_i\chi_{E_i}$$

where χ_{E_i} is the characteristic function of the set E_i. Clearly g_P and G_P are bounded measurable functions and we have the Lebesgue integrals

$$\int g_P = \sum_i m_i mE_i \; ; \; \int G_P = \sum_i M_i mE_i \; .$$

Let $\langle P_k \rangle$ be a sequence of partitions of $[a,b]$ such that $\|P_k\| \to 0$ as $k \to \infty$, and such that P_{k+1} is a refinement of P_k (i.e., all points of P_k belong to P_{k+1}). Put

$$f_k = g_{P_k} \quad , \quad F_k = G_{P_k} \; .$$

Clearly we have

$$f_k \le f_{k+1} \quad , \quad F_{k+1} \le F_k$$

and

$$-\sup_{x\varepsilon[a,b]} |f(x)| \le f_k(x) \le f(x)$$

$$\le F_k(x) \le \sup_{x\varepsilon[a,b]} |f(x)| \; .$$

Since the sequences $\langle f_k \rangle$ and $\langle F_k \rangle$ are monotonic we have

$$\lim_{k\to\infty} f_k(x) = h(x) \; ; \; \lim_{k\to\infty} F_k(x) = H(x)$$

where both h and H are bounded measurable functions. We have by Theorem 34

$$\lim_{k\to\infty} \int f_k = \int h \quad ; \quad \lim_{k\to\infty} \int F_k = \int H .$$

We now prove that

$$(*) \qquad S \subset \{x: H(x) - h(x) > 0\} \cup S_o$$

where S_o is a certain countable set. Let $x \varepsilon S$. Then since f is discontinuous at x, there is an $\varepsilon > 0$ such that for every open interval I containing x we have

$$\sup_{y\varepsilon I\cap[a,b]} f(y) - \inf_{y\varepsilon I\cap[a,b]} f(y) \geq 0 .$$

If x is not one of the points $x_1, x_2, \ldots, x_{n-1}$ in the partition P and if $x \varepsilon E_i$, then there is an open interval I containing x such that $I \cap [a,b] \subset E_i$. This implies that

$$g_P(x) = m_i \leq \inf_{y\varepsilon I\cap[a,b]} f(y) ;$$

$$\sup_{y\varepsilon I\cap[a,b]} f(y) \leq M_i = G_P(x) .$$

Denote by S_o the set of all points in the partitions $P_1, P_2, \ldots,$. Clearly S_o is countable. Then

$$x \varepsilon S - S_o \text{ implies } F_n(x) - f_n(x) \geq \varepsilon$$

so that $H(x) - h(x) \geq \varepsilon$. This proves (*). To proceed with the proof, suppose $mS = 0$. Let $\langle P_k \rangle$ be, as before, any sequence of partitions of $[a,b]$ such that $\|P_k\| \to 0$, and P_{k+1} a refinement of P_k. Let x be such that $H(x) - h(x) > 0$. Then there is an $\varepsilon > 0$ such that $H(x) - h(x) > 2\varepsilon$. We claim that f is discontinuous at x, i.e., $x \varepsilon S$. Suppose on the contrary that f is continuous at x. Then there must exist a $\delta > 0$ such that for $y \varepsilon [a,b]$ and $|y - x| < \delta$ we have $|f(y) - f(x)| < \varepsilon$. Consider an integer k such that $\|P_k\| \leq \delta$. Then x must belong to some of the closed intervals determined by P_k so that

$$f(x) - \varepsilon \leq f_k(x) \quad ; \quad F_k(x) \leq f(x) + \varepsilon .$$

This implies

$$f(x) - \varepsilon \leq h(x) \quad , \quad H(x) \leq f(x) + \varepsilon$$

or

$$H(x) - f(x) \leq 2\varepsilon$$

which contradicts the inequality $2\varepsilon < H(x) - h(x)$. Finally assume f Riemann integrable. Then $R\int_a^b h = \int_a^b h = \int_a^b H$. Since $H - h \geq 0$ it follows that $\int_a^b (H - h) = 0$, which implies that $H(x) = h(x)$ a.e. Hence the set $\{x: H(x) - h(x) > 0\}$ has measure zero. It follows that $mS = 0$ since $S \subset \{x: H(x) - h(x) > 0\} \cup S_o$ and S_o is countable.

3.9 FURTHER COMPARISON BETWEEN RIEMANN AND LEBESGUE INTEGRALS

We have mentioned in section 3.6 that the Riemann integral has certain limitations which make the Lebesgue integral desirable as a more general concept. One of the defects of the Riemann integral is related to convergent sequences of functions. Let $\langle f_n \rangle$ be a uniformly bounded sequence of real-valued functions defined on $[a,b]$, which converge uniformly to a function f, and such that all f_n are Riemann integrable. Then it is known from the theory of Riemann integral that f is Riemann integrable and

$$R \int_a^b f(x)dx = \lim_{n\to\infty} (R \int_a^b f_n(x)dx) .$$

The defect appears if the uniform convergence of the f_n is replaced by the ordinary pointwise convergence. For in the latter case it may happen, that the limit function f is not Riemann integrable. To see this, consider the following example.

Example

Let $\langle r_n \rangle$ be the rational numbers in the closed interval $[a,b]$. For any positive integer n set

$$f_n(x) = 1 \quad \text{if} \quad x \in \{r_1, r_2, \ldots, r_n\}$$

$$f_n(x) = 0 \quad \text{if} \quad x \notin \{r_1, r_2, \ldots, r_n\} .$$

Then the sequence $\langle f_n \rangle$ converges to the function f, where

$$f(x) = 1 \quad \text{if} \quad x \text{ is rational}$$

$$f(x) = 0 \quad \text{if} \quad x \text{ is irrational} .$$

For each n, f_n is Riemann integrable by Theorem 36, since the set of discontinuities is finite and so of measure zero. However, f is not Riemann integrable since the set of discontinuities is the interval $[a,b]$ and has positive measure.

Another defect is in reference to the recovery of a function by integrating its derivative. If f is a real-valued function on $[a,b]$ such that $f'(x)$ exists for every x and is bounded on $[a,b]$, then it is not always true that for every $x \in [a,b]$

$$f(x) - f(a) = \int_a^x f'(t)dt$$

if the integral is taken in the Riemann sense. To see this one can exhibit an example of a function f such that the set of discontinuities of f' is of positive measure, so that f' is not Riemann integrable. For such an example the reader is referred to Gelbaum and Olmsted ([10] p. 107). On the other hand, as we shall see in Chapter 4, the above equality holds provided the integral is taken in the Lebesgue sense.

Before we close this chapter, we wish to give the definition of the integral of a complex-valued function f defined on a set E of real numbers. For each x in E we have $f(x) = u(x) + iv(x)$ where u and v are both real-valued functions defined on the set E. We write $f = u + iv$ and call u and v respectively the *real* and *imaginary* part of f.

Definition *Let $f = u + iv$ be a complex valued function defined on a measurable subset E of the real line. Then we say that*

(i) f is measurable iff both u and v are measurable.

(ii) f is integrable iff u and v both are integrable, and when they are, then the integral $\int_E f$ is defined by

$$\int_E f = \int_E u + i \int_E v .$$

The above definition will be used in Chapter 10.

PROBLEMS

1. Show that if O_1 and O_2 are open sets, then $|O_1 \cup O_2| \leq |O_1| + |O_2|$. If in addition $O_1 \cap O_2 = \emptyset$, then show that $|O_1 \cup O_2| = |O_1| + |O_2|$ (see Remark in sec. 3.2, p. 109).
2. Show that if O is an open set, then $m^*O = |O|$.
3. Show that if $A \subset B$, then $m^*A \leq m^*B$.
4. Show that for every set A and every real number x, $m^*(A + x) = m^*A$; i.e., m^* is translation invariant.

5. (a) Prove that for any given set A and any $\varepsilon > 0$ there is an open set O containing A such that $m^*O \leq m^*A + \varepsilon$.

   (b) Show that for any given set A there is a set G of type G_δ such that $A \subset G$ and $m^*A = m^*G$.

6. The symmetric difference of the sets A, B, denoted by $A \Delta B$ is defined by $A \Delta B = (A \cup B) - (A \cap B)$. Prove that if $m^*A < +\infty$ and $m^*B < +\infty$, then

$$|m^*A - m^*B| \leq m^*(A \Delta B).$$

7. Prove that a set E is measurable if and only if there is a set G of type G_δ such that $E \subset G$ and $m^*(G - E) = 0$ (see Prop. 8).

8. Prove Proposition 7.

9. Prove that a set E of finite outer measure is measurable if and only if for every $\varepsilon > 0$ there is a finite number of open intervals $I_1, I_2, \ldots, I_n$, such that, if $U = I_1 \cup I_2 \cup \ldots \cup I_n$ we have $m^*(U \Delta E) < \varepsilon$.

10. Prove Proposition 9.

11. Prove that if E is a measurable set, then $mE = \sup\{mF\colon F \subset E,\ F = \text{bounded and closed}\}$. (Hint: Consider the cases $mE = +\infty$, $mE < +\infty$ separately.)

12. Show that the Cantor set has measure zero.

13. Show that there exists a closed nowhere-dense subset E of $[0,1]$ such that $mE > 0$. (Hint: Construct E in the same manner as the Cantor ternary set except that each of the

intervals removed at the nth step has length $\alpha 3^{-n}$ with $0 < \alpha < 1$.)

14. Show that any measurable set E of positive measure (hence any set of positive outer measure) contains a nonmeasurable set.
(Hint: If $E \subset (0,1)$ let $A_i = E \cap K_{r_n}$ as in section 3.3. If for some i, A_i is not measurable then we are done. If all A_i are measurable then $mA_i = 0$ while $\sum m^*A_i \geq m^*E > 0$.)
15. Prove that there exist measurable sets which are not Borel sets.
(Hint: Fill in the details in the proof given in the Note following Proposition 4.)
16. Show that the converse of Corollary 11 is not true in general.
17. Let E be a set. Prove that the characteristic function χ_E is measurable if and only if E is measurable.
18. Let f be an extended real-valued function on $\mathbb{R}$, such that the set $\{x: f(x) > r\}$ is measurable for each rational number r. Show that f is measurable.
19. Let $f: [0,1] \to \mathbb{R}$ such that its derivative f' exists and is finite at every point x in $[0,1]$. Prove that f' is measurable.
(Hint: Define $f(x) = f(b)$ for $x > b$. Then f' can be viewed as the limit of a certain sequence of measurable functions, see Corollary 14.)
20. Let $f: [0,1] \to \mathbb{R}$ be measurable. Show that if

O is any open set in $\mathbb{R}$, then $f^{-1}[O]$ is a measurable subset of $[0,1]$.

21. (a) Let f be a measurable function. Prove that the collection of sets S, such that $f^{-1}[S]$ is measurable, form a σ-algebra of sets.
(b) Prove that if f is a measurable function and B is a Borel set, then $f^{-1}[B]$ is a measurable set.

22. Let $\langle f_n \rangle$ be a sequence of measurable functions on a measurable set E of finite measure such that all f_n are finite almost everywhere. Suppose that $\lim_{n\to\infty} f_n(x) = f(x)$ almost everywhere on E and that $f(x)$ is finite almost everywhere. Show that $\langle f_n \rangle$ converges in measure to f. Show that the above statement is not true if $mE = +\infty$.
(Hint: Let

$$E_n(\varepsilon) = \{x \in E: |f_n(x) - f(x)| \geq \varepsilon\},$$

$$A_n(\varepsilon) = \bigcup_{k=n}^{\infty} E_k(\varepsilon), \quad S = \bigcap_{n=1}^{\infty} A_n(\varepsilon).$$

Show that $\lim_{n\to\infty} A_n(\varepsilon) = mS = 0$. The theorem follows since $E_n(\varepsilon) \subset A_n(\varepsilon)$ for each n.)

23. Let $\langle f_n \rangle$ and $\langle g_n \rangle$ be sequences of measurable functions on $[0,1]$. Suppose that f_n and g_n converge in measure to f and g respectively. Prove that $\langle f_n + g_n \rangle$ converges in measure to $f + g$.

24. Let $\langle f_n \rangle$ be a sequence of measurable functions and finite in value a.e. Let f be a function of the same kind. Suppose that for each $\varepsilon > 0$ there is a measurable set E with $mE < \varepsilon$ such that $\langle f_n(x) \rangle$ converges to $f(x)$ uniformly on $\mathbb{R} - E$. Prove that f_n converges to f in measure.

25. Give an example of a sequence of measurable functions which converges on $[0,1]$ but does not converge uniformly on any subset of measure 1.

26. Prove Lemma 23.

27. Prove the statement made in the Remark following the proof of Theorem 25. (Hint: For each positive integer n there are simple functions ϕ_n and ψ_n such that $\phi_n(x) \leq f(x) \leq \psi_n(x)$ and

$$\int \psi_n - \int \phi_n < \frac{1}{n} .$$

Set $S = \{x: \sup_n \phi_n(x) < \inf_n \psi_n(x)\}$ and show that $mS = 0$.)

28. Let f be a bounded measurable function defined on a measurable set E of finite measure.

(a) Show that if $A \leq f(x) \leq B$ $(x \in E)$, then $A \cdot mE \leq \int_E f \leq B \cdot mE$.

(b) Show that if $f(x) = c$, $(x \in E)$, then

$$\int_E f = c \cdot mE .$$

(c) Show that if $mE = 0$, then

$$\int_E f = 0 .$$

29. (a) Let f and E be as in Problem 3.28. Suppose that $E = \bigcup_{n=1}^{\infty} E_n$ where $\langle E_n \rangle$ is a sequence of pairwise disjoint measurable sets. Show that

$$\int_E f = \sum_{n=1}^{\infty} \int_{E_n} f .$$

The property of the integral expressed by the last equality is called *countable additivity*. (Hint: First prove the theorem in the case where the sequence $\langle E_n \rangle$ is finite. Set $S_n = \bigcup_{k=n+1}^{\infty} E_k$. By (a) of Problem 3.28 we have if $A \le f(x) \le B$,

$$A \cdot mS_n \le \int_{S_n} f \le mS_n .$$

Then $mS_n \to 0$ and $\int_{S_n} f \to 0$.)

(b) If g is bounded and measurable on E and if $m\{x \in E: f(x) \neq g(x)\} = 0$, then

$$\int_E f = \int_E g .$$

30. Prove Propositions (a) through (f) stated just before Proposition 26.

31. Show that the definition of the integral of a nonnegative measurable function f defined on a measurable set E, is equivalent to the definition of the integral given in section 3.6, if f is in addition bounded and if $mE < +\infty$.

32. Let $\langle f_n \rangle$ be a sequence of nonnegative measurable functions on the real line such that $f_n \to f$ a.e., and such that

$$\int f_n \to \int f < +\infty \; .$$

Prove that for each measurable set E we have

$$\lim_{n\to\infty} \int_E f_n = \int_E f \; .$$

33. Let $\langle f_n \rangle$ be a sequence of nonnegative measurable functions defined on the whole real line $\mathbb{R}$. Suppose that

$$\lim_{n\to\infty} f_n(x) = f(x) \qquad (x \in \mathbb{R})$$

and

$$f_n \le f \quad \text{for each } n.$$

Show that

$$\int f = \lim_{n\to\infty} \int f_n \; .$$

(Hint: Use Fatou's lemma.)

34. Construct a sequence of continuous functions f_n on $[0,1]$ such that $0 \leq f_n \leq 1$, such that

$$\lim_{n\to\infty} \int_0^1 f_n = 0$$

but such that the sequence $\langle f_n(x) \rangle$ converges for no $x \in [0,1]$.

35. Let $\mathcal{E}$ be the class of all extended real-valued measurable functions defined on $[0,1]$ and finite almost everywhere. If f and g are in $\mathcal{E}$ define

$$d(f,g) = \int_0^1 \frac{|f(x)-g(x)|}{1+|f(x)-g(x)|}\,dx .$$

(a) Show that $d(f,g) = 0$ iff $f \sim g$, where $f \sim g$ means $f(x) = g(x)$ a.e.
Show that $\sim$ is an equivalence relation on $\mathcal{E}$.
(b) Let $\mathcal{C}$ be the set of equivalence classes in $\mathcal{E}$ determined by $\sim$. If $[f]$ is the equivalence class containing f, define $D([f_1],[f_2]) = d(f_1,f_2)$. Show that if $\langle f_n \rangle$ is a sequence in $\mathcal{E}$, and $f \in \mathcal{E}$ then

$$\lim_{n\to\infty} D([f_n],[f]) = 0$$

if and only if f_n converges to f in measure.

(Hint: If $\varepsilon > 0$ and $A_n(\varepsilon) = \{x: |f_n(x) - f(x)| \geq \varepsilon\}$ show that $d(f_n,f) \leq m(A_n(\varepsilon)) + \varepsilon$ and

$$\frac{\varepsilon}{1+\varepsilon}\, m\,(A_n(\varepsilon)) \leq d(f_n,f)\ .)$$

36. Let $S_1, S_2, \ldots, S_n$ be measurable subsets of $[0,1]$, such that each point of $[0,1]$ belongs to at least three of these sets. Show that at least one of the sets $S_1, S_2, \ldots, S_n$ has measure greater than or equal to $3/n$.

37. Let f be a measurable function defined on a measurable set E. Show that f is integrable over E iff $|f|$ is integrable over E. Show also that if f is integrable, then

$$\left|\int_E f\right| \leq \int_E |f|\ .$$

38. Let f be a function defined on $[0,+\infty)$ and Riemann integrable on $[0,a]$ for every $a \geq 0$. The improper Riemann integral $R\int_0^\infty f$ is defined by

$$R\int_0^\infty f = \lim_{a\to\infty} \int_0^a f$$

whenever the limit exists.
Give an example of a measurable function defined on $[0,+\infty)$ whose improper Riemann integral exists and which is not integrable on $[0,+\infty)$ in the Lebesgue sense.

39. Show that if $\int |f_n - f| \to 0$ and $f_n(x) \to g(x)$

a.e., then $f(x) = g(x)$ a.e.

40. Let $f(x) = 0$ for every x in the Cantor set F, and $f(x) = n$ for x in each of the intervals of length 3^{-n} in $[0,1] - F$. Prove that f is integrable on $[0,1]$, and that $\int_0^1 f = 3$.

41. If f is an integrable function on $[0,2\pi]$ show that

$$\lim_{n\to\infty} \int_0^{2\pi} f(x) \sin nx dx$$

$$= \lim_{n\to\infty} \int_0^{2\pi} f(x) \cos nx dx = 0 .$$

(Hint: First prove the statement if f is a step function.)

42. Let f be integrable on $[0,1]$. Set

$$S_n = \{x \in [0,1]: |f(x)| \geq n\}, \quad n = 0,1,2,\ldots, .$$

Show that

$$\sum_{n=0}^{\infty} mS_n < +\infty .$$

43. Let $\langle f_n \rangle$ be a sequence of functions on $[0,1]$ defined for $n = 1,2,\ldots,$ by

$$f_n(x) = 2n \quad \text{if } \frac{1}{2n} \leq x \leq \frac{1}{n}$$

$$f_n(x) = 0 \quad \text{if } x \in (0, \frac{1}{2n}) \cup (\frac{1}{n}, 1) .$$

(a) Evaluate

$$\int_0^1 (\lim_{n\to\infty} f_n(x))dx \quad ; \quad \lim_{n\to\infty} \int_0^1 f_n(x)dx$$

(b) Show that for the sequence $\langle f_n \rangle$, Fatou's lemma applies but Lebesgue's Dominated Convergence theorem does not.

44. Let f be a function which is integrable on $[0,1]$ and such that $f(x) = 0$ for $x \notin [0,1]$. Let

$$\phi(x) = \frac{1}{2h} \int_{x-h}^{x+h} f(t)dt .$$

Show that

$$\int_0^1 |\phi(x)|dx \le \int_0^1 |f(x)|dx .$$

45. Show that if $x > 0$, then the function ϕ defined on $[0,+\infty)$ by $\phi(t) = e^{-t}t^{x-1}$ is integrable on $[0,+\infty)$.
Let

$$\Gamma(x) = \int_0^\infty e^{-t}t^{x-1}dt \quad (0 < x < \infty) .$$

This last function is known as the Gamma function.

(a) Show that

$$\Gamma(x + 1) = x\Gamma(x) \quad (0 < x < \infty) .$$

(Hint: Integrate by parts over $[\varepsilon,N]$. Then let $\varepsilon \to 0+$ and $N \to +\infty$.)

(b) Show that $\Gamma(1) = 1$; $\Gamma(n + 1) = n!$ $(n = 1,2,\ldots\ .)$

CHAPTER 4

Relation Between Differentiation and Integration

4.1 INTRODUCTION

Every real-valued integrable function f on the closed interval $[a,b]$ has the associated function

$$F(x) = \int_a^x f(t)dt \quad (a \le x \le b).$$

Two formulas with which the reader is familiar from his calculus courses are the following:

$$F'(x) = f(x)$$

and for f differentiable

$$f(x) - f(a) = \int_a^x f'(t)dt.$$

We have seen in Chapter 3 section 9, that the second formula is not always meaningful if the integration is taken in the Riemann sense. In this chapter we shall discuss these two formulas for Lebesgue integration.

However, such a discussion requires the notions of bounded variation (treated in some advanced calculus courses) and absolute continuity of real-valued functions of a real variable.

4.2 FUNCTIONS OF BOUNDED VARIATION

Definitions *If* $[a,b]$ *is a bounded and closed interval then a set of points* $P = \{x_0, x_1, \ldots, x_n\}$ *satisfying the inequalities* $a = x_0 < x_1 < \ldots < x_n = b$ *is called a* partition *of* $[a,b]$.

A real-valued function f *on* $[a,b]$ *is said to be of* bounded variation *on* $[a,b]$ *iff there exists a positive number* M *such that*

$$\sum_{k=1}^{n} |f(x_k) - f(x_{k-1})| \leq M$$

for all partitions of $[a,b]$.

Clearly if f is a monotone function (i.e., nondecreasing or nonincreasing) on a bounded and closed interval $[a,b]$, then it is of bounded variation. Also if f is continuous on $[a,b]$ and if f' exists and is bounded on (a,b) (say $|f'(x)| \leq k$ for some positive k and for all $x \in (a,b)$), then f is of bounded variation on $[a,b]$. To see this, apply the mean value theorem and write

$$f(x_k) - f(x_{k-1}) = f'(t_k)(x_k - x_{k-1})$$

with $t_k \in (x_{k-1}, x_k)$.

If f is of bounded variation on $[a,b]$, then, using the special partition $P = \{a,x,b\}$ of $[a,b]$, we find

$$|f(x) - f(a)| + |f(b) - f(x)| \leq M$$

which implies

$$|f(x) - f(a)| \leq M$$

or

$$|f(x)| \leq M + |f(a)|.$$

Thus if f *is of bounded variation, then* f *is bounded.*

The reader should carefully notice that *a bounded function or a continuous function is not necessarily of bounded variation.* For example, let

$$f(x) = x \cos(\pi/2x) \quad \text{if} \quad x \neq 0, \quad f(0) = 0.$$

Then f is bounded and continuous on $[0,1]$ but it is not of bounded variation (see Prob. 2).

Let f be of bounded variation on $[a,b]$ and let S_P denote the sum $\sum_{k=1}^{n} |f(x_k) - f(x_{k-1})|$ corresponding to the partition $P = \{x_0, x_1, \ldots, x_n\}$

of [a,b]. Let $\mathcal{P}[a,b]$ be the collection of all possible partitions P of [a,b]. Then the number

$$V_f(a,b) = \sup\{S_P : P \in \mathcal{P}[a,b]\}$$

is called the *total variation* of f on the interval [a,b]. When no confusion is possible we shall write V_f instead of $V_f(a,b)$.

1. Proposition *Let f be of bounded variation on [a,b] and let a < c < b. Then f is of bounded variation on the intervals [a,c] and [c,b]. Furthermore,*

$$V_f(a,b) = V_f(a,c) + V_f(c,b).$$

Proof Let P_1 and P_2 be any arbitrary partitions of [a,c] and [c,b] respectively. Then $P_0 = P_1 \cup P_2$ is a partition of [a,b]. If S_P denotes the sum $\Sigma|f(x_k) - f(x_{k-1})|$ corresponding to the partition P (of the appropriate interval) we have

$$S_{P_1} + S_{P_2} = S_{P_0} \leq V_f(a,b).$$

This implies that the sums S_{P_1} and S_{P_2} are bounded, so that f is of bounded variation on [a,c] and [c,b], and the last inequality shows that

$$V_f(a,c) + V_f(c,b) \leq V_f(a,b).$$

We now prove that

$$V_f(a,b) \le V_f(a,c) + V_f(c,b).$$

Let $P = \{x_0, x_1, \ldots, x_n\}$ be any arbitrary partition of $[a,b]$, and let $P' = P \cup \{c\}$ be the partition obtained by adjoining a point c. If $c \varepsilon [x_{k-1}, x_k]$ then we have

$$|f(x_k) - f(x_{k-1})| \le |f(x_k) - f(c)| + |f(c) - f(x_{k-1})|$$

so that

$$S_P \le S_{P'}.$$

The points of P' which belong to $[a,c]$ determine a partition P_1 of $[a,c]$ and the points of P' in $[c,b]$ determine a partition P_2 of $[c,b]$. Hence

$$S_P \le S_{P'} = S_{P_1} + S_{P_2} \le V_f(a,c) + V_f(c,b)$$

which says that the number $V_f(a,c) + V_f(c,b)$ is an upper bound for every sum S_P. This implies that

$$V_f(a,b) \le V_f(a,c) + V_f(c,b).$$

(*Note:* It follows from Proposition 1 that if f is of bounded variation on $[a,b]$, then f is of bounded variation on every subinterval [a',b'] of [a,b].)

2. Proposition *If f and g are both of bounded variation on [a,b], then so are fg, f+g, f-g. Also we have*

$$V_{f\pm g} \leq V_f + V_g \quad \text{and} \quad V_{fg} \leq AV_f + BV_g$$

where

$$A = \sup_{x\varepsilon[a,b]}\{|g(x)|\}; \; B = \sup_{x\varepsilon[a,b]}\{|f(x)|\}.$$

The proof is simple and is left to the reader.

3. Proposition *Let f be a function of bounded variation on [a,b]. Consider the function F defined on [a,b] as follows:*

$$F(x) = V_f(a,x) \quad \text{if} \quad a < x \leq b$$

$$F(a) = 0.$$

Then the functions F and F - f are both nondecreasing on [a,b].

Proof For $a < x < y \leq b$, we have by Proposition 1

$$V_f(a,y) = V_f(a,x) + V_f(x,y)$$

so that

$$F(y) - F(x) = V_f(x,y) \geq 0.$$

This proves that F is increasing.

Also for $a \leq x < y \leq b$ we have

$$[F(y)-f(y)]-[F(x)-f(x)] = V_f(x,y)-[f(y)-f(x)] \geq 0.$$

Hence $F - f$ is increasing.

4. Theorem *A real-valued function f defined on [a,b] is of bounded variation if and only if it can be expressed as the difference of two nondecreasing functions.*

Proof If f is of bounded variation, then $f = F - (F - f)$ with F as in Proposition 3, and both F and $F - f$ are nondecreasing. Conversely, if $f = f_1 - f_2$, where f_1 and f_2 are nondecreasing in $[a,b]$, then f_1 and f_2, as monotone functions, are of bounded variation and so is their difference, by Proposition 2.

Remark Theorem 4 also holds if "nondecreasing" is replaced by "increasing" (i.e., strictly increasing). For if $f = f_1 - f_2$ where f_1 and f_2 are nondecreasing and h is any arbitrary increasing function on $[a,b]$, we also have $f = (f_1 + h) - (f_2 + h)$ where $f_1 + h$ and $f_2 + h$ are increasing.

4.3 DIFFERENTIATION OF MONOTONE FUNCTIONS

We recall that a real-valued function f defined on an interval I is called *monotone* iff either it is nondecreasing or nonincreasing. In other words f is monotone if either

$$f(x) \leq f(y) \quad \text{whenever} \quad x < y (x \in I,\ y \in I),$$

or

$$f(y) \leq f(x) \quad \text{whenever} \quad x < y (x \in I,\ y \in I).$$

We assume that the reader is familiar with the following properties of monotone functions.

Let f be a monotone function on $[a,b]$. Then

(a) For each $x \in (a,b)$, $f(x+)$ and $f(x-)$ (i.e., the *right-hand limit* and the *left-hand limit* of f at x) both exist and are finite. Also $f(a+)$ and $f(b-)$ exist. The number $f(x+) - f(x-)$ is called the *jump* of f at x. Clearly f is continuous at x if and only if $f(x) = f(x+) = f(x-)$ (see Prob. 6).

(b) The function f has at most a countable number of jumps. This implies that the set of points of discontinuities of f is countable (see Prob. 7).

In this section we establish a much stronger property than the one given in (b). We prove that a monotone function is differentiable almost everywhere.

Definition *If S is a set and $\mathcal{J}$ is a collection of closed intervals of positive length, then $\mathcal{J}$ is said to cover S* in the sense of Vitali *iff for every $x \in S$ and every $\varepsilon > 0$ there is an interval $I \in \mathcal{J}$ such that $\ell(I) < \varepsilon$ and $x \in I$.*

5. Theorem (Vitali's Covering Theorem) *If S is a set of finite outer measure and $\mathcal{J}$ a collection of closed intervals which cover S in the sense of Vitali, then for every $\varepsilon > 0$ there is a finite number of disjoint closed intervals*

$I_1, I_2, \ldots, I_n$ in $\mathcal{J}$ which cover all of S except for a set whose outer measure is less than ε (i.e., $m^*\left(S - \bigcup_{k=1}^{n} I_k\right) < \varepsilon$).

Proof From the definition of the outer measure it follows that there exists an open set O containing S and such that $mO < +\infty$. For if c is an arbitrary positive number, there is an open set $O \supset S$ with $mO < m^*S + c < +\infty$, since $m^*S < +\infty$. Let $\mathcal{J}_o \subset \mathcal{J}$ be the collection of all intervals in $\mathcal{J}$ that are subsets of the open set O. The class $\mathcal{J}_o$ is not empty because if x is an arbitrary point in S there must be an interval I belonging to $\mathcal{J}$ that contains x and is a bubset of O. Thus for every $J \varepsilon \mathcal{J}_o$ we have $\ell(J) \leq mO < +\infty$. Now choose $I_1 \varepsilon \mathcal{J}_o$ such that for every $J \varepsilon \mathcal{J}_o$ we have $\ell(I_1) > (1/2)\ell(J)$. The existence of such I_1 follows from the simple fact that for any given bounded set of positive numbers (in our case all $\ell(J)$ for $J \varepsilon \mathcal{J}_o$) there is one whose double exceeds all the others. Choose $I_2 \varepsilon \mathcal{J}_o$ such that $I_1 \cap I_2 = \emptyset$ and such that for every $J \varepsilon \mathcal{J}_o$ with $I_1 \cap J = \emptyset$ we have $\ell(I_2) > (1/2)\ell(J)$. Suppose $I_1, I_2, \ldots, I_{k-1}$ have already been chosen, are pairwise disjoint and are such that for $i = 1,2,\ldots,k-1$ we have $\ell(I_i) > (1/2)\ell(J)$ for every $J \varepsilon \mathcal{J}_o$ with $J \cap \left(\bigcup_{j=1}^{i-1} I_j\right) = \emptyset$. Then if there is an

$x \in S - \bigcup_{j=1}^{k-1} I_j$, since x belongs to an arbitrarily small interval $J \in \mathcal{J}_o$ there are intervals $J \in \mathcal{J}_o$ such that $J \cap \left(\bigcup_{j=1}^{k-1} I_j\right) = \emptyset$. Among these intervals we pick up one, twice whose length exceeds the length of any of the others. By induction now, either the theorem holds or there is an infinite sequence $\langle I_n \rangle_{n=1}^{\infty}$ of disjoint intervals in $\mathcal{J}_o$ such that for every k, if $J \in \mathcal{J}_o$ and $J \cap \left(\bigcup_{j=1}^{k-1} I_j\right) = \emptyset$ then $\ell(I_k) > (1/2)\ell(J)$. Since the intervals I_n are pairwise disjoint and contained in the set O of finite measure, there must exist an n such that $\sum_{k=n+1}^{\infty} \ell(I_k) < \varepsilon/5$. For every $k = n+1, n+2, \ldots,$ let J_k be the closed interval with the same midpoint as I_k such that $\ell(J_k) = 5\ell(I_k)$. Let $G = \bigcup_{k=n+1}^{\infty} J_k$. Then $mG < \varepsilon$. Now suppose $x \in S$ but $x \notin \bigcup_{k=1}^{n} I_k$. There is a $J \in \mathcal{J}_o$ such that $x \in J$ but $J \cap \left(\bigcup_{k=1}^{n} I_k\right) = \emptyset$. But J is not disjoint from all the I_k since $\ell(J) > 2\ell(I_k)$ for some k. Let m be the first positive integer such that $J \cap I_m \neq \emptyset$. Then $m > n$, and for this particular m we have $\ell(J) < 2\ell(I_m)$. For $J \cap \left(\bigcup_{n=1}^{m-1} I_n\right) = \emptyset$, and if we had $\ell(J) \geq 2\ell(I_m)$ then I_m could not be chosen as a member of the sequence $\langle I_n \rangle$. Now $m > n$ and $\ell(J) < 2\ell(I_m)$ so that $J \subset J_m$. Hence

$$S - \bigcup_{k=1}^{n} I_k \subset \bigcup_{k=n+1}^{\infty} J_k$$

or

$$m^*(S - \bigcup_{k=1}^{n} I_k) < \varepsilon.$$

The theorem is proved.

(*Note:* It follows from the above proof that if S and $\mathcal{J}$ are as in Theorem 5, then there is a finite or infinite sequence $\langle I_n \rangle$ of intervals in $\mathcal{J}$ which cover S except for a set of measure zero.)

Let $f: [a,b] \to \mathbb{R}$ be a real valued function. Along with the notion of the ordinary derivative of f we consider the four *Dini derivatives* which have the advantage that they apply to functions which are not necessarily differentiable. For $x \in (a,b)$ these derivatives of f are defined as follows:

$$D^+f(x) = \overline{\lim}_{h\to0+} \frac{f(x+h)-f(x)}{h} \qquad \text{(upper right)}$$

$$D_+f(x) = \underline{\lim}_{h\to 0+} \frac{f(x+h)-f(x)}{h} \qquad \text{(lower right)}$$

$$D^-f(x) = \overline{\lim}_{h\to 0-} \frac{f(x+h)-f(x)}{h} \qquad \text{(upper left)}$$

$$D_-f(x) = \underline{\lim}_{h\to 0-} \frac{f(x+h)-f(x)}{h} \qquad \text{(lower left)}$$

The signs + and - refer to right and left respectively, and their positions to limits superior and inferior. For any f and any x the four derivatives exist as members of the extended real number system. Clearly $D^+f(x) \geq D_+f(x)$; $D^-f(x) \geq D_-f(x)$.

If $D^+f(x) = D_+f(x) = D^-f(x) = D_-f(x)$, and if these are all finite, then f is called *differentiable at* x, and its ordinary derivative $f'(x)$ equals this common value of the Dini derivatives.

6. Theorem *Let* $f\colon [a,b] \to \mathbb{R}$ *be a nondecreasing function. Then* f *is differentiable almost everywhere;* f' *is a measurable function and*

$$\int_a^b f'(x)dx \leq f(b) - f(a).$$

Proof To prove that f is differentiable almost everywhere, it suffices to prove that the Dini derivatives of f are equal almost everywhere, i.e., the set of points on which at least two of the four derivatives differ is of measure zero. We give the proof only for the set $\{x \in (a,b)\colon D^+f(x) > D_-f(x)\}$, the proof for the other sets being similar. Since f is nondecreasing each of the derivatives is nonnegative at each point of (a,b). Let $\{\langle a_n, b_n\rangle\}$

be an enumeration of the pairs of rational numbers such that $0 < a_n < b_n$ $(n = 1,2,\ldots)$. Put

$$E_n = \{x \in (a,b): D^+f(x) > b_n > a_n > D_-f(x)\},$$

$$(n = 1,2,\ldots).$$

Then

$$\{x \in (a,b): D^+f(x) > D_-f(x)\} = \bigcup_{n=1}^{\infty} E_n.$$

We next prove that $m^*E_n = 0$ for all n. Since E_n is contained in (a,b) it has finite outer measure. For a given $\varepsilon > 0$ there exists an open set O such that

$$E_n \subset O \quad \text{and} \quad mO < m^*E_n + \varepsilon.$$

For each $x \in E_n$ and any arbitrary positive number p we can find an open interval $I_x = (x - h, x)$ such that

$$0 < h < p, \quad I_x \subset O \quad \text{and} \quad f(x) - f(x-h) < a_n h.$$

This follows from the definition of $D_-f(x)$ as a limit inferior. Thus the collection $\langle \bar{I}_x \rangle$ covers E_n in the sense of Vitali so that by Theorem 5 there is a finite number $\{\bar{I}_1, \bar{I}_2, \ldots, \bar{I}_N\}$ of them with $I_i = (x_i - h_i, x_i)$ $(i = 1,2,\ldots,N)$ such that

$$m^*\left(E_n - \bigcup_{i=1}^{N} \overline{I}_i\right) = m^*\left(E_n - \bigcup_{i=1}^{N} I_i\right) < \varepsilon.$$

We have

$$E_n = \left(E_n \cap \bigcup_{i=1}^{N} I_i\right) \cup \left(E_n - \bigcup_{i=1}^{N} I_i\right)$$

$$m^*E_n \leq m^*\left(E_n \cap \bigcup_{i=1}^{N} I_i\right) + m^*\left(E_n - \bigcup_{i=1}^{N} I_i\right)$$

$$< m^*\left(E_n \cap \bigcup_{i=1}^{N} I_i\right) + \varepsilon$$

or

$$m^*E_n - \varepsilon < m^*\left(E_n \cap \bigcup_{i=1}^{N} I_i\right).$$

Next we obtain a Vitali covering of the set $E_n \cap \bigcup_{i=1}^{N} I_i$. For each $y \in E_n \cap \bigcup_{i=1}^{N} I_i$ and q an arbitrary positive number there is an open interval $J_y = (y, y+k)$, with $k < q$, contained in some of the intervals $I_1, I_2, \ldots, I_N$, and such that $f(y+k) - f(y) > b_n k$. This again follows from the definition of $D_+f(x)$ as a limit inferior. It follows that the collection $\langle \overline{J}_y \rangle$ of closed intervals is a covering of $E_n \cap \bigcup_{i=1}^{N} I_i$ in the sense of Vitali. By Theorem 5 there is a finite number $\{\bar{J}_i, \bar{J}_2, \ldots, \bar{J}_M\}$ of intervals in $\langle \bar{J}_y \rangle$ with $J_r = (y_r, y_r + k_r)$ $(r = 1, 2, \ldots, M)$, such that

$$m^*\left(E_n \cap \bigcup_{i=1}^{N} I_i - \bigcup_{r=1}^{M} \bar{J}_r\right)$$

$$- m^*\left(E_n \cap \bigcup_{i=1}^{N} I_i - \bigcup_{r=1}^{M} J_r\right) < \varepsilon.$$

We also have, as before,

$$m^*E_n - 2\varepsilon < m^*\left(E_n \cap \bigcup_{i=1}^{N} I_i \cap \bigcup_{r=1}^{M} J_r\right).$$

Summing over the intervals $I_1, I_2, \ldots, I_N$ we have

$$\sum_{i=1}^{N} [f(x_i) - f(x_i - h_i)] < a_n \sum_{i=1}^{N} h_i$$

$$< a_n mO < a_n(m^*E_n + \varepsilon).$$

Also summing over the intervals $J_1, J_2, \ldots, J_M$ we get

$$\sum_{r=1}^{M}[f(y_r+k_r) - f(y_r)] > b_n \sum_{r=1}^{M} k_r$$

$$> b_n(m^*E_n - 2\varepsilon).$$

As we pointed out, each interval J_r is contained in some interval I_i. If we sum over those intervals J_r that are contained in I_i for a fixed i, we get, since f is nondecreasing,

$$\sum_{J_r \subset I_i}[f(y_r+k_r) - f(y_r)] \leq f(x_i) - f(x_i-h_i).$$

Now, if we sum over all intervals I_i we get

$$\sum_{i=1}^{N}\sum_{J_r \subset I_i}[f(y_r+k_r)-f(y_r)] = \sum_{r=1}^{M}[f(y_r+k_r)-f(y_r)]$$

$$\leq \sum_{i=1}^{N}[f(x_i) - f(x_i-h_i)]$$

or

$$b_n(m^*E_n - 2\varepsilon) < a_n(m^*E_n + \varepsilon).$$

Since $\varepsilon > 0$ was arbitrary we have

$$b_n m^*E_n \leq a_n m^*E_n.$$

But $a_n < b_n$, so that we must have $m^*E_n = 0$. This shows that f is differentiable almost everywhere on $[a,b]$.

To proceed with the proof of the theorem let us extend the domain of definition of f by setting $f(x) = f(b)$ if $x > b$. The function f as a nondecreasing function is clearly measurable. Also its set of discontinuities being countable is of measure zero, so that f is integrable over $[a,b]$ in the Riemann sense.

Define

$$\phi_n : [a,b] \to \mathbb{R} \quad \text{by}$$

$$\phi_n(x) = \frac{f(x+(1/n)) - f(x)}{1/n} .$$

We have

$$\lim_{n\to\infty} \phi_n(x) = f'(x) \quad \text{a.e.}$$

Since f is nondecreasing we have $\phi_n(x) \geq 0$ for every x in $[a,b]$ and every positive integer n. We have, if we consider the following integrals as Riemann integrals so that the change of variable is legitimate,

$$\int_a^b \phi_n(x)dx = n\left(\int_a^b f\left(x + \frac{1}{n}\right)dx - \int_a^b f(x)dx\right)$$

$$= n\left(\int_{a+(1/n)}^{b+(1/n)} f(x)dx - \int_a^b f(x)dx\right)$$

$$= f(b) - n \int_a^{a+(1/n)} f(x)dx \leq f(b) - f(a).$$

If we apply Fatou's lemma to the sequence $\langle \phi_n \rangle$ we get

$$\int_a^b f'(x)dx \leq \underline{\lim}_{n\to\infty} \int_a^b \phi_n(x)dx \leq f(b) - f(a).$$

This shows that f' is finite almost everywhere. The theorem is proved.

7. Corollary *If* $f: [a,b] \to \mathbb{R}$ *is of bounded variation then* f' *exists and is finite almost everywhere in* $[a,b]$.

(*Note:* It is possible to give an example of a continuous nondecreasing function $f: [0,1] \to \mathbb{R}$ such that $f(0) = 0$, $f(1) = 1$, $f'(x) = 0$ a.e. This demonstrates that the equality in $\int_a^b f'(x)dx \leq f(b) - f(a)$ need not hold [see Prob. 11].)

4.4 THE DERIVATIVE OF AN INDEFINITE INTEGRAL

8. Proposition *Let* $f: [a,b] \to \mathbb{R}$ *be an integrable function. Define*

$$F(x) = \int_a^x f(t)dt.$$

Then F *is continuous and of bounded variation on*

[a,b].

Proof Let $x \in [a,b]$ and $\varepsilon > 0$. Then by Chapter 3 Proposition 32, there is a $\delta > 0$ such that for every measurable subset E of $[a,b]$ with $mE < \delta$ we have $\int_E |f| < \varepsilon$. Thus for $0 < |h| < \delta$ we have

$$|F(x+h) - F(x)| = \left| \int_x^{x+h} f(t)dt \right| < \varepsilon$$

which proves that F is continuous. To prove that F is of bounded variation let $P = \{x_0, x_1, \ldots, x_n\}$ $(a = x_0 < x_1 < \ldots < x_n = b)$ be any partition of $[a,b]$. Then

$$\sum_{k=1}^{n} |F(x_k) - F(x_{k-1})| = \sum_{k=1}^{n} \left| \int_{x_{k-1}}^{x_k} f(t)dt \right|$$

$$\leq \sum_{k=1}^{n} \int_{x_{k-1}}^{x_k} |f(t)|dt = \int_a^b |f(t)|dt .$$

Hence F is of bounded variation.

9. Proposition *Let $f\colon [a,b] \to \mathbb{R}$ be an integrable function such that for each $x \in [a,b]$ we have*

$$\int_a^x f(t)dt = 0.$$

Then $f(t) = 0$ almost everywhere in $[a,b]$.

Proof Clearly the hypothesis implies that for each subinterval I of $[a,b]$ we have $\int_I f = 0$. We prove that $\int_E f = 0$ for each measurable subset E of $[a,b]$. Suppose on the contrary that there is a measurable set $E \subset [a,b]$ for which

$$\int_E f \neq 0.$$

Let $\varepsilon = \left|\int_E f\right|$. Then by Chapter 3, Proposition 32 there is a $\delta > 0$ such that

$$\left|\int_S f\right| \leq \int_S |f| < \varepsilon$$

for every measurable set $S \subset [a,b]$ with $mS < \delta$. Choose an open set O so that $E \subset O$ and $mO < mE + \delta$. Express $O = \bigcup_{n=1}^{\infty} I_n$ where I_n are open pairwise disjoint intervals, and put

$$V_n = I_n \cap [a,b], \quad V = \bigcup_{n=1}^{\infty} V_n.$$

Then

$$\int_V f = \sum_{n=1}^{\infty} \int_{V_n} f = 0,$$

since the integral of f vanishes on every subinterval of $[a,b]$. But $E \subset V \subset O$ so that

$m(V - E) < \delta$. This implies that

$$\left|\int_{V-E} f\right| < \varepsilon.$$

We have

$$0 = \int_V f = \int_E f + \int_{V-E} f$$

so that

$$\left|\int_{V-E} f\right| = \left|-\int_E f\right| = \varepsilon$$

which contradicts the previously established inequality $\left|\int_{V-E} f\right| < \varepsilon$. Hence $\int_E f = 0$ for every measurable subset of $[a,b]$.

Consider now the sets

$$E_1 = \{x \in [a,b]: f(x) \geq 0\};$$

$$E_2 = \{x \in [a,b]: f(x) \leq 0\}.$$

We have $\int_{E_1} f = \int_{E_2} f = 0$. Hence by Chapter 3, Corollary 27, it follows that $f = 0$ a.e.

We now prove the main theorem of this section.

10. Theorem *Let* $f: [a,b] \to \mathbb{R}$ *be an integrable function and put*

$$F(x) = F(a) + \int_a^x f(t)dt.$$

Then $F'(x) = f(x)$ *a.e. in* $[a,b]$.

Proof We first prove the theorem under the additional hypothesis that f is bounded, say $|f| \leq M$. By Proposition 8, F is continuous and of bounded variation on $[a,b]$, which implies that the derivative $F'(x)$ exists and is finite almost everywhere in $[a,b]$. For each positive integer n define f_n on (a,b) by

$$f_n(x) = \frac{F(x+(1/n))-F(x)}{1/n} = n \int_x^{x+(1/n)} f(t)dt.$$

Clearly $|f_n| \leq M$ for all n and

$$\lim_{n\to\infty} f_n(x) = F'(x) \quad \text{a.e. on } [a,b].$$

By the bounded convergence theorem we get

$$\int_a^x F'(t)dt = \lim_{n\to\infty} \int_a^x f_n(t)dt$$

$$= \lim_{n\to\infty} n\left[\int_a^x (F(t+(1/n))-F(t))dt\right] =$$

$$= \lim_{n\to\infty} n\left[\int_x^{x+(1/n)} F(t)dt - \int_a^{a+(1/n)} F(t)dt\right]$$

where the integrals may be considered as Riemann integrals since F is continuous. By the mean value theorem for integrals we have

$$\int_x^{x+(1/n)} F(t)dt = F(t') \cdot \frac{1}{n}$$

with $x < t' < x + (1/n)$.

Thus

$$\lim_{n\to\infty} n \int_x^{x+(1/n)} F(t)dt = F(x).$$

Similarly

$$\lim_{n\to\infty} n \int_a^{a+(1/n)} F(t)dt = F(a).$$

Hence

$$\int_a^x F'(t)dt = F(x) - F(a) = \int_a^x f(t)dt$$

which implies that

$$\int_a^x (F' - f) = 0 \quad \text{for all} \quad x \in (a,b)$$

so that by Proposition 9

$$F'(x) = f(x) \quad \text{a.e. on} \quad [a,b].$$

We now prove the theorem for any integrable function f on $[a,b]$. Since f is integrable we have $f = f^+ - f^-$ with f^+, f^- both nonnegative and integrable. Hence without loss of generality we may assume that $f \geq 0$. Let f_n be defined by $f_n(x) = f(x)$ if $f(x) \leq n$ and $f_n(x) = n$ if $f(x) > n$. Then $f - f_n \geq 0$, and so the function g_n defined by

$$g_n(x) = \int_a^x (f(t) - f_n(t))dt$$

is an increasing function of x. Hence g_n has a nonnegative derivative almost everywhere. Since f_n is bounded, we have $\left(\int_a^x f_n\right)' = f_n(x)$ a.e. Also since $g_n'(x) \geq 0$ we have

$$\left(\int_a^x f\right)' = F'(x) = \left(\int_a^x f_n\right)' + g_n'(x)$$

$$= f_n(x) + g_n'(x) \geq f_n(x) \quad \text{a.e.}$$

By letting $n \to \infty$ we get, since $f_n \to f$, that

$$F'(x) \geq f(x) \quad \text{a.e.}$$

By integrating both members of this inequality we get

$$\int_a^b F'(x)dx \geq \int_a^b f(x)dx = F(b) - F(a).$$

On the other hand

$$\int_a^b F'(x)dx \leq F(b) - F(a)$$

by Theorem 6. Hence

$$\int_a^b F'(x)dx = F(b) - F(a) = \int_a^b f(x)dx$$

or

$$\int_a^b (F'(x) - f(x))dx = 0.$$

Since $F'(x) - f(x) \geq 0$ a.e. we conclude that

$$F'(x) = f(x) \quad \text{a.e.}$$

4.5 ABSOLUTELY CONTINUOUS FUNCTIONS

The concept of absolute continuity for functions $f: D \to \mathbb{R}$ where D is a connected subset of $\mathbb{R}$, (i.e., a finite or infinite interval of $\mathbb{R}$) plays a key role in the understanding of the relation between differentiation and integration in the

Lebesgue sense.

Definition *A function* $f\colon D \to \mathbb{R}$, *where* D *is a connected subset of* $\mathbb{R}$, *is said to be* absolutely continuous *on* D *iff to every* $\varepsilon > 0$ *there corresponds a* $\delta > 0$ *such that*

$$\sum_{i=1}^{N} (b_i - a_i) < \delta \quad \textit{implies} \quad \sum_{i=1}^{N} |f(b_i) - f(a_i)| < \varepsilon$$

whenever $(a_1, b_1), \ldots, (a_N, b_N)$ *are disjoint open intervals in* D.

Observe that *an absolutely continuous function is uniformly continuous.* This can be seen if in the definition of the absolute continuity we take $N = 1$.

11. Proposition *If* $f\colon [a,b] \to \mathbb{R}$ *is absolutely continuous then* f *is of bounded variation on* $[a,b]$.

Proof Let $\varepsilon = 1$. Then there is a $\delta > 0$ such that $\sum_{i=1}^{N} |f(b_i) - f(a_i)| < 1$ whenever $\{(a_i, b_i)\}_{i=1}^{N}$ are disjoint open intervals in $[a,b]$, with $\sum_{i=1}^{N} (b_i - a_i) < \delta$. Let now $P = \{x_0, x_1, \ldots, x_M\}$ $(a = x_0 < x_1 < \ldots < x_M = b)$ be a fixed partition of $[a,b]$ such that $x_i - x_{i-1} < \delta$ for $i = 1,2,\ldots,M$. The function f is of bounded variation on each interval $[x_{i-1}, x_i]$. For let $P_i = \{y_0, y_1, \ldots, y_k\}$ $(x_{i-1} = y_0 < y_1 < \ldots < y_k = x_i)$ be any partition of $[x_{i-1}, x_i]$. Then since $\sum_{r=1}^{k} (y_r - y_{r-1}) < \delta$ we have that

$\sum_{r=1}^{k} |f(y_r) - f(y_{r-1})| < 1$. Thus f is of bounded variation on each interval $[x_{i-1}, x_i]$ with total variation less than or equal to 1. This implies (see Prop. 1) that f is of bounded variation on $[a,b]$ with total variation not exceeding the integer M.

12. Corollary *If $f: [a,b] \to \mathbb{R}$ is absolutely continuous then the derivative $f'(x)$ exists and is finite almost everywhere on $[a,b]$.*

13. Proposition *If $f: [a,b] \to \mathbb{R}$ is absolutely continuous and $f'(x) = 0$ a.e., then f is a constant.*

Proof Let c be any point in $[a,b]$ and let $\varepsilon > 0$ be given. Since f is absolutely continuous there is a $\delta > 0$ such that

$$\sum_{i=1}^{N} (b_i - a_i) < \delta \quad \text{implies} \quad \sum_{i=1}^{N} |f(b_i) - f(a_i)| < \varepsilon$$

whenever $(a_1, b_1), \ldots, (a_N, b_N)$ are disjoint open intervals in $[a,b]$. Put $E = \{x \in (a,c) : f'(x) = 0\}$. By assumption $f'(x) = 0$ a.e. so that $mE = c - a$ = finite. For $x \in E$ since $f'(x) = 0$ we can find a closed interval $I_x = [x,y]$ of arbitrarily small positive length such that $[x,y] \subset [a,c]$ and such that $|f(y) - f(x)| < \varepsilon|y - x|$. The intervals $\langle I_x \rangle$ cover E in the sense of Vitali. Thus by Theorem 5 there is a finite number $I_{x_i} = [x_i, y_i]$

$(i = 1,2,\ldots,n)$ of these intervals which are disjoint, and such that $m^*\left(E - \bigcup_{i=1}^{n} I_{x_i}\right) = m\left(E - \bigcup_{i=1}^{n} I_{x_i}\right) < \delta$. Clearly we may assume that

$$a \leq x_1 < y_1 < x_2 < y_2 \cdots < x_n < y_n \leq c.$$

Observe now that $m\left(E - \bigcup_{i=1}^{n} I_{x_i}\right) < \delta$ implies $\sum_{k=0}^{n} |x_{k+1} - y_k| < \delta$, where $y_0 = a$ and $x_{n+1} = c$. We have

$$\sum_{k=1}^{n} |f(y_k) - f(x_k)| \leq \varepsilon \sum_{k=1}^{n} (y_k - x_k) < \varepsilon(c - a)$$

and

$$\sum_{k=0}^{n} |f(x_{k+1}) - f(y_k)| < \varepsilon$$

since f is absolutely continuous. Hence

$$\begin{aligned} |f(c) - f(a)| &= \left| \sum_{k=0}^{n} [f(x_{k+1}) - f(y_k)] \right. \\ &\qquad \left. + \sum_{k=1}^{n} [f(y_k) - f(x_k)] \right| \\ &\leq \varepsilon + \varepsilon(c - a). \end{aligned}$$

Since ε is arbitrary, it follows that $f(a) = f(c)$, so that f is constant on $[a,b]$.

14. Theorem *If* $f\colon [a,b] \to \mathbb{R}$ *is an integrable function and* $F(x) = \int_a^x f(t)dt$, *then* F *is absolutely continuous.*

Proof Let $\varepsilon > 0$ be given. Then by Cahpter 3, Proposition 32 there is a $\delta > 0$ such that

$$\left|\int_S f(t)dt\right| \le \int_S |f(t)|\, dt$$

for all measurable sets S in $[a,b]$ with $mS < \delta$. Let $\{(a_i,b_i)\}_{i=1}^n$ be any finite set of non-overlapping open intervals in $[a,b]$ such that $\sum_{i=1}^n (b_i - a_i) < \delta$. Then if we denote by O the union of these intervals we have $mO < \delta$ so that

$$\left|\int_O f(t)dt\right| \le \int_O |f(t)|dt = \sum_{i=1}^n \int_{a_i}^{b_i} |f(t)|dt$$

$$= \sum_{i=1}^n |F(b_i) - F(a_i)| < \varepsilon.$$

This proves that F is absolutely continuous.

15. Theorem *If* $f\colon [a,b] \to \mathbb{R}$ *is absolutely continuous then*

$$\int_a^x f'(t)dt = f(x) - f(a) \quad on \quad [a,b].$$

Proof By Proposition 11, f is of bounded variation, and so $f = f_1 - f_2$, where f_1 and f_2 are nondecreasing functions. By Corollary 7, f', f_1', and f_2' exist almost everywhere and $f' = f_1' - f_2'$. We have, since f_1' and f_2' are nonnegative,

$$|f'| \leq |f_1'| + |f_2'| = f_1' + f_2'.$$

By Theorem 6

$$\int_a^b f_1'(x)dx \leq f_1(b) - f_1(a);$$

$$\int_a^b f_2'(x)dx \leq f_2(b) - f_2(a).$$

Hence

$$\int_a^b |f'| \leq f_1(b) + f_2(b) - f_1(a) - f_2(a).$$

This proves that f' is integrable. Define now the function g by

$$g(x) = \int_a^x f'(t)dt.$$

By Theorem 14, g is absolutely continuous, and since f is by hypothesis absolutely continuous we have that $f - g$ is also absolutely continuous.

It follows from Theorem 10 that

$$f' - g' = f' - f' = 0 \quad \text{a.e.,}$$

so that by Proposition 13 $f - g$ is a constant, and this constant is equal to $f(a)$. Hence

$$\int_a^x f'(t)dt = f(x) - f(a) \quad \text{on} \quad [a,b].$$

(*Note:* Theorems 14 and 15 tell us that *a function* f *is an indefinite integral if and only if* f *is an absolutely continuous function.* Furthermore Theorem 15 says that *every absolutely continuous function is the indefinite integral of its derivative.)*

PROBLEMS

1. Prove that every polynomial on $[0,1]$ is a function of bounded variation.
2. Let f be a function defined on $[0,1]$ as follows:

$$f(x) = x \cos \frac{\pi}{2x} \quad \text{if} \quad x \neq 0$$

$$f(0) = 0.$$

Show that f is bounded but it is not of bounded variation on $[0,1]$.

3. Let f be a function defined on $[0,1]$ as follows:

$$f(x) = x^2 \sin \frac{1}{x} \quad \text{if} \quad x \neq 0$$

$$f(0) = 0.$$

Show that f is of bounded variation on $[0,1]$.

4. Show that if f is a function of bounded variation on $[a,c]$ and on $[c,b]$, where $a < c < b$, then f is of bounded variation on $[a,b]$.

5. A function f defined on a closed and bounded interval $[a,b]$ is said to satisfy a *Lipschitz condition* of order $\alpha > 0$ iff $|f(x) - f(y)| \leq k|x - y|^\alpha$ for some positive real number k. Show that if $\alpha > 1$ then f is a constant function. Give an example of a function satisfying a Lipschitz condition of given order $0 < \alpha < 1$, and which is not of bounded variation.

6. Let f be a nondecreasing function on the closed and bounded interval $[a,b]$. Prove that $f(x+)$ and $f(x-)$ exist for each $x \in (a,b)$ and that

$$f(x-) \leq f(x) \leq f(x+).$$

Also, show that, $f(a+)$ and $f(b-)$ exist and

$$f(a) \leq f(a+), \quad f(b-) \leq f(b).$$

(Hint: The proof involves only the basic principles of supremum and infimum of sets of real numbers.)

7. Let f and $[a,b]$ be as in Problem 6, and let $x_1, x_2, \ldots, x_n$ be n points in $[a,b]$ such that $a < x_1 < x_2 < \ldots < x_n < b$.

   (a) Show that

$$\sum_{k=1}^{n} [f(x_k+) - f(x_k-)] \le f(b-) - f(a+).$$

   (b) Use (a) to show that the set of points where f is discontinuous is at most countable.

8. Let $f\colon [0,1] \to \mathbb{R}$ be one-to-one continuous function. Show that f is either strictly increasing or strictly decreasing.

9. Let f be a function defined by $f(0) = 0$, and $f(x) = x \sin(1/x)$ if $x \neq 0$. Show that $D^+f(0) = D^-f(0) = 1$ and $D_+f(0) = D_-f(0) = -1$.

10. Let f be a function defined by

$$f(x) = x(1 + \sin(\log x)) \qquad \text{if} \quad x > 0$$

$$f(x) = x + \sqrt{-x} \cdot \sin^2(\log(-x)) \text{ if} \quad x < 0$$

$$f(0) = 0.$$

   Show that

$$D^+f(0) = 2,$$

$$D_+f(0) = 0,$$

$$D^-f(0) = 1,$$

$$D_-f(0) = -\infty.$$

11. Prove Corollary 7.

12. Show that the Cantor function k (Chap. 2,

Prob. 49) is continuous nondecreasing and such that $k(0) = 0$, $k(1) = 1$ and $k'(x) = 0$ a.e.

13. Show that the Cantor function k is not absolutely continuous, although it is uniformly continuous.

14. Let $f\colon [0,1] \to \mathbb{R}$ be an absolutely continuous function.

(a) Show that if S is a measurable subset of $[0,1]$ with $mS = 0$ then $mf[S] = 0$. (Hint: For ε and δ as in the definition of absolutely continuous function choose an open set $O \supset S$ with $mO < \delta$. Let $\langle I_n \rangle$ be the components of O with $I_n = (a_n, b_n)$. We have $\sum_{i=1}^{n} (b_i - a_i) < \delta$, $n = 1,2,\ldots$. Show that

$$m^*f[S] \le \Sigma mf[\overline{I}_i] \le \varepsilon.)$$

(b) Let E be a measurable subset of $[0,1]$. Show that $f[E]$ is also a measurable set. (Hint: Consider an increasing sequence $\langle E_n \rangle$ of closed subsets of E such that $m(E - \bigcup F_n) = 0$. Then $\langle f[E_n] \rangle$ is an increasing sequence of closed subsets of $f[E]$, and $\bigcup f[E_n]$ is a measurable subset of $f[E]$. Use (a) to prove $m[f[E] - \bigcup f[E_n]] = 0$.)

15. Show that if f and g are absolutely continuous on $[0,1]$ then so are $f + g$, fg and cf (c = constant). If $f(x) \neq 0$ for

$0 \le x \le 1$ then $1/f$ is also absolutely continuous.

16. Let $f\colon [0,1] \to \mathbb{R}$ be continuous.

(a) Show that if F is a subset of $[0,1]$ of type F_σ, then $f[F]$ is also a set of type F_σ.

(b) Show that if S is a measurable subset of $[0,1]$, then $f[S]$ is a measurable set if and only if for every subset E of $[0,1]$ with $mE = 0$, we have $mf[E] = 0$.

17. Let f be a measurable function. Then if

$$\lim_{h\to 0} \frac{1}{h} \int_x^{x+h} |f(t) - f(x)|dt = 0,$$

the point x is said to be a *Lebesgue point* for f.

(a) Let f be an integrable function on $[0,1]$. Put

$$F(x) = \int_0^x f, \quad 0 \le x \le 1.$$

Show that F is differentiable at every Lebesgue point for f in $[0,1)$.

(b) Show that if f is as in (a), then almost every point of $[0,1]$ is a Lebesgue point for f.

(Hint: Let r be a rational number. Since $|f - r|$ is integrable on $[0,1]$ we have

$$\lim_{h\to 0} \frac{1}{h} \int_x^{x+h} |f(t) - r|\,dt = |f(x) - r|$$

a.e. in $[0,1]$.

Let E_r be the set where this last equality does not hold, and let $\langle r_1, r_2, \ldots \rangle$ be an enumeration of the rational numbers. Let $A = \{x: |f(x)| = \infty\}$, and $E = A \cup \bigcup_{n=1}^{\infty} E_{r_n}$. Show that $mE = 0$, and that all points in $[0,1] - E$ are Lebesgue points.)

(c) If f is as in (a) and if f is continuous at x_0, then x_0 is a Lebesgue point for f.

CHAPTER 5

Metric Spaces

5.1 INTRODUCTION

A "space" is a nonempty set, where some additional conditions are usually imposed. In all the spaces encountered in the so-called "classical analysis," a "metric" is defined; that is, to each pair of elements of the space there is associated a nonnegative number, called "distance." But it has been noticed that a number of fundamental results obtained for a certain space depend only on the fact that the space possesses a metric. For example, many important facts from analysis do not depend on the algebraic structure of the space of real numbers (that is, on the fact that operations of addition and multiplication subject to known rules are defined) but only on those properties of real numbers which are related to the concept of distance. This situation leads to the following definition introduced by M. Frechet.

Definition *A* metric space $\langle X,d \rangle$ *is a nonempty set* X *(whose elements are called* points*) together with a real-valued function* d *(called a* metric*) defined* *on* $X \times X$ *such that for all* x, y *and* z *in* X *the following conditions are satisfied:*

(i) $d(x,y) > 0$ *if* $x \neq y$, *and* $d(x,y) = 0$ *if* $x = y$.

(ii) $d(x,y) = d(y,x)$ *(symmetry)*

(iii) $d(x,z) \leq d(x,y) + d(y,z)$ *(triangle inequality)*

Examples of Metric Spaces

1) Let X be an arbitrary nonempty set. Put $d(x,y) = 0$ if $x = y$ and $d(x,y) = 1$ if $x \neq y$. This metric space is called a *discrete* metric space.

2) The set $\mathbb{R}$ of real numbers with the metric $d(x,y) = |x - y|$.

3) The set $\mathbb{R}^n$ $(n = 2,3,\ldots)$, of ordered n-tuples of real numbers $x = \langle x_1, x_2, \ldots, x_n \rangle$ with the metric $d(x,y) = \left(\sum_{k=1}^{n} (y_k - x_k)^2\right)^{\frac{1}{2}}$. This metric space is called the *Euclidean n-space* $\mathbb{R}^n$.

4) The set $\mathbb{R}_o^n$ $(n = 2,3,\ldots)$ of ordered n-tuples of real numbers $x = \langle x_1, x_2, \ldots, x_n \rangle$ with the metric:

$$d(x,y) = \max(|y_i - x_i| : 1 \leq i \leq n) .$$

5) The set $C[a,b]$ of all continuous real-valued functions defined on the closed and bounded interval $[a,b]$ with the metric:

$$d(f,g) = \sup(|g(t) - f(t)| : a \leq t \leq b).$$

6) The set ℓ^2 of all infinite sequences $x = \langle x_i \rangle$ of real numbers such that $\sum_{i=1}^{\infty} x_i^2 < +\infty$, with the

metric: $d(x,y) = \left(\sum_{i=1}^{\infty} (y_i - x_i)^2\right)^{\frac{1}{2}}$.

7) The set $C^2[a,b]$ of all continuous real-valued functions defined on the closed and bounded interval $[a,b]$ with the metric:

$$d(f,g) = \left[\int_a^b (f(t) - f(t))^2 dt\right]^{\frac{1}{2}} .$$

8) The set ℓ^∞ of all bounded infinite sequences $x = \langle x_i \rangle$ of real numbers with the metric:

$$d(x,y) = \sup \{|x_i - y_i| : i = 1,2,\ldots,n,\ldots\} .$$

9) Let $\langle X,d \rangle$ and $\langle Y,\sigma \rangle$ be two metric spaces, and let $X \times Y$ be the Cartesian product of X and Y. For any elements $\langle x_1,y_1 \rangle$ and $\langle x_2,y_2 \rangle$ of $X \times Y$ define $\rho((x_1,y_1),(x_2,y_2)) = (d(x_1,x_2)^2 + \sigma(y_1,y_2)^2)^{\frac{1}{2}}$. Then $\langle X \times Y,\rho \rangle$ is a metric space (see Prob. 3).

The reader must always remember that *a metric space is not merely a set, but a set together with a metric.* For example, the metric spaces considered in Examples 5) and 7) (pp. 218 and 219) are distinct although they have the same points.

Metric Subspaces of a Metric Space

Let $\langle X,d \rangle$ be a metric space. Then obviously every nonempty subset A of X is also a metric space with the metric d_A defined as follows:

$$d_A(x,y) = d(x,y) \quad \text{for } x,y \text{ in } A.$$

In other words the metric d_A is the restriction of the metric d to the subset $A \times A$ of $\mathbb{R} \times \mathbb{R}$. The metric space $\langle A,d_A \rangle$ is called, by definition, a subspace of $\langle X,d \rangle$.

This method of forming subspaces of a given metric space enables us to obtain an infinity of further examples from the few described above.

5.2 ISOMETRIES

Definition *Let* $\langle X,d \rangle$ *and* $\langle X',d' \rangle$ *be two metric spaces. A function* f *from* X *onto* X' *is called an* isometry *iff* $d'(f(x),f(y)) = d(x,y)$ *for any two elements* x *and* y *in* X.

An isometry is evidently a one-to-one correspondence between X and X'. For if $x \neq y$ then $d'(f(x),f(y)) = d(x,y) \neq 0$ so that $f(x) \neq f(y)$. Also, if f is an isometry from $\langle X,d \rangle$ onto $\langle X',d' \rangle$, then the inverse mapping f^{-1} is an isometry from $\langle X',d' \rangle$ onto $\langle X,d \rangle$. Two metric spaces $\langle X,d \rangle$ and $\langle X',d' \rangle$ are called *isometric* iff there is an isometry from $\langle X,d \rangle$ onto $\langle X',d' \rangle$. Any proposition proved to be true for a metric space $\langle X,d \rangle$, and such that its proof involves only distances, yields a corresponding proposition in any metric space $\langle X',d' \rangle$ which is isometric to $\langle X,d \rangle$.

Example

Let $\langle \mathbb{R}^n,d \rangle$ be the metric space considered in Example 3) (p. 218) and $a = \langle a_1,a_2,\ldots,a_n \rangle$, $b = \langle b_1,b_2,\ldots,b_n \rangle$ be two points in $\mathbb{R}^n$, with $|a| =$

$\left(\sum_{i=1}^{n} a_i^2\right)^{\frac{1}{2}} \neq 0$. Consider the set $E = \{x \in \mathbb{R}^n : x = at + b\}$ (called a straight line in $\mathbb{R}^n$) where t is any real number and $x_i = a_i t + b_i$. Then the subspace $\langle E,d\rangle$ of $\langle \mathbb{R}^n,d\rangle$ is isometric to the metric space $\langle \mathbb{R},d'\rangle$ where $\mathbb{R}$ is the set of real numbers and $d'(x,y) = |x - y|$ (see Ex. 2). In fact define the mapping $f : E \to \mathbb{R}$ as follows: For $x = at + b$ in E, put $f(x) = |a|t$. If r is any real number then for $x = a(r/|a|) + b$ we have $f(x) = r$, which proves that the mapping f is onto $\mathbb{R}$. Also if $x = at + b$ and $y = at' + b$ are any two elements in E, we have:

$$d'(f(x),f(y)) = d'(|a|t,|a|t')$$

$$= |a||t - t'| = |x - y| = d(x,y),$$

so that $\langle E,d\rangle$ and $\langle \mathbb{R},d'\rangle$ are isometric.

1. Theorem *Let $f : \mathbb{R} \to \mathbb{R}$ be an isometry with respect to the usual metric $|x - y|$. Then f is one of the following forms:*

$$f(x) = x + f(0) \quad ; \quad f(x) = -x + f(0).$$

Proof For every $x \in \mathbb{R}$ we have $|f(x) - f(0)| = |x - 0| = |x|$. It follows that $f(x) = (-1)^{\alpha_x} \cdot x + f(0)$, where α_x may only assume the values 0 or 1. We wish to show that α_x is either zero for all x or that α_x is one for all x. We have $f(x) - f(y)$

$= (-1)^{\alpha_x} x - (-1)^{\alpha_y} y = (-1)^{\alpha_x} [x - (-1)^{\beta} y]$ where $\beta = \alpha_y - \alpha_x$. Clearly β may only be -1, 0 or 1. We have

$$|f(x) - f(y)| = |x - y| = |x - (-1)^{\beta} y| .$$

This implies that either $x - y = x - (-1)^{\beta} y$ or $y - x = x - (-1)^{\beta} y$. The case $y - x = x - (-1)^{\beta} y$ is impossible since it implies $2x = y[1 + (-1)^{\beta}]$. Then if $\beta = \pm 1$ we have $x = 0$, and if $\beta = 0$ we have $x = y$, and this contradicts our assumptions on x and y. Hence we must have $x - y = x - (-1)^{\beta} y$ which implies $\alpha_x = \alpha_y$. Therefore α_x has the same value, say α, for all $x \neq 0$, so that $f(x) = (-1)^{\alpha} x + f(0)$ for $x \neq 0$. Clearly the last relation is also satisfied for $x = 0$. The theorem is proved.

5.3 OPEN AND CLOSED SETS

In this section all sets mentioned are subsets of a given metric space $\langle X, d \rangle$. For a point $a \in X$ and a real number $r > 0$, the *open ball* (respectively *closed ball*) of *center* a and *radius* r is the set $S(a;r) = \{x \in X: d(a,x) < r\}$ (respectively $S[a;r] = \{x \in X: d(a,x) \leq r\}$). Every open or closed ball contains its center. An open ball $S(a;r)$ is also called a *neighborhood* of the point a, and is usually denoted by W or V.

If A,B are two nonempty sets, then the distance between A and B is defined to be the number $d(A,B) = \inf\{d(x,y): x \in A, y \in B\}$. When A is reduced to a single point x, then $d(A,B)$ is also written $d(x,B)$. If A is a nonempty set, then the *diameter* of A is defined to be the number $\delta(A) = \sup_{x,y \in A} d(x,y)$. A *bounded* set is a nonempty set whose diameter is finite.

Definition *A set O of X is said to be* open *iff it is empty or if for every $x \in O$ there exists $r > 0$ such that $S(x;r) \subset O$. The whole space X is an open set.*

Any open ball $S(a;r)$ is an open set: For if $x \in S(a;r)$ then $d(a,x) < r$. Put $r_1 = r - d(a,x)$. Let $y \in S(x;r_1)$, then $d(x,y) < r_1$. Also $d(a,y) \leq d(a,x) + d(x,y) < d(a,x) + r_1 = r$, so that $y \in S(a;r)$ and $S(x;r_1) \subset S(a;r)$. This proves that $S(a;r)$ is open.

Definition *A* point of closure *of a set E is a point $x \in X$ such that every open ball with center x contains a point of E. We denote by $\overline{E}$ (called the* closure *of E) the set of all points of closure of E. A point x is an* accumulation point *of a set E iff it is a point of closure of $E - \{x\}$. We denote by E' the set of all accumulation points of E.*

Since the center of every open ball $S(a;r)$ is an element of $S(a;r)$, it is clear that for every set E we have $E \subset \overline{E}$. Also $E' \subset \overline{E}$.

Definition *A set F is* closed *iff $\overline{F} = F$.*

As we mentioned in the introduction to this chapter, a number of properties of real numbers depend only on the fact that the space of real numbers is a metric space. As an example of this we have the following propositions which are true in the space of real numbers as well as in every metric space. The proofs are identical to those given in Chapter 2.

2. Proposition *The union of any collection of open sets is an open set. The intersection of two open sets is an open set.*

3. Proposition *The closure of any set is a closed set.*

4. Proposition *The empty set and the whole space* X *are closed sets. The intersection of any collection of closed sets is a closed set. The union of any two closed sets is a closed set.*

5. Proposition *The complement of an open (respectively closed) set is closed (respectively open).*

A set A is said to be *dense with respect to a set* B (or *dense in* B) iff $B \subset \bar{A}$; that is, if every point in B is a point of closure of A. A set A is called *everywhere dense or dense* if it is dense with respect to the whole space X; in this case we have $\bar{A} = X$.

A metric space $\langle X, d\rangle$ is said to be *separable* iff it contains a countable dense set. A typical example of a separable metric space is the real line $\mathbb{R}$ with the usual metric defined by $d(x,y) = |x - y|$. In fact the set of rational numbers is countable and everywhere dense in $\langle \mathbb{R}, d\rangle$.

A discrete metric space $\langle X,d\rangle$ (Ex. 1, p. 218) is separable iff X is a countable set. In fact, we first easily notice that every subset A of X is an open set (for $x \in A$, $S(x;\frac{1}{2}) = \{x\} \subset A$) which implies that A is also a closed set. Now if A is a countable everywhere dense set in X, then $A = \bar{A} = X$, i.e., X is countable. Also if X is countable, then since $\bar{X} = X$, X is separable.

6. Proposition *A metric space $\langle X,d\rangle$ is separable if and only if there exists a countable collection $\{O_i\}$ of open sets such that any open set $O \subset X$, can be expressed as the union of elements of $\{O_i\}$.*

Proof To prove that the condition is necessary suppose $\langle X,d\rangle$ separable, and let D be a countable dense set of X. The set of all open balls $S(x;\delta)$, with $x \in D$ and δ rational, is the required family $\{O_i\}$ of open sets. For if O is any open set and y any element of O, there is a rational δ such that $S(y;\delta) \subset O$. Since D is dense, there is an x in D such that $d(x,y) < \delta/2$. For $z \in S(x;\delta/2)$ we have $d(y,z) \leq d(y,x) + d(x,z) < (\delta/2) + (\delta/2) = \delta$, which means that $y \in S(x;\delta/2) \subset S(y;\delta) \subset O$. This proves that the set O can be expressed as the union of elements of $\{O_i\}$.

To prove that the condition is sufficient, suppose that there is a countable collection $\{O_i\}$ such that every open set O can be expressed as the union of elements of $\{O_i\}$. Let x_i be a point in O_i, and let D be the set of all those points x_i. We prove that D is dense. Let x be any point of X, and r any positive number. Since $S(x;r)$ is an open set it can be expressed as the union of elements of $\{O_i\}$, so that for some i we have $x_i \in O_i \subset S(x;r)$. Hence $x_i \in S$, which means that D is dense.

Before we close this section we wish to point out that the open and closed balls in an arbitrary metric space generally do not have the same properties as the open or closed balls in the Euclidean n-space $\mathbb{R}^n$. For example, it is true that in $\mathbb{R}^n$ the closure of any open ball $S(a;r)$ equals the closed ball $S[a;r]$. Here is an example of a metric space where this is not true, Artémiadis [2]: Let $\langle X,d\rangle$ be a subspace of $\mathbb{R}^2$ with $X = \{x: 0 \leq x \leq 1\} \cup \{x: x = e^{it}, 0 < t \leq (\pi/2)\}$. For $r = 1$ we have:

$$\overline{S}(0;1) = [0,1], \quad S[0;1] = X,$$

$$\text{so that } \overline{S}(0;1) \neq S[0,1].$$

Numerous other examples could be given. The question now arises of finding all metric spaces in which the closure of any open ball $S(a;r)$ equals $S[a;r]$. The following theorem answers this question.

Given any two distinct points a and b put: $E(a,b) = \{z \in X: d(z,a) < d(a,b), d(z,b) < d(a,b)\} \cup \{a\} \cup \{b\}$.

7. Theorem [2] *For each* $x \in X$ *and for each* $r > 0$, $\overline{S}(x;r) = S[x;r]$ *if and only if, for every* a *and* b *in* X, $a \neq b$, *the set* $E(a,b)$ *is dense with respect to itself; i.e.,* $E(a,b) \subset E'(a,b)$.

Proof Suppose that for every $x \in X$ and every $r > 0$ we have $\overline{S}(x;r) = S[x;r]$. Let a and b, $a \neq b$ be any two elements of X and $z \in E(a,b)$. Since $a \neq b$ we may suppose $z \neq a$. We have:

$$d(a,z) \leq d(a,b) \quad , \quad d(b,z) < d(a,b) .$$

By assumption $\overline{S}(a;d(a,z)) = S[a;d(a,z)]$. Since $z \in S[a;d(a,z)]$ and $z \notin S(a;d(a,z))$, z must be an accumulation point of $S(a;d(a,z))$. Let $0 < \ell < d(a,b) - d(z,b)$. There is a point y such that $y \neq z$, $y \in S(z;\ell) \cap S(a;d(a,z))$. We have:

$$d(y,b) \leq d(y,z) + d(z,b) < \ell + d(z,b) < d(a,b),$$

$$d(y,a) < d(a,z) \leq d(a,b).$$

Therefore, $y \in E(a,b)$ and $z \in E'(a,b)$. This proves that $E(a,b)$ is dense in itself.

To prove the second half of the theorem suppose that for every a and b in X, $a \neq b$, $E(a,b)$ is dense in itself. We prove that for every $x \in X$ and every $r > 0$, we have $\overline{S}(x,r) = S[x,r]$. We easily see that $\overline{S}(x,r) \subset S[x,r]$. In fact let $z \in \overline{S}(x,r)$. If $z \in S(x,r)$ then $z \in S[x,r]$. Suppose $z \in S'(x,r)$, then $d(z,x) \leq r$, because if $d(z,x) > r$ then $S(z;(d(z,x)-r)/2)$ would contain no point of $S(x,r)$, which would contradict $z \in S'(x;r)$. We prove now $S[x;r] \subset \overline{S}(x;r)$. Let $z \in S[x;r]$. If $d(z,x) < r$ then $z \in \overline{S}(x;r)$. Suppose $d(z,x) = r$. Since $E(z,x) \subset E'(z,x)$, z is an accumulation point of $E(z,x)$. Therefore, for $\ell > 0$ there is a $y \in S(z;\ell)$, $y \neq z$, such that, either $y = x$ or

$d(x,y) < d(x,z) = r$, $d(z,y) < d(x,z)$. In both cases we see that $y \in S(x;r)$. In other words in every open ball with center z, there is a point of $S(x;r)$ different from z, which means that z is an accumulation point of $S(x;r)$ and $z \in \overline{S}(x;r)$.

Definition *The metric d of a metric space $\langle X,d \rangle$ is said to be* convex *iff for each x and y in X there exists an element z in X such that*

$$d(x,z) = d(z,y) = \frac{d(x,y)}{2} .$$

The metric of the space $\mathbb{R}^n$ is an example of a convex metric. Here are some consequences of Theorem 7. The proofs are left to the reader. Also see Artémiadis [2].

8. Proposition *If $\langle X,d \rangle$ is a metric space and d is a convex metric, then for each x in X, $r > 0 : S[x;r] = \overline{S}(x;r)$.*

9. Proposition *If $\langle X,d \rangle$ is a metric space and for all x in X, $r > 0$ it is true that $S[x;r] = \overline{S}(x;r)$, and O is an open subset of X, then for each t in O, $r > 0$ it is true that $S[t;r] = \overline{S}(t;r)$ in the subspace $\langle O,d \rangle$.*

10. Proposition *If $\langle X,d \rangle$ is a metric space and for each x in X, $r > 0$ it is true that $\overline{S}[x;r] = S(x;r)$ and Y is a dense subset of X, then for each $y \in Y$, $r > 0$ it is true that $S[y,r] = \overline{S}(y,r)$ in the subspace $\langle Y,d \rangle$.*

5.4 CONTINUOUS MAPPINGS--HOMEOMORPHISMS

Let $\langle X,d\rangle$ and $\langle Y,\sigma\rangle$ be two metric spaces and x_0 a point in X. A mapping f from X to Y is said to be *continuous* at x_0 iff for every $\varepsilon > 0$ there is a $\delta > 0$ such that if $d(x_0,x) < \delta$ then $\sigma(f(x_0),f(x)) < \varepsilon$; f is said to be *continuous in* X (or simply continuous) iff it is continuous at every point of X. In terms of neighborhoods, the above definition is restated as follows: A mapping f from X to Y is said to be continuous at x_0 iff for every neighborhood W of $f(x_0)$ in Y, there is a neighborhood V of x_0 such that $f[V] \subset W$.

11. Theorem *Let f be a mapping from a metric space $\langle X,d\rangle$ to a metric space $\langle Y,\sigma\rangle$. The following statements are equivalent:*

(a) f is continuous.

(b) For every open set O in Y, $f^{-1}[O]$ is an open set in X.

(c) For every closed set F in Y, $f^{-1}[F]$ is a closed set in X.

(d) For every set A in X, $f[\overline{A}] \subset \overline{f[A]}$.

Proof (a) ⇒ (d). Let $x_0 \in \overline{A}$. We want to prove that every neighborhood W of $f(x_0)$ contains a point of $f[A]$. Since f is continuous at x_0 there is a neighborhood V of x_0 such that $f[V] \subset W$. Since $x_0 \in \overline{A}$, there is a point y such that $y \in V \cap A$, so that $f(y) \in f[A] \cap W$. This proves that $f[\bar{A}] \subset \overline{f[A]}$.

(d) ⇒ *(c)*. Let F be a closed set in Y and $A = f^{-1}[F]$. Then $f[\bar{A}] \subset \overline{f[A]} \subset \bar{F} = F$, hence $\bar{A} \subset f^{-1}[F] = A$; this implies that $\bar{A} = A$ so that $f^{-1}[F]$ is closed in X.

(c) ⇒ *(b)*. Let O be an open set in Y. Then $\complement O$ is a closed set in Y and $f^{-1}[\complement O]$ is a closed set in X. But it is easily seen that $f^{-1}[\complement O] = \complement f^{-1}[O]$, so that $f^{-1}[O]$ is open in X.

(b) ⇒ *(a)*. Let $x_0 \in X$ and W be a neighborhood of $f(x_0)$. Then $f^{-1}[W]$ is an open set in X containing x_0. Hence there is a neighborhood V of x_0 contained in $f^{-1}[W]$. It follows that $f[V] \subset W$, so that f is continuous at x_0.

It should be observed that *the direct image of an open* (respectively closed) *set by a continuous mapping is not necessarily an open set.* For example, if O is an open set in $\mathbb{R}$ and if $f(x) = 1$ for every x in O, the direct image of O by f is the set $\{1\}$ which is not open. Also the mapping g defined by $g(x) = \arctan(x)$ shows that the direct image of the closed set $\mathbb{R}$ is the open interval $(-(\pi/2),(\pi/2))$ which is not a closed set.

12. Proposition *Let f be a mapping from a metric space X into a metric space Y, and g a mapping from Y to a metric space Z. If f is continuous at $x_0 \in X$ and g continuous at $f(x_0)$ then $h = g\circ f$ is continuous at x_0.*

Proof Let W be a neighborhood of $h(x_0) = g(f(x_0))$. Then since g is continuous at $f(x_0)$ there is a neighborhood V of $f(x_0)$ such that $g[V] \subset W$. Also, since f is continuous at x_0, there is a neighborhood S of x_0 such that $f[S] \subset V$. Therefore $h[S] = g[f[S]] \subset W$, which proves that h is continuous at x_0.

13. Corollary *Let f be a mapping from a metric space X into a metric space Y, continuous at* $x_0 \in X$, *and F a subset of X containing* x_0. *Then the restriction of f to F is continuous at* x_0.

Proof Let I_F be the identity mapping from F into X (i.e., for $x \in F$, $I_F(x) = x$). Then obviously I_F is continuous in F and since the restriction of f to F is the mapping $f \circ I_F$, it follows from Proposition 12 that this restriction is continuous in F, hence at x_0. It is left to the reader to verify that the converse of the corollary is not necessarily true.

Definition *A mapping f from a metric space X into a metric space Y is called a* homeomorphism *if and only if:*

(a) f is a one-to-one continuous mapping from X onto Y (so that the inverse mapping f^{-1} *exists).*

(b) f^{-1} *is continuous on Y.*

Two metric spaces X,Y are said to be homeomorphic *if and only if there exists a homeomorphism from X onto Y.*

The word homeomorphic is derived from the Greek word ὁμοιόμορφος, meaning of similar form or structure. The study of topology is essentially the study of those properties, called *topological properties*, which remain invariant under every homeomorphism. For example the property for a subset O of a metric space X to be an open set is a topological property. For let f be a homeomorphism from X onto a metric space Y. Then since f^{-1} is continuous on Y, by virtue of Theorem 11 the inverse image of the open set O by f^{-1}, that is $f[O]$, must be open. It can be proved in the same way that the property for a set in X to be closed is a topological property. In other words the direct image of an open (respectively closed) set by a homeomorphism is an open (respectively closed) set. This leads to the following equivalent definition of a homeomorphism and homeomorphic spaces: *Two metric spaces X and Y are called homeomorphic if and only if there is a one-to-one continuous mapping from X onto Y such that the direct image of every open set in X is an open set in Y.*

An isometry f from $\langle X,d\rangle$ to $\langle Y,\sigma\rangle$ is a homeomorphism. For we already proved in section 5.2 that f is one-to-one. To prove that f is continuous let $x_0 \in X$ and $S(f(x_0);r)$ be any neighborhood of $f(x_0)$. Then there is a neighborhood $S(x_0;r)$ of x_0 such that $f[S(x_0;r)] \subset S(f(x_0);r)$. In fact for $x \in S(x_0;r)$ we have $\sigma(f(x), f(x_0)) = d(x,x_0) < r$, i.e., $f(x) \in S(f(x_0);r)$. It remains to prove that the direct image of any open set O in X is an open set in Y. To prove this, let y be a point of f[O]. Then there is an $x \varepsilon O$ with $f(x) = y$. Since O is open there is a $\delta > 0$ such that $S(x;\delta) \subset O$. We prove that $S(y;\delta) \subset f[O]$. Let $y_1 \varepsilon S(y;\delta)$. Since f maps X onto Y there is an x_1 such that $f(x_1) = y_1$. We have $\sigma(y,y_1) = d(x,x_1) < \delta$. Therefore $x_1 \varepsilon S(x;\delta)$ so that $x_1 \varepsilon O$. This means that $y_1 \varepsilon f[O]$, hence f[O] is open.

5.5 EQUIVALENT METRICS

Let $\langle X,d\rangle$, $\langle X,\sigma\rangle$ be two metric spaces. As we pointed out in section 5.1, although these metric spaces have the same underlying set X they must be considered as distinct if the metrics d and σ are distinct. If the identity mapping $x \to x$ from $\langle X,d\rangle$ onto $\langle X,\sigma\rangle$ is a homeomorphism then d and σ are called *equivalent metrics*. It follows from this definition that if d and σ are equivalent, then every set which is open in $\langle X,d\rangle$ is also open in $\langle X,\sigma\rangle$ and vice versa. In other words the two metric spaces have the same open sets. Conversely, if $\langle X,d\rangle$ and $\langle X,\sigma\rangle$ have the same open sets then, d and σ are equivalent metrics. For in that case the identity mapping is obviously a one-to-one continuous mapping such that the direct image of any open set in $\langle X,d\rangle$ is an open set in $\langle Y,\sigma\rangle$.

5.6 LIMITS--CONVERGENCE--COMPLETENESS

A sequence $\langle x_n\rangle$ from a metric space $\langle X,d\rangle$ is said to *converge* to a point a in X iff for every neighborhood V of a there exists an integer N such that the relation $n \geq N$ implies $x_n \in V$ (that is all x_n, except a finite number of them, are in V). We say in this case that a is *limit* of the sequence $\langle x_n \rangle$ and we write $\lim_{n\to\infty} x_n = a$, or $x_n \to a$.

A restatement of the above definition is the following. *The sequence* $\langle x_n \rangle$ *converges to* a *iff for every* $\varepsilon > 0$ *there exists an integer* N *such that the relation* $n \geq N$ *implies* $d(a,x_n) < \varepsilon$.

It follows from the last definition that $\lim_{n\to\infty} x_n = a$ iff $\lim_{n\to\infty} d(a,x_n) = 0$.

We can prove in the same way as in the case of real numbers, that a sequence can converge at most to one point. Also if $\lim_{n\to\infty} x_n = a$ then $\lim_{k\to\infty} x_{n_k} = a$ for any subsequence $\langle x_{n_k} \rangle$ of $\langle x_n \rangle$ (see Prob. 29).

A point a is called a *cluster point* of a sequence $\langle x_n \rangle$ iff given $\varepsilon > 0$ and given N, there is an $n \geq N$ such that $d(x_n,a) < \varepsilon$. This is equivalent to saying that a is a cluster point of $\langle x_n \rangle$ iff every neighborhood of a contains infinitely many terms of $\langle x_n \rangle$. If a is the limit of $\langle x_n \rangle$ then a is a cluster point of $\langle x_n \rangle$, but, as in the case of real numbers, the converse is not true.

A sequence $\langle x_n \rangle$ from a metric space $\langle X,d \rangle$ is called a *Cauchy sequence* iff given $\varepsilon > 0$ there exists an integer N such that the relations $p \geq N$, $q \geq N$ imply $d(x_p, x_q) < \varepsilon$.

Every convergent sequence is a Cauchy sequence. For let $\langle x_n \rangle$ be a sequence converging to a point a. Then for any $\varepsilon > 0$ there exists an integer N such that $n \geq N$ implies $d(a, x_n) < \varepsilon/2$. For $p \geq N$, $q \geq N$ we have $d(x_p, x_q) \leq d(x_p, a) + d(a, x_q) < \varepsilon/2 + \varepsilon/2 = \varepsilon$. The converse of this statement is not true in an arbitrary metric space. That is every Cauchy sequence is not always convergent to some point of the space. To show this consider the subspace X of the real line consisting of all positive real numbers. The sequence $\langle (1/n) \rangle$ is a Cauchy sequence but it is not convergent to any point of X.

Definition *Let $\langle X, d \rangle$ and $\langle Y, \sigma \rangle$ be metric spaces; suppose $E \subset X$, f maps E into Y, and a is an accumulation point of E. We write*

$$\lim_{x \to a} f(x) = y \tag{1}$$

iff there is a point $y \in Y$ with the following property: For every $\varepsilon > 0$ there exists a $\delta > 0$ such that

$$\sigma(f(x), y) < \varepsilon$$

for all points $x \in E$ for which

$$0 < d(x, a) < \delta .$$

In terms of limits of sequences this definition

can be reformulated as follows:
(1) holds iff

(2) $$\lim_{n\to\infty} f(x_n) = y$$

for every sequence $\langle x_n \rangle$ in E such that

(3) $$x_n \neq a \quad , \quad \lim_{n\to\infty} x_n = a \ .$$

For suppose (1) holds. Let $\langle x_n \rangle$ be a sequence in E satisfying (3) and let $\varepsilon > 0$ be given. Then there exists $\delta > 0$ such that $\sigma(f(x),y) < \varepsilon$ if $x \in E$ and $0 < d(x,a) < \delta$. Also there exists N such that $n > N$ implies $0 < d(x_n,a) < \delta$. Hence for $n > N$ we have $\sigma(f(x_n),y) < \varepsilon$ which shows that (2) holds.

Next suppose (1) is false. Then there exists some $\varepsilon > 0$ such that for every $\delta > 0$ there exists a point $x \in E$ (depending on δ) such that $\sigma(f(x),y) \geq \varepsilon$ and $0 < d(x,a) < \delta$. Taking $\langle \delta_n \rangle$ to be a strictly decreasing sequence of positive real numbers such that $\lim_{n\to\infty} \delta_n = 0$, we find a sequence in E satisfying (3) for which (2) does not hold. A contradiction.

It is clear that *if* f *has a limit at* a, *this limit is unique.*

Definition *A metric space* X *is called* complete *iff every Cauchy sequence from* X *is convergent to a point in* X.

Examples

All metric spaces given in the examples considered in section 5.1 are complete, except the one given in Example 7. In fact:

1) In a discrete space since the distance of two distinct points is one, only those sequences in which all the terms, except a finite number of them, are equal, are Cauchy sequences. Obviously such a sequence is convergent, i.e., the space is complete.

2) The completeness of the space $\mathbb{R}$ of real numbers is known from calculus.

3) We prove the completeness of the Euclidean n-space $\mathbb{R}^n$. Let $\langle x_i \rangle$ (with $x_i = \langle x_i^{(1)}, x_i^{(2)}, \ldots, x_i^{(n)} \rangle$) be a Cauchy sequence. Then for every $\varepsilon > 0$ there exists a positive integer N such that $\sum_{k=1}^{n} (x_p^{(k)} - x_q^{(k)})^2 < \varepsilon^2$ for $p \geq N$, $q \geq N$. It follows that for each $k = 1,2,\ldots,n$ and for all $p,q \geq N$ we have $|x_p^{(k)} - x_q^{(k)}| < \varepsilon$, i.e., $\langle x_p^{(k)} \rangle$ is a Cauchy sequence of real numbers. We set $x^{(k)} = \lim_{p\to\infty} x_p^{(k)}$ and $x = \langle x^{(1)}, x^{(2)}, \ldots, x^{(n)} \rangle$. Then it is easy to prove that $\lim_{i\to\infty} x_i = x$.

4) The completeness of the space $\mathbb{R}_o^n$ can be proved in exactly the same way as in the case of the space $\mathbb{R}^n$.

5) We prove the completeness of the space $C[a,b]$. Let $\langle x_n \rangle$ be a Cauchy sequence from $C[a,b]$. Then for each $\varepsilon > 0$ there is an N such that for

$n,m \geq N$, $d(x_n,x_m)<\varepsilon$. From the definition of the metric in $C[a,b]$ it follows that $|x_n(t) - x_m(t)| < \varepsilon$ for all $t \in [a,b]$. This means that the sequence $\langle x_n \rangle$ converges uniformly in $[a,b]$ to a function x which is continuous in $[a,b]$, i.e., $x \in C[a,b]$. Therefore, for given $\varepsilon > 0$ there is an integer N such that for $n \geq N$ $|x_n(t) - x(t)| < \varepsilon$ for all $t \in [a,b]$. This implies that $\sup_{t\in[a,b]} \{|x_n(t) - x(t)|\}$ $= d(x_n,x) < \varepsilon$, which means that $\lim_{n\to\infty} d(x_n,x) = 0$ so that $\lim_{n\to\infty} x_n = x$, i.e., the space $C[a,b]$ is complete.

6) The space ℓ^2 is complete. Let $\langle x^{(n)} \rangle$ be a Cauchy sequence from ℓ^2, with $x^{(n)} = \langle x_1^{(n)}, x_2^{(n)}, \ldots, x_m^{(n)}, \ldots \rangle$. Then for given $\varepsilon > 0$ there is a positive integer N such that $d^2(x^{(p)},x^{(q)}) =$ $\sum_{k=1}^{\infty} (x_k^{(p)} - x_k^{(q)})^2 < \varepsilon$ for $p,q \geq N$. It follows that for arbitrary k, $(x_k^{(p)} - x_k^{(q)})^2 < \varepsilon$ which means that for each k the sequence $\langle x_k^{(n)} \rangle$ is a Cauchy sequence of real numbers so that it converges to some real number, say x_k. Denote the sequence $\langle x_k \rangle$ by x. We show that:

a) $$\sum_{k=1}^{\infty} x_k^2 < +\infty ;$$

b) $$\lim_{n\to\infty} d(x^{(n)},x) = 0$$

In fact for an arbitrary positive integer M, we have

$$\sum_{k=1}^{\infty}(x_k^{(p)} - x_k^{(q)})^2 =$$

$$\sum_{k=1}^{M}(x_k^{(p)} - x_k^{(q)})^2 + \sum_{k=M+1}^{\infty}(x_k^{(p)} - x_k^{(q)})^2 < \varepsilon.$$

It follows that $\sum_{k=1}^{M}(x_k^{(p)} - x_k^{(q)})^2 < \varepsilon$. If in this inequality we fix q and pass to the limit as $p \to \infty$ we get $\sum_{k=1}^{M}(x_k - x_k^{(q)})^2 \leq \varepsilon$. Since this inequality is valid for arbitrary M we have $\sum_{k=1}^{\infty}(x_k - x_k^{(q)})^2 \leq \varepsilon$. From this last inequality and the fact that the series $\sum_{k=1}^{\infty} x_k^{(q)^2}$ is convergent it follows that the series $\sum_{k=1}^{\infty} x_k^2$ is also convergent, which means that x is a point of ℓ^2. Further, since ε is arbitrary it follows that:

$$\lim_{q\to+\infty} d(x^{(q)},x) =$$

$$\lim_{q\to\infty}\left(\sum_{k=1}^{\infty}(x_k - x_k^{(q)})^2\right)^{\frac{1}{2}} = 0, \quad \text{i.e., } x^{(n)} \to x.$$

7) The space $C^2[a,b]$ is not complete. We leave to the reader to verify that the sequence $f_n(t) = \arctan(nt)$ $(-1 \leq t \leq 1)$ is a Cauchy sequence from $C^2[a,b]$ but it does not converge to any point of this space (see Prob. 34).

8) The space ℓ^∞ of all bounded infinite sequences is complete. The proof is left to the reader

9) Let X be an arbitrary nonempty set and let $\langle Y,d\rangle$ be a complete metric space. Let $\mathcal{B}(X,Y)$ be the set of all bounded mappings from X into Y. For f and g in $\mathcal{B}(X,Y)$ define

$$D(f,g) = \sup\{d(f(x), g(x)) : x \in X\} .$$

Then D is a metric on $\mathcal{B}(X,Y)$, and $\langle \mathcal{B}(X,Y),D\rangle$ is a complete metric space.
(*Note:* A mapping $f : X \to Y$ is called *bounded* iff $f[X]$ is a bounded subset of Y.)
Proof The proof that D is a metric is left to the reader. Let $\langle f_n\rangle$ be a Cauchy sequence in $\mathcal{B}(X,Y)$. For every $x \in X$ the inequality $d(f_p(x), f_q(x)) \leq D(f_p,f_q)$ shows that the sequence $\langle f_n(x)\rangle$ is a Cauchy sequence in Y. Since Y is complete, this sequence has a limit, say $f(x)$. Since $\langle f_n\rangle$ is a Cauchy sequence, for every $\varepsilon > 0$ there exists an integer n_ε such that for all $p,q \geq n_\varepsilon$ and every $x \in X$

$$d(f_p(x), f_q(x)) \leq \varepsilon .$$

If in this inequality we fix x and p, and let

$q \to \infty$, then $f_q(x)$ tends to $f(x)$, and we obtain the inequality

$$d(f_p(x), f(x)) \leq \varepsilon \quad \text{for all} \quad p \geq n_\varepsilon .$$

This implies $D(f_p, f) \leq \varepsilon$ which proves that the sequence $\langle f_n \rangle$ converges to f in the space $\langle \mathcal{B}(X,Y), D\rangle$. Hence this space is complete.

10) Let $\langle X,d\rangle$ be a metric space and let $\mathcal{C}(X,\mathbb{R})$ be the collection of all bounded continuous real-valued functions on X. For $f,g \in \mathcal{C}(X,\mathbb{R})$ define $D(f,g)$ as in Example 9. Then $\mathcal{C}(X,\mathbb{R})$ is a complete subspace of $\mathcal{B}(X,\mathbb{R})$.

14. Theorem *A metric space $\langle X,d\rangle$ is complete if and only if every sequence of closed nested balls $\langle S_n[x_n;r_n]\rangle$ (i.e., $S_{n+1} \subset S_n$ for all n) with radii tending to zero, have a nonempty intersection.*

Proof Suppose $\langle X,d\rangle$ is complete and let $\langle S_n[x_n;r_n]\rangle$ be a sequence of closed nested balls such that $r_n \to 0$. Then $\langle x_n\rangle$ is a Cauchy sequence. In fact let $\varepsilon > 0$ be given. Since by hypothesis $\lim_{n\to\infty} r_n = 0$, there is an N such that for $n \geq N$, $r_n < \varepsilon$. If $q > p > N$ we have $S_q \subset S_p$ so that $d(x_p, x_q) < r_p < \varepsilon$. Since X is complete there is an $x \in X$ such that $\lim_{n\to\infty} x_n = x$. We next prove that $x \in \bigcap_{n=1}^{\infty} S_n$. Each S_n contains all the x_n except, perhaps, the points $x_1, x_2, \ldots, x_{n-1}$. This means that x is a point of closure of S_n and since S_n is closed we have $x \in S_n$, for all n. Suppose now that $\langle X,d \rangle$ is not complete: i.e., suppose there is a Cauchy sequence from X which does not converge to any point of X. We prove that it is possible to construct a sequence of closed nested balls in X whose radii tend to zero and whose intersection is empty. In fact let $\langle x_n \rangle$ be a nonconvergent Cauchy sequence. Let n_1 be such that $d(x_{n_1},x_m) < 1/2$ for all $m > n_1$; n_2 be such that $n_2 > n_1$ and $d(x_{n_2},x_m) < 1/4$ for all $m > n_2$, and so forth. We obtain, by induction, a subsequence of integers $\langle n_k \rangle$ such that $d(x_{n_k},x_m) < 1/2^k$ for all $m > n_k$. Consider now the sequence $S_k[x_{n_k};1/(2^k-1)]$ of closed balls whose radii obviously tend to zero. We have $S_{k+1} \subset S_k$ $(k = 1,2,\ldots)$. For if $z \in S_{k+1}$, then $d(z,x_{n_{k+1}}) \leq 1/2^k$, and $d(z,x_{n_k}) \leq d(z,x_{n_{k+1}}) + d(x_{n_{k+1}},x_{n_k}) \leq (1/2^k) + (1/2^k) = 1/2^{k-1}$. We have $\bigcap_{k=1}^{\infty} S_k = \emptyset$. For suppose on the contrary that there is a point $x \in \bigcap_{k=1}^{\infty} S_k$. Then the ball S_k contain all points x_n beginning x_{n_k}, and consequently $d(x,x_n) < 1/2^{k-1}$ for all $n > n_k$. This implies that $\lim_{n\to\infty} x_n = x$, which contradicts the assumption that

$\langle x_n \rangle$ is not convergent. Therefore $\bigcap_n S_n = \emptyset$.

The great advantage that one has in dealing with complete metric spaces is that to prove a sequence is convergent it suffices only to prove that it is a Cauchy sequence. In other words the limit itself of the sequence does not have to be known in advance.

15. Theorem *If* $\langle X,d \rangle$ *is an incomplete metric space then it is possible to find a complete metric space* $\langle X^*,d^* \rangle$ *such that* X *is isometric to a dense subspace of* X^*.

Proof Let E be the set of all Cauchy sequences from X. We define the binary relation $\sim$ on E as follows: For $\langle x_n \rangle \in E$, $\langle x_n' \rangle \in E$ we put $\langle x_n \rangle \sim \langle x_n' \rangle$ iff $\lim_{n\to\infty} d(x_n, x_n') = 0$. It is easy to prove that $\sim$ is an equivalence relation on E. Therefore $\sim$ partitions E into equivalence classes. Put $E/\sim\ = X^*$. The elements of X^* are the equivalence classes. Next, let x^*, y^* be any two elements in X^*, $\langle x_n \rangle$ any Cauchy sequence in x^*, and $\langle y_n \rangle$ any Cauchy sequence in y^*. We prove that the sequence $\langle d(x_n,y_n) \rangle$ is convergent and that its limit does not depend on the choice of the sequences $\langle x_n \rangle$, $\langle y_n \rangle$. In fact since $\langle x_n \rangle$, $\langle y_n \rangle$ are Cauchy sequences, using the triangle inequality, we have for all sufficiently large p and q and for a given $\varepsilon > 0$:

$$|d(x_p,y_p) - d(x_q,y_q)|$$

$$= |d(x_p,y_p) - d(x_p,y_q) + d(x_p,y_q) - d(x_q,y_q)|$$

$$\leq |d(x_p,y_p) - d(x_p,y_q)| + |d(x_p,y_q) - d(x_q,y_q)|$$

$$\leq d(y_p,y_q) + d(x_p,x_q) < \frac{\varepsilon}{2} + \frac{\varepsilon}{2} = \varepsilon \ .$$

Thus the sequence $\langle d(x_n,y_n)\rangle$ is a Cauchy sequence of real numbers and consequently it has a limit. To prove that this limit does not depend on the choice of $\langle x_n\rangle \; \varepsilon \; x^*$ and $\langle y_n\rangle \; \varepsilon \; y^*$ let $\langle x_n'\rangle \; \varepsilon \; x^*$, $\langle y_n'\rangle \; \varepsilon \; y^*$. We have $\langle x_n\rangle \sim \langle x_n'\rangle$, $\langle y_n\rangle \sim \langle y_n'\rangle$. This implies that

$$|d(x_n,y_n) - d(x_n',y_n')|$$

$$= |d(x_n,y_n) - d(x_n',y_n) + d(x_n',y_n) - d(x_n',y_n')|$$

$$\leq |d(x_n,y_n) - d(x_n',y_n)| + |d(x_n',y_n) - d(x_n',y_n')|$$

$$\leq d(x_n,x_n') + d(y_n,y_n') \to 0 \ .$$

i.e., $\lim_{n\to\infty} d(x_n,y_n) = \lim_{n\to\infty} d(x_n',y_n')$.

We now define the metric d^* in X^* as follows: For any two elements x^*, y^* we set

$$d^*(x^*,y^*) = \lim_{n\to\infty} d(x_n,y_n)$$

where $\langle x_n\rangle \; \varepsilon \; x^*$, $\langle y_n\rangle \; \varepsilon \; y^*$.

We show that $\langle X^*,d^*\rangle$ is a metric space. In

fact let x^*, y^*, z^* be any three elements in X^*. Condition (i) of the definition of metric space follows from the definition of the equivalence relation $\sim$. Condition (ii) is obvious. To prove condition (iii) let $\langle x_n \rangle \varepsilon x^*$, $\langle y_n \rangle \varepsilon y^*$, $\langle z_n \rangle \varepsilon z^*$. Since the triangle inequality is satisfied in $\langle X,d \rangle$ we have for all n:

$$0 \leq d(x_n,y_n) + d(y_n,z_n) - d(x_n,z_n) .$$

Passing to the limit as $n \to \infty$ we obtain:

$$\lim_{n\to\infty} d(x_n,z_n) \leq \lim_{n\to\infty} d(x_n,y_n) + \lim_{n\to\infty} d(y_n,z_n)$$

$$\text{i.e.,} \quad d^*(x^*,z^*) \leq d^*(x^*,y^*) + d(y^*,z^*) .$$

Now define the mapping $f : X \to X^*$ as follows: For $x \varepsilon X$ let $f(x)$ be the equivalence class which contains the Cauchy sequence $\langle x_n \rangle$ where $x_n = x$ for all n. Put $f[X] = Y$. We want to prove that:

(a) $\langle X,d \rangle$ and $\langle Y,d^* \rangle$ are isometric.

(b) $\overline{Y} = X^*$.

(c) $\langle X^*,d^* \rangle$ is complete.

To prove (a) we notice that the mapping f is one-to-one. In fact if $x \varepsilon X$, $y \varepsilon X$, $x \neq y$ then $f(x) \neq f(y)$; for if $f(x) = f(y)$ the Cauchy sequences $\langle x_n \rangle$ with $x_n = x$, and $\langle y_n \rangle$ with $y_n = y$ for all n, would be equivalent, so that we would have $\lim_{n\to\infty} d(x_n,y_n) = d(x,y) = 0$, i.e., $x = y$ which contradicts $x \neq y$. Also $d^*(f(x),f(y)) = \lim_{n\to\infty} d(x_n,y_n) = d(x,y)$.

To prove (b) let $x^* \varepsilon X^*$ and $\langle x_n \rangle$ be an element in x^*. For every $\varepsilon > 0$ there is an integer N such that, $p \geq N$, $q \geq N$ imply $d(x_p, x_q) < \varepsilon/2$. Let x_p^* be the equivalence class which contains the Cauchy sequence $\langle x_m \rangle$ where $x_m = x_p$ for all m. We have $x_p^* = f(x_p) \varepsilon Y$. Also $d^*(x_p^*, x^*) = \lim_{m\to\infty} d(x_p, x_m) \leq \varepsilon/2 < \varepsilon$. This last inequality shows that in an arbitrary neighborhood of x^* there is an element of Y, namely x_p^*, i.e., $x^* \varepsilon \overline{Y}$. This proves (b).

We finally prove (c). Let $\langle x_n^* \rangle$ be a Cauchy sequence in X^*. Since Y is dense in X^*, for each n there is a $y_n^* \varepsilon Y$ such that $d^*(y_n^*, x_n^*) < 1/n$. The sequence $\langle y_n^* \rangle$ is a Cauchy sequence in X^*. In fact for $\varepsilon > 0$ and for p, q large enough, using the triangle inequality and the fact that $\langle x_n^* \rangle$ is a Cauchy sequence we have:

$$d^*(y_p^*, y_q^*)$$

$$\leq d^*(y_p^*, x_p^*) + d^*(x_p^*, x_q^*) + d^*(x_q^*, y_q^*)$$

$$< \frac{\varepsilon}{3} + \frac{\varepsilon}{3} + \frac{\varepsilon}{3} = \varepsilon .$$

Since X and Y are isometric, the sequence $\langle y_m \rangle$ where $f^{-1}(y_m^*) = y_m \varepsilon X$, is also a Cauchy sequence. Let $x^* \varepsilon X^*$ be the equivalence class containing $\langle y_m \rangle$. Then it follows easily from the definition of the metric d^* that $\lim_{n\to\infty} d^*(x_n^*,x^*) = 0$, i.e., $\langle X^*,d^*\rangle$ is complete.

5.7 UNIFORMLY CONTINUOUS MAPPINGS

A *uniformly continuous mapping* from a metric space $\langle X,d\rangle$ to a metric space $\langle Y,\sigma\rangle$ is a mapping f such that for every $\varepsilon > 0$ there exists a $\delta > 0$ such that for all x and x' in X with $d(x,x') < \delta$ we have $\sigma(f(x),f(x')) < \varepsilon$. It follows from this definition that *a mapping f from a metric space* $\langle X,d\rangle$ *to a metric space* $\langle Y,\sigma\rangle$ *is not uniformly continuous iff there is an* $\varepsilon_o > 0$ *such that for every* $\delta > 0$ *there are* x,x' *in* X, *with* $d(x,x') < \delta$ *such that* $\sigma(f(x),f(x')) \geq \varepsilon_o$.

It is obvious that a uniformly continuous mapping is also continuous, while the converse is not true. For example, the function h defined on $(0,1]$ by $h(x) = 1/x$ is continuous but not uniformly continuous on $(0,1]$. In fact let $\varepsilon_o = 1$. If δ is given put $\delta_1 = \min(1,\delta)$ and choose $x = \delta_1/2$, $x' = \delta_1$. Then x,x' belong to $(0,1]$, $|x - x'| < \delta$, and

$$|h(x) - h(x')| = |(2/\delta_1) - (1/\delta_1)| = (1/\delta_1) \geq 1 .$$

The same example shows that the image of a Cauchy sequence under a continuous mapping is not necessarily a Cauchy sequence. For the image of the Cauchy sequence $\langle(1/n)\rangle$ $(n = 1,2,\ldots)$ under h is the sequence $\langle n\rangle$ which is not a Cauchy sequence.

The following proposition shows that uniformly continuous mappings have the property to preserve Cauchy sequences.

16. Proposition *Let* $\langle X,d\rangle$ *and* $\langle Y,\sigma\rangle$ *be metric spaces and* $f : X \to Y$ *a uniformly continuous mapping. If* $\langle x_n\rangle$ *is a Cauchy sequence in* X, *then* $\langle f(x_n)\rangle$ *is a Cauchy sequence in* Y.

Proof For every $\varepsilon > 0$ there is a $\delta > 0$ such that for all x, x' in X, with $d(x,x') < \delta$ we have $\sigma(f(x),f(x')) < \varepsilon$. Since $\langle x_n\rangle$ is a Cauchy sequence for this δ there is an integer N such that p,q $\geq$ N imply $d(x_p,x_q) < \delta$. Thus p,q $\geq$ N imply $\sigma(f(x_p),f(x_q)) < \varepsilon$. This proves that $\langle f(x_n)\rangle$ is a Cauchy sequence.

A mapping f from a metric space $\langle X,d\rangle$ into a metric space $\langle Y,\sigma\rangle$ is called a *uniform homeomorphism* iff f is a homeomorphism and f and f^{-1} are both uniformly continuous.

Obviously if f is a uniform homeomorphism from X onto Y then f^{-1} is a uniform homeomorphism from Y onto X.

A property of a metric space which is preserved under every uniform homeomorphism is called a *uniform property*.

17. Proposition *If* X *is a complete metric space and* f *a uniform homeomorphism from* X *to a metric space* Y, *then* Y *is also complete.*

Proof Let $\langle y_n\rangle$ be a Cauchy sequence in Y. Since f^{-1} is uniformly continuous, we have by Proposition 16 that $\langle f^{-1}(y_n)\rangle$ is a Cauchy sequence in X. Since

X is complete, there is an $x \in X$ such that $\lim_{n\to\infty} f^{-1}(y_n) = x$. Put $f(x) = y$. Since f is continuous at x we have $\lim_{n\to\infty} f(f^{-1}(y_n)) = \lim_{n\to\infty} y_n = f(x) = y$. Therefore Y is complete.

It follows from Propositions 16 and 17 that the properties of being Cauchy sequence and of completeness are uniform properties.

It is left to the reader to prove that uniform continuity is another uniform property. By this we mean that if f is a uniformly continuous mapping from a metric space X to a metric space Y, and if h is a uniform homeomorphism from Y onto a metric space Z, then the composite mapping $h \circ f$ is uniformly continuous from X to Z. Also if h' is a uniform homeomorphism from a metric space X' onto X then $f \circ h'$ is uniformly continuous from X' to Y. However these three uniform properties are not topological properties. To see this let h be the mapping, considered earlier, defined on $(0,1]$ with $h(x) = 1/x$. Then h is a homeomorphism from the incomplete space $(0,1]$ onto the complete space $[1,+\infty)$, and it does not preserve Cauchy sequences. Furthermore, if f is the identity mapping on $(0,1]$, then f is uniformly continuous, while $h \circ f$ is not.

Let $\langle X,d\rangle$, $\langle X,\sigma\rangle$ be two metric spaces with the same underlying set X. Then the metrics d and σ are said to be *uniformly equivalent* iff the identity mapping from $\langle X,d\rangle$ to $\langle X,\sigma\rangle$ is a uniform homeomorphism. It follows from this definition that d and σ are uniformly equivalent iff for given $\varepsilon > 0$ there is a $\delta > 0$ such that:

(i) $d(x,y) < \delta$ implies $\sigma(x,y) < \varepsilon$.

(ii) $\sigma(x,y) < \delta$ implies $d(x,y) < \varepsilon$.

In fact, suppose d and σ uniformly equivalent. Then for a given $\varepsilon > 0$ there are $\delta_1 > 0$, $\delta_2 > 0$ such that $d(x,y) < \delta_1$ implies $\sigma(x,y) < \varepsilon$, and $\sigma(x,y) < \delta_2$ implies $d(x,y) < \varepsilon$. For $\delta = \min(\delta_1,\delta_2)$ conditions (i) and (ii) are both satisfied. Conversely, it is obvious that for a given $\varepsilon > 0$ conditions (i) and (ii) imply that d and σ are uniformly equivalent.

5.8 EXTENSION OF MAPPINGS

Let E be a subset of a metric space X and f be a mapping from E into a metric space Y. A mapping g from X into Y is called an *extension* of f (from E to X) iff $f(x) = g(x)$ for every $x \varepsilon E$.

The following few lines will help the reader to understand what we mean by an extension problem. We recall from our high school mathematics that if n is a positive integer then 3^n is by definition the number obtained by multiplying 3 by itself n-times. Also, there was no great difficulty in understanding the definition of 3^x for x rational. Then the question was how to define 3^x for any real number x, without destroying any definition or property previously obtained for powers with rational exponents. In other words the problem that the student was facing at that moment was an extension problem that can be formulated as follows:

Construct a continuous mapping f from the set $\mathbb{R}$ of real numbers into $\mathbb{R}$ such that for every rational number x, $f(x) = 3^x$. Prove that the mapping f is unique.

The following theorem provides another example of the extension problem concerning uniformly continuous mappings.

18. Proposition *Let* $\langle X,d\rangle$ *be a metric space and* E *be a dense subset of* X. *Let* f *be a uniformly continuous mapping from* E *into a complete metric space* $\langle Y,\sigma\rangle$. *Then there exists a unique continuous extension* g *of* f *from* E *to* X. *Moreover,* g *is uniformly continuous.*

Proof Let $x \in X$. Since E is dense, there is a sequence $\langle x_n\rangle$ from E which converges to x. It follows from Proposition 16 that $\langle f(x_n)\rangle$ is a Cauchy sequence and since Y is complete $\langle f(x_n)\rangle$ converges to an element $y \in Y$. The point y does not depend on the choice of the sequence $\langle x_n\rangle$. In fact let $\langle x_n'\rangle$ be another sequence from E converging to x. Then $d(x_n,x_n') \leq d(x_n,x) + d(x,x_n') \to 0$, i.e., $d(x_n,x_n') \to 0$. This implies that $\sigma(f(x_n),f(x_n'))$ $\to 0$. But $\sigma(f(x_n'),y) \leq \sigma(f(x_n'),f(x_n)) + \sigma(f(x_n),y)$ so that $\sigma(f(x_n'),y) \to 0$, i.e., $\lim_{n\to\infty} f(x_n') = y$. Next, for every $x \in X$ define $g(x) = y$. We prove that the mapping g is the extension sought. To show that g is uniformly continuous on X, let $\varepsilon > 0$ be given. Then there is a $\delta_1 > 0$ such that for $z,z' \in E$ with $d(z,z') < \delta_1$ we have

$\sigma(f(z),f(z')) < \varepsilon/3$. Let $x,x' \in X$ with $d(x,x') < \delta$, where $\delta_1 = 2\delta$. Because of the way that g is constructed we can find x_o,x_o' in E such that $d(x,x_o) < \delta/2$, $d(x',x_o') < \delta/2$ imply $\sigma(g(x),g(x_o)) < \varepsilon/3$, $\sigma(g(x'),g(x_o')) < \varepsilon/3$. We have

$$d(x_o,x_o') \leq d(x_o,x) + d(x,x') + d(x',x_o')$$

$$< \frac{\delta}{2} + \delta + \frac{\delta}{2} = \delta_1$$

which implies $\sigma(f(x_o),f(x_o')) < \varepsilon/3$. Also

$$\sigma(g(x),g(x'))$$

$$\leq \sigma(g(x),g(x_o)) + \sigma(g(x_o),g(x_o')) + \sigma(g(x_o'),g(x'))$$

$$< \frac{\varepsilon}{3} + \frac{\varepsilon}{3} + \frac{\varepsilon}{3} = \varepsilon .$$

Finally if h is any other extension of f, then $h = g$. The proof is left to the reader.

5.9 THE METHOD OF SUCCESSIVE APPROXIMATIONS

A powerful method of proving the existence of a solution of an equation (either algebraic or differential or integral equation) and sometimes effectively calculating this solution, is the so-called *method of successive approximations*. An outline of this method is given by the following theorem.

A mapping f from a metric space $\langle X,d\rangle$ into a metric space $\langle Y,\sigma\rangle$ is called a *contraction* iff there is a real number $0 \leq \lambda < 1$ such that for every x and y in X, $\sigma(f(x),f(y) \leq \lambda d(x,y)$. Every contraction mapping is continuous. For if $x_n \to x$ then $\sigma(f(x_n),f(x)) \to 0$, i.e., $f(x_n) \to f(x)$.

19. Theorem *Let $\langle X,d\rangle$ be a complete metric space and f a contraction mapping from X into itself. Then f has one and only one fixed point (i.e., the equation $f(x) = x$ has one and only one solution).*

Proof Let x_0 be an arbitrary point in X. Set $x_1 = f(x_0)$ and $x_n = f(x_{n-1}) = f^n(x_0)$ for n = 2, 3,... . The sequence $\langle x_n\rangle$ is a Cauchy sequence. In fact, for m > n we have:

$$d(x_n,x_m) = d(f^n(x_0),f^m(x_0)) \leq \lambda^n d(x_0,x_{m-n})$$

$$\leq \lambda^n\{d(x_0,x_1) + d(x_1,x_2) + \ldots + d(x_{m-n-1})x_{m-n})\}$$

$$\leq \lambda^n d(x_0,x_1)\{1 + \lambda + \lambda^2 + \ldots + \lambda^{m-n-1}\}$$

$$\leq \lambda^n d(x_0,x_1)(1/(1 - \lambda)) .$$

Since $0 \leq \lambda < 1$ the last member of this inequality can be made arbitrarily small, which proves that $\langle x_n\rangle$ is a Cauchy sequence. Since X is complete, there is an $x \in X$ such that $\lim_{n\to\infty} x_n = x$. By virtue of the continuity of f we have:

$$f(x) = f(\lim_{n\to\infty} x_n)$$

$$= \lim_{n\to\infty} f(x_n) = \lim_{n\to\infty} x_{n+1} = x .$$

This proves that the point x is a fixed point. To prove the uniqueness of x let us suppose that for some point $y \in X$ we also have $f(y) = y$. Then $d(x,y) \leq \lambda d(f(x),f(y)) = \lambda d(x,y)$ which implies that $d(x,y) = 0$, i.e., $x = y$.

Here are some applications of Theorem 19.

1) Let f be a mapping defined on the closed interval $[a,b]$ satisfying the Lipschitz condition $|f(x_2) - f(x_1)| \leq k|x_2 - x_1|$ with $k < 1$, and taking $[a,b]$ into itself. Then f is a contraction mapping. If x_0 is any point in $[a,b]$ the sequenc $x_0, x_1 = f(x_0)$, $x_2 = f(x_1), \ldots$ converges to the singl root of the equation $f(x) = x$.

Using the mean value theorem we see that the Lipschitz condition is satisfied if the function f is differentiable and satisfies the condition $|f'(x)| \leq k < 1$ in $[a,b]$.

2) Let

(i) $dy/dx = f(x,y)$

be a given differential equation with the initial condition

(ii) $y(x_0) = y_0$,

where $f(x,y)$ is defined and continuous in some oper set G of the plane containing the point $\langle x_0, y_0 \rangle$, and satisfies a Lipschitz condition with respect to y

$$|f(x,y_1) - f(x,y_2)| \leq M|y_1 - y_2| .$$

Then there exists a unique solution $y = \phi(x)$ of (i) defined on a certain closed interval $[x_0 - c, x_0 + c]$, and such that $y_0 = \phi(x_0)$ (Picard's Theorem).

Proof Equation (i) together with the initial condition (ii) is equivalent to the integral equation:

$$\text{(i')} \qquad \phi(x) = y_0 + \int_{x_0}^{x} f(t,\phi(t))dt .$$

In other words every solution which satisfies (i) and (ii) satisfies also (i') and vice versa. Since f is continuous in G we can find a neighborhood G' of $\langle x_0,y_0\rangle$ and a number k such that $G' \subset G$ and $|f(x,y)| \leq k$ in G'. Next, choose a $c > 0$ such that the following conditions are satisfied:

(a) $\langle x,y\rangle \in G'$ if $|x - x_0| \leq c$, $|y - y_0| \leq k\,c$

(b) $M\,c < 1$.

Let $\langle C^*,d\rangle$ be the metric space whose points are all continuous functions ϕ defined on the closed interval $[x_0 - c, x_0 + c]$ such that $|\phi(x) - y_0| \leq kc$, with the metric $d(\phi_1,\phi_2) = \max\{|\phi_1(x) - \phi_2(x)|: x \in [x_0 - c, x_0 + c]\}$. It is easy to see that C^* is a closed subset of the complete metric space C[a,b] (Ex. 5, p. 218). This implies that $\langle C^*,d\rangle$ is complete. Let us now consider the mapping $F : C^* \to C^*$ defined by:

$$F(\phi) = \psi \quad \text{with} \quad \psi(x) = y_0 + \int_{x_0}^{x} f(t,\phi(t))dt$$

for $x \in [x_0 - c, x_0 + c]$. We have

$$|\psi(x) - y_0| = |\int_{x_0}^{x} f(t,\phi(t)dt| \leq kc \ ,$$

which means that $F[C^*] \subset C^*$. Moreover we have

$$|\psi_1(x) - \psi_2(x)| \leq \int_{x_0}^{x} |f(t,\phi_1(t)) - f(t,\phi_2(t))|dt$$

$$\leq M\,c \;\; d(\phi_1,\phi_2).$$

Since $Mc < 1$, the mapping F is a contraction. Therefore the equation $F(\phi) = \phi$, and consequently equation (i'), has a unique solution.

3) As a third application we prove the existence and uniqueness of the solution of the Fredholm nonhomogeneous linear integral equation of the second kind:

$$f(x) = \lambda \int_a^b k(x,y)f(y)dy + \phi(x)$$

where k is a continuous function on $[a,b] \times [a,b]$, ϕ a continuous function on $[a,b]$, f the unknown continuous function on $[a,b]$, and λ a certain parameter to be determined later. Since k is continuous on a compact set there is an M such that $|k(x,y)| \leq M$. Consider the mapping: $F : C[a,b] \to C[a,b]$ defined by:

$$F(f) = g \quad \text{where} \quad g(x) = \lambda \int_a^b k(x,y)f(y)dy + \phi(x) .$$

For $g_1, g_2 \in C[a,b]$ we have:

$$d(g_1,g_2) = \sup_{x\in[a,b]} |g_1(x) - g_2(x)|$$

$$\leq |\lambda| M(b - a)\cdot d(f_1,f_2) .$$

This proves that for $|\lambda| < 1/(M\cdot(b - a))$, F is a contraction mapping on the complete metric space $C[a,b]$ into $C[a,b]$. It follows from Theorem 19 that the equation $F(f) = f$, i.e., the Fredholm equation, has a unique solution. The successive approximations to this solution are: $f_0, f_1, \ldots, f_n$ where

$$f_n(x) = \int_a^b k(x,y)f_{n-1}(y)dy + \phi(x) .$$

5.10 COMPACT METRIC SPACES AND THE BOLZANO-WEIERSTRASS THEOREM

A collection $\mathcal{C}$ of open sets in a metric space $\langle X,d\rangle$ is said to be an *open covering* of a subset $S \subset X$, iff S is contained in the union of the sets in $\mathcal{C}$.

A metric space $\langle X,d \rangle$ is said to be *compact* iff every open covering $\mathcal{E}$ of X has a finite subcovering, that is, if there exists a finite collection $\{O_1, O_2, \ldots, O_N\} \subset \mathcal{E}$, such that $X = \bigcup_{i=1}^{N} O_i$. A subset K of a metric space $\langle X,d \rangle$ is said to be compact iff the subspace $\langle K,d \rangle$ of X is compact. This is equivalent to saying that K is compact if and only if every open covering of K by open sets in X has a finite subcovering.

In Chapter 2 we proved the Bolzano-Weierstrass Theorem which asserts that every infinite and bounded set of real numbers has at least one accumulation point. Unfortunately this theorem, one of the earliest in point-set theory, is not valid in an arbitrary metric space. An example of such space is the space of rational numbers with the usual metric. The set $\{.01, .01001, .010010001, \ldots\}$ is obviously infinite and bounded, but has no accumulation point. For if such a point existed it would be a number with non-repeating decimals, that is an irrational number. A second example is the space ℓ^2 given in section 5.1, Example 6). The set E whose elements are $\xi_1 = \langle 1,0,\ldots \rangle$, $\xi_2 = \langle 0,1,0,\ldots \rangle$, $\xi_3 = \langle 0,0,1,0,\ldots \rangle$, ... is infinite and bounded since the distance of any two elements in E is $\sqrt{2}$. However E has no accumulation point. Suppose on the contrary that $x = \langle x_n \rangle$ is an accumulation point of E. Then there is a point ξ_m in E which belongs to the neighborhood $S(x;(1/2))$ of x. We have $d(x,\xi_m) < 1/2$, which implies that $x_m > 1/2$. Hence for any $k \neq m$,

$d(x,\xi_k) > 1/2$, which means that x is not an accumulation point of E; a contradiction.

These two examples show that the consequences of boundedness in an arbitrary metric space are not the same as in the space $\mathbb{R}$ of real numbers.

The notion of total boundedness, introduced below, is equivalent to boundedness in the space of real numbers.

Definition *A metric space* $\langle X,d\rangle$ *is called* totally bounded *(or precompact) if and only if for every* $\varepsilon > 0$ *there is a finite subset* F *of* X *such that* $d(x,F) < \varepsilon$ *for every* $x \in X$. *This is equivalent to saying that* $\langle X,d\rangle$ *is totally bounded if and only if for every* $\varepsilon > 0$ *there is a finite subset* $\{x_1,x_2,\ldots,x_n\}$ *of* X *such that the union of the open balls* $S(x_i;\varepsilon)$ $(i = 1,2,\ldots,n)$ *covers* X.

A subset T of $\langle X,d\rangle$ is called *totally bounded* if and only if for every $\varepsilon > 0$ there exists a finite subset $\{x_1,x_2,\ldots,x_n\}$ of X such that the union of the open balls $S(x_i;\varepsilon)$ $(i = 1,\ldots,n)$ covers T.

Clearly if X is totally bounded then X is bounded. The converse is not true. The set $\{\xi_n\}$ considered above is a counterexample.

20. Theorem *For a metric space* $\langle X,d\rangle$ *the following conditions are equivalent:*

(a) X *is compact.*

(b) Any infinite sequence in X *has at least one cluster point.*

(c) X *is totally bounded and complete.*

Proof *(a)* $\Rightarrow$ *(b)*: Let X be compact and $\langle x_n\rangle$ an infinite sequence from X. Denote by F_n the closure of the set $\{x_n, x_{n+1}, \ldots,\}$. We have $F_{n+1} \subset F_n$ for all n. We prove that $\bigcap_{n=1}^{\infty} F_n \neq \emptyset$. In fact, suppose on the contrary that $\bigcap_{n=1}^{\infty} F_n = \emptyset$. Then $\complement\left(\bigcap_{n=1}^{\infty} F_n\right) = \bigcup_{n=1}^{\infty}\left(\complement F_n\right) = X$, i.e., the collection $\{\complement F_n\}$ is an open covering of X. Hence, a finite number of $\complement F_n$, say $\complement F_{n_1}, \complement F_{n_2}, \ldots, \complement F_{n_k}$ cover X, i.e., $\complement F_{n_1} \cup \complement F_{n_2} \cup \ldots \cup \complement F_{n_k} = X$. It follows that $F_{n_1} \cap F_{n_2} \cap \ldots \cap F_{n_k} = \emptyset$. Let $M = \max\{n_1, n_2, \ldots, n_k\}$. In virtue of the relation $F_{n+1} \subset F_n$, we have $F_M \subset F_{n_1} \cap F_{n_2} \cap \ldots \cap F_{n_k}$, i.e., $F_M = \emptyset$; a contradiction. Hence $\bigcap_{n=1}^{\infty} F_n \neq 0$. Let now $a \in \bigcap_{n=1}^{\infty} F_n$, V be any neighborhood of a, and N be any positive integer. We have $a \in F_N$, i.e., a is a point of closure of the set $\{x_N, x_{N+1}, \ldots\}$, which implies that there is an x_n with $n \geq N$, such that $x_n \in V$. Hence a is a cluster point of $\langle x_n \rangle$.

(b) ⇒ (c): Let $\langle x_n \rangle$ be a Cauchy sequence from X. Then by assumption $\langle x_n \rangle$ has a cluster point,

say a. We prove that $\lim_{n\to\infty} x_n = a$. Let $\varepsilon > 0$ be given. Then there is an N such that $n,m \geq N$ imply $d(x_m,x_n) < \varepsilon/2$. We have, for $n \geq N$ and some $m \geq N$, $d(a,x_m) < \varepsilon/2$ and $d(a,x_n) \leq d(a,x_m) + d(x_m,x_n) < (\varepsilon/2) + (\varepsilon/2) = \varepsilon$; this proves that $\langle x_n \rangle$ converges to a, i.e., X is complete. To prove that X is totally bounded, suppose on the contrary that there exists an ε such that X cannot be covered by a finite number of open balls with radius ε. Then we can define the sequence $\langle x_n \rangle$ by induction as follows. Let x_1 be any point in X. Since $S(x_1;\varepsilon)$ does not cover X there is a point $x_2 \in X - S(x_1;\varepsilon)$. Similarly the union $S(x_1;\varepsilon) \cup S(x_2;\varepsilon)$ does not cover X, therefore there is an $x_3 \in X - S(x_1;\varepsilon) \cup S(x_2;\varepsilon)$ and so on. From the way that $\langle x_n \rangle$ is defined we have that all the x_n are distinct and for $p \neq q$ $d(x_p,x_q) \geq \varepsilon$. By assumption $\langle x_n \rangle$ has at least a cluster point, say a. If $\langle x_{n_k} \rangle$ is a subsequence of $\langle x_n \rangle$ converging to a, then there is a k_o such that for $k \geq k_o$, we have $d(a,x_{n_k}) < \varepsilon/2$, which implies $d(x_{n_h},x_{n_k}) \leq d(x_{n_h},a) + d(a,x_{n_k}) < (\varepsilon/2) + (\varepsilon/2) = \varepsilon$ for $h \geq k_o$, $k \geq k_o$; a contradiction. Hence X is totally bounded.

(c) ⇒ (a): To prove that X is compact suppose again that there is an open covering $\mathcal{C}$ of X such

that no finite number of sets in $\mathcal{C}$ cover X. Since X is totally bounded it can be covered by a finite number of open balls with radius $1/2^{n-1}$. At least one of these balls, say B_{n-1}, is not covered by a finite number of sets in $\mathcal{C}$; otherwise there would exist a finite subcovering of X. Next consider a finite covering of X by open balls with radius $1/2^n$. Then among the balls which have a nonempty intersection with B_{n-1}, there is at least one, call it B_n, which is not covered by a finite number of sets in $\mathcal{C}$. We obtain in this way, by induction, a sequence $\langle B_n \rangle$ of open balls such that $B_{n-1} \cap B_n \neq \emptyset$ for all n, and such that no one of the B_n is covered by a finite number of sets in $\mathcal{C}$. Let x_n be the center of B_n. For $z \in B_{n-1} \cap B_n$ we have:

$$d(x_{n-1}, x_n) \le d(x_{n-1}, z) + d(z, x_n) < \frac{1}{2^{n-1}} + \frac{1}{2^n} < \frac{1}{2^{n-2}} .$$

Hence, if $q > p \geq n$, we have:

$$d(x_p, x_q) \le d(x_p, x_{p+1}) +, \ldots, + d(x_{q-1}, x_q)$$

$$< \frac{1}{2^{p-1}} + \ldots + \frac{1}{2^{q-2}} < \frac{1}{2^{n-2}} .$$

This implies that $\langle x_n \rangle$ is a Cauchy sequence, hence it converges to some point a of X. Let O be an open set in $\mathcal{C}$ containing a. There is an open ball $S(a;\varepsilon)$ such that $S(a;\varepsilon) \subset O$. Since a is the limit of $\langle x_n \rangle$, there is an integer n such that $d(x_n,a) < \varepsilon/2$ and $(1/2^n) < (\varepsilon/2)$. For $y \in B_n$ we have

$$d(a,y) \leq d(a,x_n) + d(x_n,y) < \frac{\varepsilon}{2} + \frac{1}{2^n} < \frac{\varepsilon}{2} + \frac{\varepsilon}{2} = \varepsilon.$$

This proves that $B_n \subset S(a;\varepsilon) \subset O$, which contradicts the fact that no one of the B_n is covered by a finite number of sets in $\mathscr{C}$.

21. Corollary *If* X *is a complete metric space, every totally bounded infinite set in* X *has at least one accumulation point.*

22. Corollary *Let* E *be a subset of a metric space* X. *Then* $\overline{E}$ *is compact if and only if every infinite sequence of points of* E *has a subsequence which converges to a point of* X.

23. Proposition *Any continuous mapping f from a compact metric space* $\langle X,d \rangle$ *into a metric space* $\langle Y,\sigma \rangle$ *is uniformly continuous.*

Proof Let $\varepsilon > 0$ and $x \in X$ be given. There is a $\delta_x > 0$ such that $d(x,y) < \delta_x$ implies $\sigma(f(x),f(y)) < \varepsilon/2$. The collection $\{S(x;(1/2)\delta_x): x \in X\}$ is an open covering of X. Since X is compact, there is a finite subcovering $S(x_1;(1/2)\delta_{x_1}),\ldots,$ $S(x_n;(1/2)\delta_{x_n})$. Put $\delta = \min\{(\delta_{x_1}/2),\ldots,(\delta_{x_n}/2)\}$, and let y and z be any two points in X with $d(y,z) < \delta$. We want to prove that $\sigma(f(y),f(z)) < \varepsilon$.

In fact, we first notice that y and z belong to the same open ball, say $S(x_i;\delta_{x_i})$. For if $y \in S(x_i;(1/2)\delta_{x_i}) \subset S(x_i;\delta_{x_i})$ then $d(y,x_i) < (1/2)\delta_{x_i}$, so that $d(x_i,z) \le d(x_i,y) + d(y,z) < (1/2)\delta_{x_i} + \delta \le \delta_{x_i}$, i.e., $z \in S(x_i;\delta_{x_i})$. Next, we have:

$$\sigma(f(z),f(y)) \le \sigma(f(z),f(x_i)) + \sigma(f(x_i),f(y)) < \frac{\varepsilon}{2} + \frac{\varepsilon}{2} = \varepsilon.$$

This proves that f is uniformly continuous on X.

5.11 EQUICONTINUOUS SPACES OF FUNCTIONS

In Theorem 20 of the preceding section we proved that a metric space X is compact if and only if it is complete and totally bounded. Since a closed subspace of a complete metric space is also complete we conclude that a closed subspace of a complete metric space is compact if and only if it is totally bounded. This gives total boundedness a prominent part in determining whether a metric space is compact or not. Hence a closed subspace of the space $\mathcal{C}(X,\mathbb{R})$ of all continuous real-valued bounded functions on the metric space X (see Ex. 10, p. 241) is compact if and only if it is totally bounded. Unfortunately, this information is not of great value in most applications to analysis. A criterion for compactness in terms of the individual functions in the subspace of $\mathcal{C}(X,\mathbb{R})$

is needed. The notion of equicontinuity plays a central role in such kinds of questions.

Definition *Let* $\mathcal{F}$ *be a family of mappings of a metric space* $\langle X,d\rangle$ *into a metric space* $\langle Y,\sigma\rangle$. *The family* $\mathcal{F}$ *is said to be* equicontinuous *iff for every* $\varepsilon > 0$ *there exists an* $\eta > 0$ *such that for every* $f \in \mathcal{F}$ *and for all* $x_1, x_2 \in X$ *with* $d(x_1,x_2) \leq \eta$ *we have* $\sigma(f(x_1),f(x_2)) \leq \varepsilon$.

It follows from this definition that every function in $\mathcal{F}$ is uniformly continuous on X. The equicontinuity of $\mathcal{F}$ means that the number η in the definition depends only on ε and not on f.

The sequence $\langle f_n\rangle$ of functions defined on [0,1] by $f_n(x) = x^n$ is not equicontinuous although each of these functions is uniformly continuous on [0,1].

Let $\langle X,d\rangle$ and $\langle Y,\sigma\rangle$ be compact metric spaces. We shall denote by $\mathcal{C}(X,Y)$ the set of all continuous functions from X into Y. For $f,g \in \mathcal{C}(X,Y)$ put

$$D(f,g) = \sup\{\sigma(f(x),g(x)) : x \in X\}.$$

Since Y is compact it is bounded, so that D defines a metric on $\mathcal{C}(X,Y)$ (see Ex. 10 p. 241).

24. Theorem (Ascoli) *Let* $\langle X,d\rangle$ *and* $\langle Y,\sigma\rangle$ *be compact metric spaces and let* $\mathcal{F}$ *be a subset of the metric space* $\langle \mathcal{C}(X,Y),D\rangle$. *Then* $\mathcal{F}$ *is equicontinuous if and only if the closure* $\overline{\mathcal{F}}$ *of* $\mathcal{F}$ *is a compact subspace of* $\mathcal{C}(X,Y)$.

Remark *Recalling the definition of relatively compact sets (Prob. 52) we see that* $\mathcal{F}$ *is equicontinuous iff* $\mathcal{F}$ *is relatively compact in* $\mathcal{C}(X,Y)$.

Proof Suppose that $\mathcal{F}$ is equicontinuous. By Corollary 22, to prove that $\bar{\mathcal{F}}$ is compact, we need only to show that every infinite sequence from $\mathcal{F}$ has a subsequence which converges to a point of $\mathcal{C}(X,Y)$. But since Y is compact it is complete and so $\mathcal{C}(X,Y)$ is complete (see Ex. 10, p. 241). Hence it suffices to prove that every infinite sequence in $\mathcal{F}$ contains a Cauchy sequence. So let S be an infinite sequence from $\mathcal{F}$. Since $\mathcal{F}$ is equicontinuous, for every $\varepsilon > 0$ there is an $\eta(\varepsilon) > 0$ such that for each $f \in \mathcal{F}$ and all $x', x'' \in X$ with $d(x', x'') \leq \eta(\varepsilon)$ we have $\sigma(f(x'), f(x'')) \leq \varepsilon$. Since X is compact there exists a finite covering of X by open balls of radius $\eta(\varepsilon)$, with centers x_i $(i = 1, 2, \ldots, n)$. Consider the subset $\{f(x_1) : f \in S\}$ of Y. Since Y is compact we can find a subsequence $S_1(\varepsilon)$ of S such that for all $f, g \in S_1(\varepsilon)$ we have $\sigma(f(x_1), g(x_1)) \leq \varepsilon$. Also we can find a subsequence $S_2(\varepsilon)$ of $S_1(\varepsilon)$ such that for all $f, g \in S_2(\varepsilon)$ we have $\sigma(f(x_2), g(x_2)) \leq \varepsilon$. After n operations we will have constructed a sequence $S_n(\varepsilon)$ such that for all $f, g \in S_n(\varepsilon)$ we have $\sigma(f(x_n), g(x_n)) \leq \varepsilon$. But by construction, for each x we have $d(x, x_i) \leq \eta$ for some i $(1 \leq i \leq n)$, which implies that $\sigma(f(x), f(x_i)) \leq \varepsilon$ for every $f \in S$. Denote the subsequence $S_n(\varepsilon)$ by $S(\varepsilon)$. Hence for all f and g in $S(\varepsilon)$ we have

$$\sigma(f(x), g(x)) \leq \sigma(f(x), f(x_i)) +$$

$$+ \sigma(f(x_i),g(x_i)) + \sigma(g(x_i),g(x)) \leq 3\varepsilon \ .$$

Recalling the definition of the metric D (Ex. 9, p. 240), we see that the last inequality implies $D(f,g) \leq 3\varepsilon$.

What we have established so far is that for given $\varepsilon > 0$ there is a subsequence $S(\varepsilon)$ of the sequence S such that for every two elements f and g of $S(\varepsilon)$ we have $D(f,g) \leq 3\varepsilon$. Consider now a sequence $\langle \varepsilon_k \rangle$ of positive real numbers such that $\lim_{k\to\infty} \varepsilon_k = 0$. Let $S(\varepsilon_1)$ be the subsequence of S for $\varepsilon = \varepsilon_1$. Let $S(\varepsilon_1,\varepsilon_2)$ be the subsequence of $S(\varepsilon_1)$ obtained in the same way as $S(\varepsilon_1)$ was obtained from S, and so on. In this way we obtain successive sequences

$$S(\varepsilon_1), S(\varepsilon_1,\varepsilon_2), S(\varepsilon_1,\varepsilon_2,\varepsilon_3), \ldots, S(\varepsilon_1,\varepsilon_2,\varepsilon_3,\ldots,\varepsilon_n), \ldots$$

each of which is a subsequence of the preceding. Let us now write down these sequences explicitly.

$$S(\varepsilon_1) : a_{11}, a_{12}, a_{13}, \ldots$$

$$S(\varepsilon_1,\varepsilon_2) : a_{21}, a_{22}, a_{23}, \ldots$$

$$S(\varepsilon_1,\varepsilon_2,\varepsilon_3) : a_{31}, a_{32}, a_{33}, \ldots$$

....

....

....

Consider the diagonal sequence $\langle a_{11}, a_{22}, a_{33}, \ldots \rangle$. It is clear that this is the desired sequence, that is the Cauchy subsequence of S. Hence $\bar{\mathcal{F}}$ is compact.

Assume now that $\bar{\mathcal{F}}$ is a compact subset of $\mathcal{C}(X,Y)$. For every $\varepsilon > 0$ there exists a finite sequence $\langle g_1, g_2, \ldots, g_n \rangle$ of elements of $\bar{\mathcal{F}}$ such that the open balls $S_\varepsilon(g_i)$ form an open covering of $\bar{\mathcal{F}}$. Since X is compact, the g_i are uniformly continuous. So for each g_i there exists an $\eta_i > 0$ such that the inequality $d(x_1, x_2) < \eta_i$ implies $\sigma(g_i(x_1), g_i(x_2)) < \varepsilon$. Let η be the smallest of the η_i. Since every function $f \in \mathcal{F}$ belongs to some ball $S_\varepsilon(g_i)$ for some i, we have

$$\sup\{\sigma(f(x), g_i(x)) : x \in X\} = D(f, g_i) < \varepsilon$$

and so

$$\sigma(f(x), g_i(x)) < \varepsilon \quad \text{for all} \quad x \in X.$$

Hence for every $f \in \mathcal{F}$ and for all $x_1, x_2 \in X$ with $d(x_1, x_2) < \eta$ we have

$$\sigma(f(x_1), f(x_2)) \leq \sigma(f(x_1), g_i(x_1))$$

$$+ \sigma(g_i(x_1), g_i(x_2)) + \sigma(g_i(x_2), f(x_2)) < 3\varepsilon$$

This proves that $\mathcal{F}$ is equicontinuous.

(*Note*: Theorem 24 is also sometimes called Arzela's Theorem.)

In analysis it is often useful to know when a sequence of continuous functions has a uniformly convergent subsequence. Theorem 25, below, provides such conditions.

The reader should be very careful in dealing with subsequences of a given sequence of functions. The situation is not as simple as it is in the case of sequences of numbers. We know that any bounded sequence of real (or complex) numbers always has a convergent subsequence. The following example, however, provides a uniformly bounded sequence $\langle f_n \rangle$ of functions on $[0,2\pi]$ such that no subsequence converges:

$$f_n(x) = \sin nx, \quad 0 \leq x \leq 2\pi, \quad n = 1,2,\ldots .$$

We claim there is no convergent subsequence of $\langle f_n \rangle$. For suppose on the contrary that there is a subsequence $\langle f_{n_k} \rangle_{k=1}^{\infty}$ such that $\langle \sin n_k x \rangle$ converges for each $x \in [0,2\pi]$. Then

$$\lim_{k\to\infty} \sin n_k\, k = \lim_{k\to\infty} \sin n_{k+1}\, x$$

or

$$\lim_{k\to\infty} (\sin n_k x - \sin n_{k+1} x)^2 = 0$$

By applying Lebesgue bounded convergence theorem to

the sequence $\langle (f_{n_k} - f_{n_{k+1}})^2 \rangle$ we obtain

$$\lim_{k\to\infty} \int_0^{2\pi} (\sin n_k x - \sin n_{k+1} x)^2 dx = 0 .$$

But this last equality is impossible because an easy calculation shows that for every positive integer k we have

$$\int_0^{2\pi} (\sin n_k x - \sin n_{k+1} x)^2 dx = 2\pi .$$

Hence no subsequence of $\langle f_n \rangle$ is convergent.

For real-valued functions defined on a compact interval we have the following theorem which can be easily obtained from Theorem 24.

The reader should observe that a sequence $\langle f_n \rangle$ in $\langle \mathcal{C}(X,Y), D \rangle$ is convergent to f if and only if $\langle f_n(x) \rangle$ is uniformly convergent to $f(x)$ in X.

25. Theorem *Let* $\langle f_n \rangle$ *be a sequence of real-valued continuous functions on a compact interval* $[a,b]$.

(a) If $\langle f_n \rangle$ *is uniformly convergent on* $[a,b]$ *then* $\langle f_n \rangle$ *is equicontinuous.*

(b) If $\langle f_n \rangle$ *is uniformly bounded on* $[a,b]$ *(i.e.,* $|f_n(x)| < M$ *for all* n *and all* x *in* $[a,b]$*), and equicontinuous, then* $\langle f_n \rangle$ *has a uniformly convergent subsequence.*

5.12 CATEGORY

Let X be a metric space. A subset E of X is said to be *nowhere dense* iff $\complement\overline{E}$ is everywhere dense. Another way of saying this is that $\overline{E}$ does not contain an open ball. Clearly this is also equivalent to saying that E is nowhere dense iff E is empty or if every open ball in X contains an open ball that is disjoint from E. Thus a closed set is nowhere dense if and only if its complement is dense. Any finite set in $\mathbb{R}$, and the set of natural numbers are nowhere dense sets.

If E is a subset of metric space X then any point of E which is not an accumulation point of E is called an *isolated point of* E (see Prob. 10). If the space X does not have isolated points, then the set consisting of a single point of X is nowhere dense. But if X has isolated points and x_o is one of them, then the set $\{x_o\}$ fails to be nowhere dense.

A subset E of a metric space is called *perfect* iff $E' = E$. An example of nowhere dense perfect set is the Cantor set considered in Chapter 2. In fact, it is closed and it contains no interval, since the total length of intervals removed is one.

A set E is said to be of *first category* iff it is the union of a countable collection of nowhere dense sets. If E is not of first category it is said to be of *second category*.

Clearly any countable union of sets of first category is a set of first category. Thus *if the union of two sets is a set of second category then at least one of them is of second category.*

Since the empty set is nowhere dense and hence of first category it follows that *a set of the second category cannot be empty*. This fact suggests the use of category arguments for existence proofs. By this we mean that if we prove that the set of all things of a particular kind is a set of the second category, then there must exist things of that kind. Theorem 27, below, is a remarkable example of an existence proof based on category arguments.

One of the most useful consequences of completeness of a metric space is expressed by the following Baire's theorem.

26. Theorem *Every complete metric space is of second category.*

Proof Let X be a complete metric space. To prove that X is of second category it suffices to prove that there is no set of first category which contains all the points of X. Let E be an arbitrary set of first category; i.e., $E = \bigcup_{n=1}^{\infty} E_n$ where for each n, E_n is nowhere dense. Since $\complement\overline{E}_1$ is dense and open, there is a point $x_1 \in \complement\overline{E}_1$ and a neighborhood $S(x_1;r_1) = S_1$ of x_1, contained in $\complement\overline{E}_1$. Since $\complement\overline{E}_2$ is dense there is a point $x_2 \in \complement\overline{E}_2 \cap S(x_1;r_1)$. Since $\complement\overline{E}_2 \cap S(x_1;r_1)$ is open there is a neighborhood $S(x_2;r_2)$ contained in $\complement\overline{E}_2 \cap S(x_1;r_1)$. If we take $r_2 < \min\{(1/2)r_1, r_1 - d(x_1,x_2)\}$ then we also have $\overline{S}(x_2;r_2) \subset S_1$. We obtain by induction, a sequence $S(x_n;r_n) = S_n$ of open balls such that

$\overline{S}(x_n;r_n) \subset S(x_{n-1};r_{n-1})$ and $\lim_{n\to\infty} r_n = 0$. For $n,m \geq N$ we have $x_n \in S_N$, $x_m \in S_N$, so that $d(x_n,x_m) \leq d(x_n,x_N) + d(x_N,x_m) < 2r_N$. Hence $\langle x_n \rangle$ is a Cauchy sequence, since $r_n \to 0$. The space X being complete there is a point $x \in X$ such that $x_n \to x$. Since for $n > N$ $x_n \in S(x_{N+1};r_{N+1})$ we have $x \in \overline{S}(x_{N+1};r_{N+1}) \subset S(x_N;r_N) \subset \complement\overline{E}_N$ which means that

$$x \in \bigcap_{n=1}^{\infty} \complement\overline{E}_n, \text{ i.e., } x \notin E.$$ This proves that X is of second category. The theorem is proved.

Let S be the set of all continuous functions on $\mathbb{R}$, with period 1 (i.e., $f(t + 1) = f(t)$ for all $t \in \mathbb{R}$). For f and g in S define $d(f,g) = \sup\{|f(t) - g(t)|: -\infty < t < +\infty\}$. Then, as in the case of the space $C[a,b]$, one can show that $\langle S,d \rangle$ is a complete metric space, so that by Theorem 26, S is of second category. Let Δ denote the subset of S consisting of all functions which are differentiable at least at one point. If we prove that Δ is a set of first category then $S - \Delta$ will be of second category since S is of second category. This will prove that there are functions in S whose derivative exists nowhere. We have the following theorem.

27. Theorem *There exist continuous functions on the real line which are nowhere differentiable.*
Proof Let $\mathcal{F}$ be the subset of S consisting of all functions whose four Dini derivatives (see section 4.3) are finite at least at one point (not necessarily the same for all functions in $\mathcal{F}$). Clearly

$\mathcal{F} = \bigcup_{n=1}^{\infty} \mathcal{F}_n$ where $\mathcal{F}_n$ is the set of all functions f in S for which there is an α and a positive integer n such that

$$\left|\frac{f(\alpha+h) - f(\alpha)}{h}\right| \leq n$$

for every $h > 0$. We first note that $S - \mathcal{F}_n$ is everywhere dense in S. Let $\varepsilon > 0$ and $f \varepsilon S$. Using the uniform continuity of f, one can show that there is a function $g \varepsilon S$ whose graph is a broken line each of whose segments has slope either greater than n or less than $-n$, such that $|f(t) - g(t)| < \varepsilon$ for every t (see Prob. 64). Then $g \varepsilon S - \mathcal{F}_n$ and $d(f,g) < \varepsilon$, so that $S - \mathcal{F}_n$ is everywhere dense in S. Next we show that $\mathcal{F}_n$ is a closed subset of S. Let f be a point of $\bar{\mathcal{F}}_n$. Then there is a sequence $\langle f_n \rangle$ in $\mathcal{F}_n$ such that $\lim_{n\to\infty} d(f_n,f) = 0$. For every m there is an α_m such that for every $h > 0$

$$\left|\frac{f_m(\alpha_m+h) - f_m(\alpha_m)}{h}\right| \leq n .$$

Since f_m is periodic of period 1, α_m may be taken to be in the interval $[0,1]$. Furthermore, we may assume, without loss of generality, that the sequence $\langle \alpha_m \rangle$ is convergent. (Why?) Let $\alpha = \lim_{m\to\infty} \alpha_m$. We prove that for every $h > 0$ we have

$$\left|\frac{f(\alpha+h) - f(\alpha)}{h}\right| \leq n .$$

Let $h > 0$ and $\varepsilon > 0$. There is an N such that for every $m > N$ and for every t, $|f(t) - f_m(t)| < \varepsilon h/4$. But there is an $m > N$ such that

$$|f(\alpha) - f(\alpha_m)| < \varepsilon h/4$$

$$\text{and} \quad |f_m(t+h) - f_m(\alpha_m+h)| < \varepsilon h/4.$$

Then

$$\left|\frac{f(\alpha+h) - f(\alpha)}{h}\right| \leq \left|\frac{f(\alpha+h) - f_m(\alpha+h)}{h}\right|$$

$$+ \left|\frac{f_m(\alpha+h) - f_m(\alpha_m+h)}{h}\right| + \left|\frac{f_m(\alpha_m+h) - f_m(\alpha_m)}{h}\right|$$

$$+ \left|\frac{f_m(\alpha_m) - f(\alpha_m)}{h}\right| + \left|\frac{f(\alpha_m) - f(\alpha)}{h}\right|$$

$$< \frac{\varepsilon}{4} + \frac{\varepsilon}{4} + n + \frac{\varepsilon}{4} + \frac{\varepsilon}{4} = n + \varepsilon .$$

Since ε was arbitrary it follows that $f \in \mathcal{F}_n$. Hence $\mathcal{F}_n$ is a closed set whose complement is everywhere dense. Therefore $\mathcal{F}_n$ is nowhere dense so that $\mathcal{F}$ is of first category. Since clearly Δ is a subset of $\mathcal{F}$ it follows that Δ is of first category. The theorem is proved.

28. Theorem (Uniform Boundedness Principle) *Let $\mathcal{F}$ be a family of real-valued continuous functions on a complete metric space X, and suppose that for each $x \in X$ there is a number M_x such that $|f(x)| \leq M_x$ for all $f \in \mathcal{F}$. Then there is a nonempty open set $O \subset X$ and a constant M such that $|f(x)| \leq M$ for all $f \in \mathcal{F}$ and all $x \in O$.*

Proof For each natural number m and each f in $\mathcal{F}$, set $E_{m,f} = \{x: |f(x)| \leq m\}$, and $E_m = \bigcap_{f \in \mathcal{F}} E_{m,f}$. Since f is continuous, $E_{m,f}$ is a closed set and consequently E_m is closed. For each $x \in X$ let m be an integer greater or equal to M_x. Then for every $f \in \mathcal{F}$ we have $|f(x)| \leq m$. In other words each point x of X belongs to some E_m; i.e., $X = \bigcup_{m=1}^{\infty} E_m$. The space X being complete, it is of second category, hence at least one of the E_m is not nowhere dense. This implies that for some integer $m = m_o$ the set $\overline{E}_{m_o} = E_{m_o}$ contains an open ball S. Therefore, for every $x \in S$ and every $f \in \mathcal{F}$, $|f(x)| \leq m_o$. The theorem is proved.

PROBLEMS

1. Show that the spaces described in section 5.1, examples 1 through 9 are metric spaces.
2. Let $\langle X,d\rangle$ be a metric space. For any x,y in X define

$$d_1(x,y) = \frac{d(x,y)}{1 + d(x,y)} .$$

Show that $\langle X,d_1\rangle$ is a metric space.

3. Let $\langle X,\rho\rangle$ and $\langle Y,\sigma\rangle$ be metric spaces and let $X \times Y = \{(x,y): x \in X,\ y \in Y\}$ be the Cartesian product of X and Y. Show that $\langle X \times Y,\ D_i\rangle$ is a metric space, $i = 1,2,3$, where

(a) $D_1((x_1,y_1),(x_2,y_2))$

$$= [\rho(x_1,x_2)^2 + \sigma(y_1,y_2)^2]^{\frac{1}{2}}$$

(b) $D_2((x_1,y_1),(x_2,y_2))$

$$= \max[\rho(x_1,x_2),\sigma(y_1,y_2)]$$

(c) $D_3((x_1,y_1),(x_2,y_2))$

$$= \rho(x_1,x_2) + \sigma(y_1,y_2) .$$

4. Let p be a fixed prime number, and let Q denote the set of all rational numbers. For any positive integer n define $f(n)$ to be the exponent of p in the decomposition of n into prime numbers (e.g., if $p = 3$, then $f(360) = f(2^3 \cdot 3^2 \cdot 5) = 2$). If $x = \pm r/s$ is any nonzero rational number with r and s positive integers, define $f(x) = f(r) - f(s)$.

(a) If m and n are any positive integers show that $f(mn) = f(m) + f(n)$.

(b) Show that if $x = \pm r/s$ is a nonzero rational number, then the definition of $f(x)$ given by $f(x) = f(r) - f(s)$ does not depend on the particular expression of x as a fraction.

(c) Show that if x and y are any nonzero rational numbers, then $f(xy) = f(x) + f(y)$.

(d) For any x and y in Q define

$$\rho(x,y) = p^{-f(x-y)} \quad \text{if} \quad x \neq y$$

$$\rho(x,x) = 0$$

Show that $\langle Q,\rho\rangle$ is a metric space; ρ is called the *p-adic distance*.

(e) Show that for any x,y and z in Q we have

$$\rho(x,z) \leq \max\{\rho(x,y),\rho(y,z)\} .$$

(Hint: Suppose x,y,z distinct. Then for x and y in Q with $x \neq 0$, $y \neq 0$ and $x - y \neq 0$ we have to prove that $f(x - y) \geq \min(f(x),f(y))$. Suppose $f(x) \geq f(y)$ then the relation to prove becomes $f(z - 1) \geq 0$, for any $z \in Q$ such that $z \neq 0$, $z \neq 1$ and $f(z) \geq 0$. By definition $z = \pm p^h r/s$ with $h \geq 0$, r and s not divisible by p, and the relation $f(z - 1) \geq 0$ follows.)

5. Let $f: \mathbb{R} \to \mathbb{R}$ be an isometry.

(a) Show that the mapping f is one-to-one and onto $\mathbb{R}$.

(b) Let O and F be an open and a closed set respectively. Prove that $f[O]$ is open and $f[F]$ is closed.

(c) If O is a bounded open set show that $mO = mf[O]$.

(d) If E is any bounded set show that $m^*E = m^*f[E]$.

(e) If E is any measurable set show that $mE = mf[E]$.

6. Prove Propositions 2, 3, 4, and 5.

7. Show that in a discrete metric space every set is both closed and open.

8. (a) Let O be an open set in a metric space X. Show that for any subset $B \subset X$ we have $O \cap \bar{B} \subset \overline{O \cap B}$.

(b) If X is the real line with the usual metric, show that there are open sets O_1, O_2 such that the sets $O_1 \cap \bar{O}_2$, $O_2 \cap \bar{O}_1$, $\overline{O_1 \cap O_2}$ and $\bar{O}_1 \cap \bar{O}_2$ are all different.

9. Let E be any set in a metric space X. Then the *interior* of E (sec. 7.3) written $\overset{\circ}{E}$, is defined to be the largest open set contained in E, and it may be empty.

(a) Show that if O is an open set then the set $O \cup (X \overset{\circ}{-} O)$ is everywhere dense.

(b) Show that if $A \subset B$ then $\overset{\circ}{A} \subset \overset{\circ}{B}$.

(c) Show that for any two sets A and B, $A \overset{\circ}{\cap} B = \overset{\circ}{A} \cap \overset{\circ}{B}$.

(d) Let A be any subset on a metric space X. Show that a point $x \in X$ is in the interior of the complement of A if and only if the distance $d(x,A)$ between $\{x\}$ and A is positive.

10. If E is a subset of a metric space X, a point $x \in E$ is said to be an *isolated point* of E iff there is an open ball $S(x;r)$ such that

$S(x;r) \cap E = \{x\}$.

(a) Let X be a separable metric space and let E be a subset of X such that all points of E are isolated points of E. Show that E is countable.

(b) Let X be a separable metric space and let $\{O_\alpha\}_{\alpha \varepsilon \Lambda}$ be any nonempty collection of open sets such that $O_\alpha \cap O_\beta = \emptyset$ if $\alpha \neq \beta$. Show that $\{O_\alpha\}_{\alpha \varepsilon \Lambda}$ is at most denumerable.

11. If E is a subset of a metric space X, a point $x \varepsilon E$ is called a *condensation point* of E iff every open ball containing x contains an uncountable set of points of E.

(a) Let X be a separable metric space, and let E be a subset of X. Show that if E has no condensation point it is countable. (Hint: Consider the intersection of E with the sets $\{O_i\}$ provided by Proposition 6.)

(b) Let X and E be as in part (a), and let F be the set of condensation points of E. Show that every point of F is a condensation point of F. Show that $E \cap (\complement F)$ is countable. (Hint: Observe that F is a closed set. Use (a).)

12. Prove Propositions 8, 9, and 10.

13. Prove that every metric space $\langle X,d \rangle$ in which every infinite set has an accumulation point is separable.

(Hint: Fix $\varepsilon > 0$ and pick $x_1 \varepsilon X$. Then choose x_2 such that $d(x_1,x_2) \geq \varepsilon$, choose x_3 such that $d(x_2,x_3) \geq \varepsilon$, and so on. Show that this process has to stop after a finite number of steps. Take $\varepsilon = 1/n$ $(n = 1,2,\ldots)$.)

14. Let $\langle X,d\rangle$, $\langle Y,\sigma\rangle$ be metric spaces and E a nonempty subset of X. Let f and g be continuous mappings from X into Y such that $f(x) = g(x)$ for every $x \in E$. Show that $f(x) = g(x)$ for every $x \in \overline{E}$.
15. Let $\langle X,d\rangle$ be a metric space, and a be a fixed element in X. Show that the function $f\colon X \to \mathbb{R}$ defined by $f(x) = d(a,x)$, is continuous.
16. Show that if $\langle X,d\rangle$ is a discrete metric space then any function defined on X is continuous.
17. Let $\langle X,d\rangle$, $\langle Y,\sigma\rangle$ be metric spaces, and $f\colon X \to Y$ be a mapping from X into Y. Show that f is continuous if and only if for every subset E of Y we have

$$f^{-1}[\overset{\circ}{E}] \subset \overset{\circ}{f^{-1}[E]}$$

where $\overset{\circ}{E}$ denotes the interior of E (see Prob. 9).
18. Let $\langle X,d\rangle$ and $\langle Y,\sigma\rangle$ be metric spaces. Show that a mapping $f\colon X \to Y$ is continuous at a point $x_o \in X$ if and only if $\lim_{n\to\infty} d(x_n,x_o) = 0$ implies $\lim_{n\to\infty} \sigma(f(x_n),f(x_o)) = 0$.
19. Show that the function $f(x) = \log x$ is a homeomorphism of $(0,+\infty)$ onto $\mathbb{R}$.
20. Show that the metric spaces $(0,1)$ and $[0,1]$ are not homeomorphic. In fact, prove that there is no one-to-one continuous mapping of either of the intervals $(0,1)$ or $[0,1]$ onto the other.

21. Let X_1, X_2, X_3 be metric spaces. Show that if X_1 and X_2 are homeomorphic and if X_2 and X_3 are homeomorphic, then X_1 and X_2 are homeomorphic.

22. On the real line $\mathbb{R}$ define two metrics d and σ such that the identity mapping $x \to x$ from $\langle \mathbb{R}, d \rangle$ onto $\langle \mathbb{R}, \sigma \rangle$ is not a homeomorphism.

23. Show that the three metrics described in Problem 3 are equivalent.

24. Let d and d_1 be two metrics for a set X. Show that these metrics are equivalent if and only if: whenever $\lim_{n\to\infty} d(x_n, x) = 0$, then $\lim_{n\to\infty} d_1(x_n, x) = 0$, and whenever $\lim_{n\to\infty} d_1(x_n, x) = 0$, then $\lim_{n\to\infty} d(x_n, x) = 0$.

25. Let d and d_1 be metrics for a set X. Show that these metrics are equivalent if and only if to each pair ε, y where $\varepsilon > 0$ and $y \in X$, there corresponds a $\delta > 0$ and an $\eta > 0$ such that

$$\{x: d_1(x,y) < \delta\} \subset \{x: d(x,y) < \varepsilon\}$$

and

$$\{x: d(x,y) < \eta\} \subset \{x: d_1(x,y) < \varepsilon\} .$$

26. Let $\langle X, d \rangle$ be a metric space and define

$$d_1(x,y) = \frac{d(x,y)}{1 + d(x,y)} .$$

Show that the metrics d and d_1 are equivalent (see Prob. 2).

27. Let $C[0,1]$ be the class of all real-valued continuous functions on $[0,1]$. If $x,y \in C[0,1]$ let

$$d(x,y) = \sup\{|x(t) - y(t)| : 0 \leq t \leq 1\}$$

and

$$d_1(x,y) = \int_0^1 |x(t) - y(t)|dt .$$

We know (see Prob. 1) that d is a metric on $C[0,1]$. Show that d_1 is also a metric on $C[0,1]$ and that the metrics d and d_1 are not equivalent.
(Hint: Let x_0 be the zero function, and let $S = \{x: d(x,x_0) < 1\}$. If d and d_1 were equivalent metrics, S would contain some open ball $\{x: d_1(x,x_0) < \varepsilon\}$. But this is impossible: For any given ε choose n such that $1 < \varepsilon(n + 1)$, and let x_n be defined by $x_n(t) = t^n$. Then $d(x_n,x_0) = 0$, so $x_n \not\in S$. But $d_1(x_n,x_0) < \varepsilon$.)

28. Let $C^1[0,1]$ be the class of all continuous functions $x: [0,1] \to \mathbb{R}$ having continuous derivative x' on $[0,1]$. For x,y in $C^1[0,1]$ let

$$D(x,y) = \max_{0 \le t \le 1} |x(t) - y(t)|$$

$$+ \max_{0 \le t \le 1} |x'(t) - y'(t)| .$$

(a) Show that $C^1[0,1]$ is a metric space with D as metric.

(b) We see that $C^1[0,1]$ is a subset of $C[0,1]$ considered in Problem 27. Hence $C^1[0,1]$ is also a metric space with respect to the metric d of $C[0,1]$. Show that d and D are not equivalent metrics on $C^1[0,1]$.

29. Let $\langle x_n \rangle$ be an infinite sequence in a metric space $\langle X,d \rangle$, and $a \in X$. Show that if $\lim_{n\to\infty} x_n = a$, then $\lim_{k\to\infty} x_{n_k} = a$ for any subsequence $\langle x_{n_k} \rangle$ of $\langle x_n \rangle$.

30. Let $\langle x_n \rangle$ be an infinite sequence in a metric space X. Show that if the three subsequences $\langle x_{2n} \rangle$, $\langle x_{2n+1} \rangle$, and $\langle x_{3n} \rangle$ are convergent, then $\langle x_n \rangle$ is convergent.

31. Let X be a separable metric space and let f be an arbitrary mapping of X into $\mathbb{R}$. Let S be the set of all points a in X such that

$$\lim_{x\to a,\ x\neq a} f(x)$$

exists and is different from $f(a)$. Show that S is a countable set.

(Hint: For every pair of rational numbers p,q

with $p < q$ consider the set of points a such that $f(a) \leq p < q \leq \lim_{x \to a,\ x \neq a} f(x)$, and show that this set is countable. Use Problem 10 (a). Consider also the set of points a such that

$$\lim_{x \to a,\ x \neq a} f(x) \leq p < q \leq f(a)$$

32. Let $\langle X,d \rangle$ be a metric space and let $\langle S,d \rangle$ be a complete subspace of X. Show that S is a closed set in $\langle X,d \rangle$
33. Let $\langle X,d \rangle$ be a complete metric space and let $\langle S,d \rangle$ be a subspace of X such that S is a closed set in $\langle X,d \rangle$. Show that $\langle S,d \rangle$ is complete.
34. Show that the space $C^2[a,b]$ (sec. 5.6) is not complete.
35. Show that the space ℓ^∞ is complete (sec. 5.6).
36. Show that the spaces $\mathcal{F}(X,Y)$ and $C(X,\mathbb{R})$ described in the Examples 9 and 10 of section 5.6 are complete metric spaces.
37. Show that the three metrics considered in Problem 3 are uniformly equivalent (see Prob. 23).
38. (a) Show that the metrics d and d_1 considered in Problem 26 are uniformly equivalent.

    (b) Show that boundedness is not a uniform property.
39. Let E be a nonempty subset of a metric space $\langle X,d \rangle$. Show that the mapping $x \to d(x,E)$ is uniformly continuous on X.
40. Let X, Y, and Z be metric spaces. Let f be a uniformly continuous mapping of X into

Y and g a uniformly continuous mapping of Y into Z. Show that the composite mapping $h = g \circ f$ is uniformly continuous on X.

41. Let E be a dense subset of a metric space X, and let f be a mapping of E into a metric space Y. Show that there exists a continuous mapping F of X into Y which coincides with f on the set E, if and only if for any $x \in E$, the limit

$$\lim_{y \to x,\ y \in E} f(y) \quad \text{exists in } Y.$$

Show that if such an F exists it is unique. (Hint: *Necessity and uniqueness*: For $x \in X = \overline{E}$, we must have

$$F(x) = \lim_{y \to x,\ y \in E} F(y).$$

Hence

$$F(x) = \lim_{y \to x,\ y \in E} f(y)$$

Sufficiency: Define F by

$$F(x) = \lim_{y \to x,\ y \in E} f(y) .$$

If $x \in E$, then $F(x) = f(x)$, so that F extends f. It remains to prove that F is continuous. Let $x \in X$, and G be a neighborhood of F(x) in Y. There is a closed ball B of center F(x) contained in G. By assumption there is a neighborhood V of x in X such that $f(V \cap E) \subset B$. For any $y \in V$, $F(y)$ is the limit of f at the point y with respect to E, hence also with respect to $V \cap E$. Hence $F(y) \in \overline{f(V \cap E)}$, and therefore $F(y) \in B$ since B is closed.)

Does Problem 41 provide a solution to the problem of extending 3^x, as stated just before Proposition 18?

42. Prove the uniqueness of mapping g, given by Proposition 18.
43. Let $f: [a,b] \to [a,b]$ be a differentiable function on the bounded and closed interval $[a,b]$. Suppose $|f'(x)| \leq k < 1$ for $x \in [a,b]$. Show that the equation $f(x) = x$ has a unique solution in $[a,b]$.
44. Let f be a contraction mapping of a metric space $\langle X,d \rangle$ into itself. Then clearly for all x,y in X with $x \neq y$ we have $d(f(x),f(y)) < d(x,y)$. Show that the converse is false, by giving an example where the last inequality is satisfied but f is not a contraction mapping.
    (Hint: Consider the mapping $x \to \sqrt{1 + x^2}$ on the real line.)
45. Give an example of a noncomplete metric space where a contraction mapping may have no fixed point.
46. Prove Corollaries 21 and 22.
47. Give an example of a metric space which is bounded but not totally bounded.
48. Let X be a metric space. Show that any two

of the following properties imply the third:

(a) X is compact.

(b) X is homeomorphic to a discrete metric space.

(c) X is finite.

49. Let X be a compact metric space, and let $\langle x_n \rangle$ be a sequence with only one cluster point at x_o. Show that $\langle x_n \rangle$ converges to x_o.

50. Show that in a metric space any compact set is closed.

51. Show that every closed subset of a compact metric space is compact.

52. A subset E of a metric space X is said to be *relatively compact* iff the closure $\bar{E}$ is compact.

(a) Show that any subset of a relatively compact set is relatively compact.

(b) Show that a relatively compact set is totally bounded.

(c) Show that in a complete metric space a totally bounded set is relatively compact.

(d) Show that a subset of the real line is relatively compact if and only if it is bounded.

53. Show that a subset E of a metric space X is relatively compact if and only if every infinite sequence of points of E have a cluster point in X.

54. Let $f: X \to Y$ be a continuous mapping of a metric space X into a metric space Y. Show that if E is a compact (respectively relatively compact) set of X, then $f[E]$ is a compact (respectively relatively compact) set of Y.

55. Let X be a compact metric space, and let $f: X \to Y$ be a one-to-one continuous mapping of X onto a metric space Y. Show that f is a homeomorphism of X onto Y.
(Hint: Show that for every closed set $F \subset X$, $f[F]$ is closed. Use Problems 51 and 54.)

56. Let $\langle X,d \rangle$ be two metric spaces. We know that the Cartesian product $X \times Y$ is a metric space with respect to either one of the metrics given in Problem 3. We also know that all three metrics are uniformly equivalent metrics on $X \times Y$. Show that in order for $X \times Y$ to be a metric space of one of the following types:
(a) discrete
(b) bounded
(c) separable
(d) complete
(e) compact
(f) totally bounded
it is necessary and sufficient that both X and Y be of the same type.
The proofs of these facts can be found in Dieudonné [9] page 72.

57. Prove Theorem 25.

58. Is the sequence $\langle \sin nx \rangle$ equicontinuous on $[0,2\pi]$? Why or why not?

59. Let $\langle f_n \rangle_{n=1}^{\infty}$ be any sequence of differentiable functions on $[0,1]$ such that

$$|f_n'(x)| < M \quad (0 \leq x \leq 1,\ n = 1,2,\ldots)$$

for some $M > 0$. Show that the sequence $\langle f_n \rangle$

has a uniformly convergent subsequence.

60. Let $\langle X,d\rangle$ and $\langle Y,\sigma\rangle$ be compact metric spaces. Let f be a mapping of $X \times Y$ into another metric space $\langle Z,\rho\rangle$. For each $x \in X$ let f_x denote the mapping $y \to f(x,y)$. Show that if f is continuous, then the mapping $x \to f_x$ of X into $\langle \mathcal{C}(Y,Z),D\rangle$ is continuous and that the family $\langle f_x\rangle_{x\in X}$ is equicontinuous.

(Hint: Since $X \times Y$ is compact, f is uniformly continuous. Hence for $\varepsilon > 0$ there is $\eta > 0$ such that $d(x_1,x_2) < \eta$ and $\sigma(y_1,y_2) < \eta$ imply $\rho(f(x_1,y_1), f(x_2,y_2)) < \varepsilon$. Also $d(x_1,x_2) < \eta$ implies, for every y, that

$$\rho(f(x_1,y), f(x_2,y)) < \varepsilon$$

which means that $D(f_{x_1},f_{x_2}) < \varepsilon$. It follows that $\langle f_x\rangle_{x\in X}$ is a compact set in $\mathcal{C}(Y,Z)$, hence equicontinuous, by Theorem 24.)

61. (a) Show that a subset E of a metric space is nowhere dense iff $\bar{E}$ contains no open ball.

(b) Show that if a metric space $\langle X,d\rangle$ has no isolated points, then the set consisting of a single point of X is nowhere dense.

(c) Show that if x_0 is an isolated point, then $\{x_0\}$ is nowhere dense.

62. Show that a countable union of sets of first category is of first category.

63. Show that if the four Dini derivatives of f

are finite at some point of $(0,1)$, then there is a positive integer n, and an $\alpha \varepsilon (0,1)$ such that

$$\left|\frac{f(\alpha+h) - f(\alpha)}{h}\right| \leq n$$

for every $h > 0$.

64. Show that if $f: [0,1] \to \mathbb{R}$ is continuous, then for every $\varepsilon > 0$ there is a continuous function g whose graph is a broken line, each of whose segments has slope either greater than n or less than $-n$, such that $|f(t) - g(t)| < \varepsilon$ for every $t \varepsilon [0,1]$.

65. Let $f: [0,1] \to \mathbb{R}$ be continuous. Let f_1 be any indefinite integral of f, f_2, any indefinite integral of f_1, and so on.
Consider the sequence $\langle f_n \rangle_{n=1}^{\infty}$ where f_{n+1} is an indefinite integral of f_n, $n = 1,2,\ldots$.
Suppose that for each $x \varepsilon [0,1]$ there is a positive integer n such that $f_n(x) = 0$. Show that $f(x) = 0$ for all x in $[0,1]$.
(Hint: Let $E_n = \{x \varepsilon [0,1]: f_n(x) = 0\}$. Then $[0,1] = \bigcup_{n=1}^{\infty} E_n$. By Theorem 26 not every E_n is nowhere dense. Hence for some n, E_n contains an open interval, say I_n, such that $f_n(x) = 0$ for every $x \varepsilon I_n$. If I_n is not all of $[0,1]$ repeat this argument. Conclude

that $f(x) = 0$ on a dense subset of $[0,1]$. Hence $f(x) = 0$ for every $x \in [0,1]$.

CHAPTER 6

L^p Spaces

6.1 INTRODUCTION

The spaces considered in this chapter form a special subclass of the class of Banach spaces which will be studied in Chapter 8. An L^p space (where p is a positive real number) essentially consists of all measurable functions f defined on the closed interval $[0,1]$ (or any other bounded and closed interval) such that

$$\int_0^1 |f(x)|^p \, dx$$

is finite. In Chapter 10 we shall see that one can study the L^p spaces in a more general and abstract framework, where the interval $[0,1]$ is replaced by some abstract set X, the measurable subsets of $[0,1]$ by a certain family of subsets of X, and the Lebesgue integration by a more general process. Most of the theorems and proofs given here follow in exactly the same way in the abstract case. However, this duplication of proofs and results is intentional, and it is done in order to help the reader to make the transition from the classical case to the abstract case. The L^p spaces are of particular interest in the theory of Fourier transforms and integral equations. The case $p = 2$ is especially important.

6.2 THE HÖLDER AND MINKOWSKI INEQUALITIES

Definition *Let* p *be a positive real number, and let* f *be a measurable function defined on* [0,1] *such that* $\int_0^1 |f(x)|^p dx < +\infty$. *Then we say that* f *belongs to the class* $L^p[0,1]$ *or in short* L^p. *The norm* $\|f\|_p$ *is defined by*

$$\|f\|_p = \left(\int_0^1 |f(x)|^p \, dx\right)^{1/p} .$$

Remark If two functions in L^p coincide except on a set of measure zero, we shall consider them to represent the same element of L^p. In other words, if we write $f \sim g$, iff $f = g$ a.e., then clearly $\sim$ is an equivalence relation and the elements of L^p are actually equivalence classes of functions, the functions in any one class differing from one another only on sets of measure zero. We shall refer to a function f in L^p instead of the equivalence class containing f. Henceforth we shall be concerned with L^p for values of $p \geq 1$ exclusively. A measurable function f defined on $[0,1]$ is called *essentially bounded* iff there is some $M \geq 0$ such that the set $\{x: |f(x)| \geq M\}$ has measure zero. In other words, f is essentially bounded iff the inequality $|f(x)| \leq M$ holds almost everywhere on $[0,1]$ for some M. The norm $\|f\|_\infty$ of an essentially bounded function is defined by

$$\|f\|_\infty = \inf\{M: m\{x: |f(x)| \geq M\} = 0\}.$$

The number $\|f\|_\infty$ is also called the *essential least upper bound* or the *essential supremum* of f. A direct consequence of the definition of $\|f\|_\infty$ is that for every $\varepsilon > 0$ the set $\{x: |f(x)| > \|f\|_\infty - \varepsilon\}$ has positive measure.

We denote by L^∞ the class of all measurable and essentially bounded functions on $[0,1]$. Here also, as in the case of the L^p spaces, we identify two functions in L^∞ if they are equal almost everywhere.

If α is a real number and $f, g \in L^p$ we define αf to be the function $(\alpha f)(x) = \alpha f(x)$ and $f + g$ to be the function $(f + g)(x) = f(x) + g(x)$. The inequalities $f \leq g$, $f \geq g$ and $f \geq 0$ are defined in a similar manner. Also $|f|$ is defined by $|f|(x) = |f(x)|$, and for $f \geq 0$, f^p is the function $f^p(x) = [f(x)]^p$.

A set X of real-valued functions is called a real vector space iff for any real number α and any $f, g \in X$ we have $\alpha f \in X$ and $f + g \in X$. For the general definition of vector space see

Chapter 8. We now prove that $L^p (p \geq 1)$ is a vector space. Clearly $\alpha f \in L^p$ if $f \in L^p$. It is also true that $f + g \in L^p$ if $f, g \in L^p$. This fact follows from the inequality

$$|f + g|^p \leq 2^p[|f|^p + |g|^p].$$

To prove this inequality we may assume for the moment that f and g are any real or complex numbers and observe that

$$\max\{|f|,|g|\} \leq |f| + |g| \leq 2\max\{|f|,|g|\},$$

$$|f + g|^p \leq (|f| + |g|)^p \leq (2\max\{|f|,|g|\})^p$$

$$= \max\{2^p|f|^p, 2^p|g|^p\} \leq 2^p|f|^p + 2^p|g|^p$$

or

$$|f + g|^p \leq 2^p|f|^p + 2^p|g|^p.$$

Clearly this inequality holds if f and g are functions, so that

$$\int_0^1 |f + g|^p < +\infty.$$

Hence $f + g \in L^p$ whenever both f and $g \in L^p$. We shall now prove two fundamental inequalities for functions in L^p.

1. Theorem (Hölder's Inequality) *Let p and q be real numbers such that $1 < p < +\infty$ and*

$(1/p) + (1/q) = 1$. Let $f \in L^p$ and $g \in L^q$. Then $fg \in L^1$ and

$$(1) \qquad \int_0^1 |fg| \le \|f\|_p \|g\|_q .$$

Equality holds if and only if there exist constants α and β not both zero such that $\alpha|f(x)|^p = \beta|g(x)|^q$ a.e.

Proof If $\|f\|_p = 0$ or $\|g\|_q = 0$, then respectively (see Chap. 3, Corol. 27) $f = 0$ a.e. or $g = 0$ a.e., and inequality (1) holds trivially. Thus suppose $\|f\|_p \neq 0$, $\|g\|_q \neq 0$. Consider the function

$$h(t) = t - \frac{t^p}{p} - \frac{1}{q}$$

defined for $t \geq 0$. We have $h(1) = h'(1) = 0$. Furthermore, $h'(t) > 0$ for $0 < t < 1$, and $h'(t) < 0$ for $t > 1$. It follows that for $t = 1$ the function has an absolute maximum equal to zero. Hence

$$(2) \qquad t \le \frac{t^p}{p} + \frac{1}{q} \qquad \text{for } t \ge 0$$

the equality holding iff $t = 1$.

Setting $t = |a||b|^{1-q}$ with $a \neq 0$ and $b \neq 0$ and multiplying both members of (2) by $|b|^q$ we obtain

$$|a \cdot b| \leq \frac{|a|^p}{p} + \frac{|b|^q}{q}.$$

The last inequality has been proved for $a \neq 0$ and $b \neq 0$, but obviously holds also when a or b or both a and b are zero. If we choose $a = f(x)/\|f\|_p$, $b = g(x)/\|g\|_q$ we get

$$(3) \qquad \frac{|f(x)g(x)|}{\|f\|_p \ \|g\|_q} \leq \frac{|f(x)|^p}{p\|f\|_p^p} + \frac{|g(x)|^q}{q\|g\|_q^q}.$$

Hence integrating inequality (3) we get (1).

It is often useful to know the conditions under which equality holds in (1). We obtain this information by examining the proof. So equality holds in (1) iff equality holds in (3) a.e. In (3) equality holds iff

$$\frac{|f(x)|^p}{\|f\|_p^p} = \frac{|g(x)|^q}{\|g\|_q^q} \quad \text{a.e.}$$

The theorem is proved, for we can take

$$\alpha = \frac{1}{\|f\|_p^p}, \qquad \beta = \frac{1}{\|g\|_q^q}.$$

(*Note:* It is easily seen that (1) holds if either p or $q = 1$ and the other $= +\infty$ [see Prob. 1].)

2. Theorem (Minkowski's Inequality) *If* f *and*

g belong to L^p, $p \geq 1$, *then*

$$\|f + g\|_p \leq \|f\|_p + \|g\|_p.$$

Proof We proved earlier that L^p is a vector space, so that $f + g \in L^p$. If $p = 1$ or $p = \infty$, then the theorem follows from the inequality $|f + g| \leq |f| + |g|$. Hence we assume $1 < p < \infty$. We have

$$\int_0^1 |f + g|^p \leq \int_0^1 |f + g|^{p-1}|f| + \int_0^1 |f + g|^{p-1}|g|.$$

Applying Hölders' inequality to each term of the second member of this inequality we get, if $q = p/(p-1)$

$$\int_0^1 |f+g|^p \leq \|f\|_p \|(|f+g|^{p-1})\|_q + \|g\|_p \|(|f+g|^{p-1})\|_q.$$

We also have

$$\|(|f+g|^{p-1})\|_q = \left[\int_0^1 |f+g|^{(p-1)q}\right]^{1/q} = \left[\|f+g\|_p\right]^{p/q},$$

since

$$\frac{1}{p} + \frac{1}{q} = 1.$$

Hence the last inequality becomes

$$\|f+g\|_p^p \leq \left(\|f\|_p + \|g\|_p\right)\left(\|f+g\|_p\right)^{p/q}$$

or

(1) $$\|f+g\|_p \leq \|f\|_p + \|g\|_p.$$

The discussion of the case where equality holds is left to the reader (see Prob. 2).

6.3 CONVERGENCE IN THE MEAN OF ORDER p ($1 \leq p < +\infty$)

Let us consider the mapping $d: L^p \times L^p \to \mathbb{R}$ defined by $d(f,g) = \|f-g\|_p$ for all f and g in L^p. We claim that d has all the properties of a metric so that $\langle L^p, d\rangle$ is a metric space. Clearly $0 \leq \|f-g\|_p = d(f,g) < +\infty$. If $d(f,g) = 0$ then $\int_0^1 |f-g|^p = 0$, so that by the Corollary 27 of Chapter 3 we have $f - g = 0$ a.e., and by the Remark in section 6.2, f and g represent the same element of L^p, so that $f = g$. Thus $d(f,g) = 0$ if and only if $f = g$.

Next we observe that the symmetry property trivially holds, for

$$d(f,g) = \|f-g\|_p = \|g-f\|_p = d(g,f).$$

Finally if $f, g, h \in L^p$ then $-g$ and $-h$ also belong to L^p, so that by the Minkowski inequality we have

$$d(f,h) = \|f-h\|_p \le \|f-g\|_p + \|g-h\|_p = d(f,g) + d(g,h)$$

which proves that the triangle inequality holds. Hence $\langle L^p, d\rangle$ is a metric space. The following definitions are as in a metric space.

Definitions *Let* $\langle f_n \rangle$ *be a sequence of functions in* L^p, $(1 \le p < \infty)$ *and let* $f \in L^p$. *We say that* $\langle f_n \rangle$ converges to f in the mean of order p *(or that* $\langle f_n \rangle$ *converges to* f *in* L^p *or that* $\langle f_n \rangle$ *is* L^p*-convergent to* f*) if and only if*

$$\lim_{n\to\infty} \|f_n - f\|_p = 0.$$

A sequence $\langle f_n \rangle$ *in* L^p *is said to be a* Cauchy sequence in L^p *if and only if to every* $\varepsilon > 0$ *there corresponds an integer* N *such that* $n \ge N$, $m \ge N$ *imply* $\|f_n - f_m\|_p < \varepsilon$.

The reader should be careful to distinguish the above notion of convergence in L^p, from the notion of *pointwise convergence* (see Prob. 9). We recall that a sequence $\langle f_n \rangle$ is said to converge pointwise to f iff for each x we have $\lim_{n\to\infty} f_n(x) = f(x)$. The next theorem exhibits the very important fact that the L^p spaces considered as metric spaces are complete, i.e., every Cauchy sequence in L^p converges in the mean of order p to an element of L^p.

3. Theorem *The space* L^p *is a complete metric space for* $1 \leq p \leq \infty$.

Proof We first consider the case $p = +\infty$. Let $\langle f_n \rangle$ be a Cauchy sequence in L^∞. Put

$$S_k = \{x: |f_k(x)| > \|f_k\|_\infty\};$$

$$S_{r,n} = \{x: |f_n(x) - f_r(x)| > \|f_n - f_r\|_\infty\}.$$

Then it follows from the definition of the norm in L^∞ that $mS_k = mS_{r,n} = 0$ for all k,r and n.

Let S be the union of all sets $S_k, S_{r,n}$ for $k,r,n = 1,2,3,\ldots$. Clearly S is the union of a countable number of sets of measure zero so that $mS = 0$. Then the sequence $\langle f_n \rangle$ converges uniformly on the set $[0,1] - S$ to a bounded function f. If we define f to be zero on S, since $mS = 0$, it follows that f belongs to L^∞ and

$$\lim_{n\to\infty} \|f_n - f\|_\infty = 0.$$

Hence L^∞ is complete.

Next assume that $1 \leq p < +\infty$ and let $\langle f_n \rangle$ be a Cauchy sequence in L^p. Then for any given positive integer i there is a positive integer n_i such that $m \geq n_i$, $n \geq n_i$ imply $\|f_n - f_m\|_p < 2^{-i}$. Clearly we may take the n_i's such that $n_i < n_{i+1}$, $i = 1,2,\ldots$. Consider the sequence $\langle f_{n_i} \rangle$. We have

$$\|f_{n_{i+1}} - f_{n_i}\|_p < 2^{-i}.$$

Put

$$g_k = \sum_{i=1}^{k} \left|f_{n_{i+1}} - f_{n_i}\right| ,$$

$$g = \sum_{i=1}^{\infty} \left|f_{n_{i+1}} - f_{n_i}\right| .$$

By applying Minkowski's inequality a finite number of times to the equality that defines g_k we get $\|g_k\|_p < 1$ or

$$\int_0^1 g_k{}^p < 1 \quad \text{for } k = 1,2,\ldots.$$

If we apply Fatou's lemma to the sequence $\langle g_k{}^p \rangle$ of nonnegative functions, we obtain

$$\int_0^1 g^p = \int_0^1 \underline{\lim}_{k\to\infty} g_k{}^p \leq \underline{\lim}_{k\to\infty} \int_0^1 g_k{}^p \leq 1$$

or

$$\|g\|_p \leq 1.$$

In particular, the last inequality asserts that g^p

is integrable, and hence g is finite a.e. Thus the series

$$f_{n_1}(x) + \sum_{i=1}^{\infty} (f_{n_{i+1}}(x) - f_{n_i}(x))$$

converges absolutely for almost all values of x in $[0,1]$. Let us denote by $f(x)$ the sum of this series for the values of x at which it converges, and set $f(x) = 0$ on the remaining set which has measure zero. The function f as the limit of a sequence of measurable functions is also measurable. To proceed with the proof, let $\varepsilon > 0$ be given. Since $\langle f_n \rangle$ is a Cauchy sequence there must exist an N such that $n \geq N$, $m \geq N$ imply $\|f_n - f_m\|_p < \varepsilon$. For every $m > N$ we have by Fatou's lemma

$$\int_0^1 |f - f_m|^p \leq \underline{\lim}_{i\to\infty} \int_0^1 |f_{n_i} - f_m|^p \leq \varepsilon^p.$$

This means that $f - f_m$ belongs to L^p, which implies that f also belongs to L^p since $f = (f - f_m) + f_m$. Furthermore, the last inequality shows that $\langle f_n \rangle$ converges in L^p to f.

6.4 BOUNDED LINEAR FUNCTIONALS ON L^p

Let F be a real-valued function defined on L^p such that

$$F(\alpha f + \beta g) = \alpha F(f) + \beta F(g)$$

for all f and g in L^p and for all real numbers α and β. Then F is called a *linear functional* on L^p.

Linear functionals are of interest in analysis because they are intimately connected with the theory of integration.

A linear functional F on L^p is said to be *bounded* if and only if there is a positive constant M such that

$$|F(f)| \leq M\|f\|_p \qquad \text{for all } f \in L^p.$$

The norm $\|F\|$ of a bounded linear functional F is defined to be the infimum of the set of all the numbers M satisfying the above inequality. That is

$$\|F\| = \inf\{M: |F(f)| \leq M\|f\|_p, \quad f \in L^p\}.$$

From this we see that

$$|F(f)| \leq \|F\| \, \|f\|_p \qquad \text{for all } f \in L^p.$$

4. Proposition *Let p and q be extended real numbers such that $1 \leq p \leq +\infty$ and $(1/p) + (1/q) = 1$. If $g \in L^q$, and if F_g is defined by*

$$F_g(f) = \int_0^1 f(x)g(x)dx$$

for all $f \in L^p$, then F_g is a bounded linear

functional on L^p. We also have

$$\|F_g\| = \|g\|_q.$$

Proof By Hölder's inequality we have

$$|F_g(f)| = \left|\int_0^1 f(x)g(x)dx\right| \leq \int_0^1 |f(x)g(x)|dx \leq \|f\|_p \|g\|_q.$$

We conclude that F_g is a well-defined bounded linear functional on L^p with $\|F_g\| \leq \|g\|_q$. We shall now establish equality $\|F_g\| = \|g\|_q$. There are three cases.

Case 1: $1 < p < +\infty$. Since $g \in L^q$ we have that $g(x)$ is finite for almost all values of x. Set $S = \{x: g(x) = +\infty\}$ and consider the function f defined by

$$f(x) = 0 \qquad \text{if } x \in S$$

$$f(x) = |g(x)|^{q/p} \cdot \operatorname{sgn}(g(x)) \qquad \text{if } x \notin S$$

where $\operatorname{sgn}(g(x))$ is equal to $-1, 0$ or 1 if $g(x)$ is respectively negative, zero, or positive. One can easily verify that $f \in L^p$ with $\|f\|_p = \|g\|_q^{q/p}$. We have

$$F_g(f) = \int_0^1 fg = \int_0^1 |g|^q = \|g\|_q^q = \|g\|_q \|f\|_p.$$

This equality and the definition of the norm of a bounded linear functional imply that $\|F_g\|$ cannot be smaller than $\|g\|_q$. Hence $\|F_g\| = \|g\|_q$.

Case 2: $p = 1$. We have $g \in L^\infty$. If $\|g\|_\infty = 0$ then we are done, for $\|F_g\| = \|g\|_\infty = 0$. Suppose $\|g\|_\infty > 0$, and let ε be an arbitrary positive number. It follows from the definition of $\|g\|_\infty$ that at least one of the sets

$$A = \{x\colon g(x) \geq \|g\|_\infty - \varepsilon\},$$

$$B = \{x\colon -g(x) \geq \|g\|_\infty - \varepsilon\}$$

must have positive measure. Suppose $mA > 0$, and define the function f to be the characteristic function of A, that is $f = \chi_A$. Then

$$F_g(f) = \int_0^1 fg = \int_A g \geq (\|g\|_\infty - \varepsilon)mA = \|g\|_\infty \cdot mA - \varepsilon \cdot mA$$

$$= \|g\|_\infty \|f\| - \varepsilon \cdot mA.$$

Since ε is arbitrary we have

$$F_g(f) \geq \|g\|_\infty \|f\|_1,$$

which implies that $\|F_g\| = \|g\|_\infty$. The proof is similar if $mB > 0$.

Case 3: $p = \infty$. We have $g \in L^1$. Consider the function f defined by

$$f(x) = 0 \qquad \text{if} \quad g(x) = \pm\infty$$

$$f(x) = \operatorname{sgn}(g(x)) \qquad \text{if} \quad g(x) \text{ is finite.}$$

Clearly $f \in L^\infty$ and $\|f\|_\infty = 1$. We have

$$F_g(f) = \int_0^1 fg = \int_0^1 g \operatorname{sgn} g = \int_0^1 |g| = \|g\|_1 = \|g\|_1 \|f\|_\infty.$$

Hence

$$\|F_g\| = \|g\|_1.$$

It is natural to ask the question whether all bounded linear functionals on L^p have the form as given in Proposition 4, and whether that representation is unique. We shall see in Chapter 8, section 4, that for $p = +\infty$ the answer to this question is negative: L^1 *does not provide all bounded linear functionals on* L^∞. For $1 \leq p < +\infty$ the answer is affirmative and is given by the following theorem known as the *Riesz representation theorem.* We shall not prove this theorem here, for it will be established in a more general setting in Chapter 10.

5. Theorem (F. Riesz) *Let p and q be extended real numbers such that $1 \leq p < \infty$ and $(1/p) + (1/q) = 1$, and let F be a bounded linear functional on L^p. Then there is a unique $g \in L^q$ such that*

$$F(f) = \int_0^1 fg \quad \text{for all} \quad f \in L^p.$$

We have also $\|F\| = \|g\|_q$.

Let us denote by $(L^p)^*$ the set of all bounded linear functionals on L^p where $1 \leq p < \infty$. $(L^p)^*$ is called the *dual* or the *conjugate space of* L^p. If F and G are in $(L^p)^*$, and if α is a real number, then we define $F + G$ and αF by

$$(F + G)(f) = F(f) + G(f) \quad \text{for all} \quad f \in L^p$$

$$(\alpha F)(f) = \alpha F(f) \quad \text{for all} \quad f \in L^p.$$

One easily sees that $F + G$ and αF are linear functionals, so that $(L^p)^*$ is a real vector space. Define now the distance $d(F,G)$ between F,G by

$$d(F,G) = \|F - G\|.$$

We now prove, as in the case of L^p, that $\langle (L^p)^*, d \rangle$ is a metric space.

Let F,G and H be in $(L^p)^*$. We have $d(F,G) \geq 0$. If $d(F,G) = \|F - G\| = 0$, then for every $f \in L^p$ we have $|(F - G)(f)| \leq \|F - G\| \|f\|_p = 0$, so that $F = G$. Clearly $d(F,G) = d(G,F)$. Finally we prove that the triangle inequality is satisfied. Put $F - G = A$, $G - H = B$. We have

$$d(F,H) = \|F-H\| = \|A+B\| = \inf\{M: |(A+B)(f)| \le M\|f\|_p, \ f \in L^p\}$$

$$\le \inf\{M: |A(f)| + |B(f)| \le M\|f\|_p, \ f \in L^p\}$$

$$\le \|A\| + \|B\|$$

$$= \|F-G\| + \|G-H\| = d(F,G) + d(G,H)$$

Hence $\langle (L^p)^*, d \rangle$ is a metric space.

Proposition 4 together with Theorem 5 tell us that there is a one-to-one correspondence between L^q and $(L^p)^*$ which is an isomorphism between vector spaces, and an isometry between metric spaces. Thus, we say that L^q and $(L^p)^*$ are *isometrically isomorphic*, hence identical from an abstract viewpoint. The isomorphism simply amounts to a relabeling of the elements of these two spaces.

PROBLEMS

1. Prove Hölder's inequality for $p = 1$.
2. Discuss the case where equality holds in Minkowski's inequality.
   (Hint: If $1 < p < \infty$, then equality holds iff there are numbers α and β not both zero such that $\alpha f(x) = \beta g(x)$ a.e. Consider also $p = 1$, $p = +\infty$.)
3. Let $1 \le p \le q$. Show that if $f \in L^q$, then

$f \in L^p$.

4. Let p and q be real numbers such that $1 < p < \infty$ and $(1/p) + (1/q) = 1$. Show that for any real or complex numbers $x_1, x_2, \ldots, x_n$ and $y_1, y_2, \ldots, y_n$ we have

$$\sum_{i=1}^{n} |x_i y_i| \le \left(\sum_{i=1}^{n} |x_i|^p\right)^{1/p} \left(\sum_{i=1}^{n} |y_i|^q\right)^{1/q}.$$

This is Hölder's inequality for sums. The special case $p = q = 2$ is called Cauchy's inequality.

5. If $1 \le p < \infty$, and $x_1, x_2, \ldots, x_n$, $y_1, y_2, \ldots, y_n$ are real or complex numbers show that

$$\left(\sum_{i=1}^{n} |x_i + y_i|^p\right)^{1/p} \le \left(\sum_{i=1}^{n} |x_i|^p\right)^{1/p} + \left(\sum_{i=1}^{n} |y_i|^p\right)^{1/p}.$$

This is Minkowski's inequality for sums.

6. For $1 \le p < \infty$ let ℓ^p be the set of all infinite sequences of complex numbers.

$$x = \langle \alpha_1, \alpha_2, \ldots, \alpha_n, \ldots \rangle$$

such that

$$\sum_{i=1}^{\infty} |\alpha_i|^p < +\infty.$$

The norm $\|x\|_p$ of $x \in \ell^p$ is defined to be

$$\|x\|_p = \left(\sum_{i=1}^{\infty} |\alpha_i|^p\right)^{1/p} .$$

We denote by ℓ^∞ the set of all infinite bounded sequences of complex numbers

$$y = \langle \beta_1, \beta_2, \ldots, \beta_n, \ldots \rangle .$$

The norm $\|y\|_\infty$ of $y \in \ell^\infty$ is defined by

$$\|y\|_\infty = \sup\{|\beta_n| : n = 1,2,\ldots\}.$$

(a) Let $1 \leq p < \infty$ and $(1/p) + (1/q) = 1$. Let $x = \langle \alpha_n \rangle_{n=1}^{\infty}$ be in ℓ^p and $y = \langle \beta_n \rangle_{n=1}^{\infty}$ be in ℓ^q. Show that $\langle \alpha_n \beta_n \rangle_{n=1}^{\infty}$ is in ℓ^1 and

$$\sum_{n=1}^{\infty} |\alpha_n \beta_n| \leq \|x\|_p \|y\|_q .$$

This is Hölder's inequality for sequences. (Hint: Use Problem 4.)

(b) Let $1 \leq p \leq \infty$ and $x = \langle \alpha_n \rangle_{n=1}^{\infty}$ and $y = \langle \beta_n \rangle_{n=1}^{\infty}$ in ℓ^p. Put $z = \langle \alpha_n + \beta_n \rangle_{n=1}^{\infty}$. Show that $z \in \ell^p$ and

$$\|z\|_p \leq \|x\|_p + \|y\|_p.$$

This is Minkowski's inequality for sequences. (Hint: Use Problem 5.)

7. Let $f \varepsilon L^\infty$. Clearly for all $0 < p < \infty$ we have $f \varepsilon L^p$. Show that

$$\lim_{p\to\infty} \|f\|_p = \|f\|_\infty.$$

(Hint: Let $0 < \alpha < \|f\|_\infty$. Put $E = \{x: |f(x)| > \alpha\}$. Then $mE > 0$. We have

$$\alpha(mE)^{1/p} \leq \|f\|_p \leq \|f\|_\infty.$$

By letting $p \to \infty$ we get

$$\alpha \leq \underline{\lim}_{p\to\infty} \|f\|_p \leq \overline{\lim}_{p\to\infty} \|f\|_p \leq \|f\|_\infty.)$$

8. Let $\langle f_n \rangle$ be a sequence in L^∞ and let $f \varepsilon L^\infty$. Show that

$$\lim_{n\to\infty} \|f_n - f\|_\infty = 0$$

if and only if there is a set E of measure zero such that given $\varepsilon > 0$ there is a positive integer N such that $n \geq N$ implies

$$|f_n(x) - f(x)| < \varepsilon \quad \text{for all} \quad x \varepsilon [0,1] - A.$$

9. Let $1 \le p < \infty$, and let $f \in L^p$ and $\langle f_n \rangle$ a sequence in L^p, such that

$$\lim_{n\to\infty} f_n(x) = f(x) \quad \text{a.e.}$$

Show that

$$\lim_{n\to\infty} \|f_n - f\|_p = 0$$

iff

$$\lim_{n\to\infty} \|f_n\|_p = \|f\|_p.$$

10. Let p and q be real numbers such that $1 < p < \infty$ and $(1/p) + (1/q) = 1$. Let $\langle f_n \rangle$ be a sequence in L^p such that $\|f_n\| \le M$ for all n, and let g be in L^q. Show that

$$\lim_{n\to\infty} \int_0^1 f_n g = \int_0^1 fg.$$

11. If f and g are the functions considered in the proof of Proposition 4, Case 1, show that

$$f \in L^p, \text{ and } \|f\|_p = \|g\|_q^{q/p}.$$

12. Show that the space $\langle (L^p)^*, d \rangle$ as defined in the lines following Theorem 5, is a complete metric space.

CHAPTER 7

Topological Spaces

7.1 INTRODUCTION

In section 2.1 we already considered some of the fundamental algebraic structures of mathematics, such as groups, rings, and fields. Another fundamental structure, called *topological structure*, which appears in all parts of analysis and plays a role of vital importance, is the one that gives precise mathematical meanings to the intuitive notions of limit, continuity, nearness, and so forth.

Topology is the branch of mathematics that studies topological structures. Etymologically the word *topology* means the "science of the place" and derives from the Greek word τόπος meaning a "place." An older synonym, not in use anymore, is "analysis situs." The notions of limit and continuity appeared first in geometry and they are closely related with the notion of approximation of the exact value of a magnitude in a physical experiment. Thus, in the first place, limit and continuity played a role in the theory of functions of a real variable in the Euclidean plane and especially in the theory of real numbers. In all these cases it "seems" that the notions of limit and continuity are intrinsically related to the notion of distance, which can be expressed in terms of one or several numbers. Take an example on the real line. Roughly speaking we say that the limit of x^2, when x approaches a, is a^2 because the number $|x^2 - a^2|$, measuring the distance between the points x^2 and a^2, becomes "sufficiently" small. That is, the nearness of two points is measured here by the distance between the two points. It was the discovery by the modern mathematician that one can speak of nearness of two points of a space, and consequently of limits and continuity in a space, independently of the notion of distance. The basic observation was that the spaces where the notion of distance is defined (so that we can speak of limit and continuity) have some properties which are independent of the notion of the distance itself which gave birth to the notions of limit and continuity. To be more explicit consider the real line and define a neighborhood at a point a to be any set that contains an open interval containing a. Then any set containing a neighborhood of a is also a neighborhood of a. Also the intersection of any two neighborhoods of a is a neighborhood of a. These two properties and some others, if they are possessed by a space, entail a great number of consequences which are independent of the notion of distance, even though it was this notion of distance which allowed us in the first place to define a neighborhood. Thus we obtain in these spaces theorems in which the notion of distance is not mentioned at all. Generally speaking, an arbitrary nonempty set X is said to become a *topological space* whenever one assigns to every element a in X a collection of subsets of X, called the neighborhoods of a, such that these collections satisfy certain axioms. However, it took a long time in history to distinguish these axioms and formulate them in their simplest form. In the lines to follow (see Prop. 2) we shall define a topological space in a different way, which, however, is equivalent to the one given above.

7.2 OPEN SETS AND CLOSED SETS

Definition *A topological space* $\langle X, \mathcal{O} \rangle$ *is an ordered pair consisting of a nonempty set* X *together with a collection* $\mathcal{O}$ *of subsets of* X, *called open sets, satisfying the following properties:*

O_1: *Every union (finite or infinite) of open sets is open.*

O_2: *Every* finite *intersection of open sets is open.*

O_3: *The set* X *and the empty set* $\emptyset$ *are open.*

The collection $\mathcal{O}$ is called a *topology* for the set X. The elements of X are also called *points*. If the set X contains more than one point, then it is possible to define several topologies on X. For example, the *discrete topology* is the one for which $\mathcal{O}$ is the collection of all subsets of X. Obviously this topology contains the largest possible number of open sets. Another is the *trivial topology*, in which $\mathcal{O}$ contains only the empty set $\emptyset$ and the set X. This is the topology that contains the fewest number of open sets. Both the discrete and trivial topology are probably the least interesting topologies that we can define on X. More generally, if $\mathcal{O}_1$ and $\mathcal{O}_2$ are two different topologies on X, then if $\mathcal{O}_1 \subset \mathcal{O}_2$ we say that $\mathcal{O}_1$ is weaker (coarser) than $\mathcal{O}_2$ or that $\mathcal{O}_2$ is stronger (finer) than $\mathcal{O}_1$. If $\langle \mathcal{O}_i \rangle$ is any collection of topologies on X we easily see that the intersection of this collection of topologies is also a topology on X.

We observe that the axioms O_1, O_2, O_3 are the same as those mentioned in the study of the topology of the real line in Chapter 2. These three simple axioms are sufficient to guarantee a number of results.

Let $\langle X,d \rangle$ be a given metric space and define a set O in X to be open iff for every $x \in O$, there is an $r > 0$ such that the open ball $S_r = \{y \in X: d(x,y) < r\}$ belongs to the set O. Then we can easily verify that the axioms O_1, O_2, O_3 are satisfied, so that with the open sets defined in the above manner X is a topological space. The topology so derived from a metric d is called a *metric topology* for X. The reader should have no difficulty in constructing examples, in the Cartesian plane for instance, where two or more distinct metrics determine the same topology. Hence the distinction between the metric space and its associated topological space by its metric is essential. Two metrics are said to be *equivalent* iff they generate the same metric topology.

If, for a given topological space $\langle X, \mathcal{O} \rangle$, there is a metric d such that the metric topology generated by d is the topology $\mathcal{O}$, then $\langle X, \mathcal{O} \rangle$ is said to be a *metrizable* topological space.

Example 1

It follows from the above discussion that every metric space $\langle X,d \rangle$ provides a topological space $\langle X, \mathcal{O} \rangle$ where $\mathcal{O}$ is its metric topology.

Example 2

Let X be an arbitrary infinite set. We define a subset O to be open iff $O = \emptyset$ or if $\complement O$ is finite. It is easy to verify that this collection of open sets is a topology for X, called the *co-finite* topology.

Example 3

Let $\langle X,< \rangle$ be an arbitrary totally ordered set. An *open interval* (a,b) is the set of all $x \in X$ such that $a < x$, $x < b$, $a \neq x$, $b \neq x$. Define a set O in X to be *open* if and only if $O = \emptyset$ or O is a union of open intervals. (Obviously we repeat here the procedure used in the topology of the real line.) Then clearly X is a topological space, for axioms O_1, O_2, O_3 are easily seen to be satisfied. The topology defined in the above manner is called the *order topology* of X.

Definition *Let* X *be a topological space. Then a subset* $F \subset X$ *is said to be* closed *if and only if* $\complement F$ *is open.*

As in the case of the real line the statements O_1, O_2, O_3 imply three statements F_1, F_2, F_3

which are equivalent to them.

F_1: Every intersection (finite or infinite) of closed sets is closed.

F_2: Every *finite* union of closed sets is closed.

F_3: The empty set $\emptyset$ and the set X are closed.

The proof of these statements follows immediately if we apply De Morgan's laws.

Definition *A* neighborhood *of a point* x *of a topological space* X *is any subset of* X *which contains an open set containing* x. *We denote by* $\mathcal{V}(x)$ *the collection of all neighborhoods of a point* x. *If* E *is a subset of* X, *then a neighborhood of* E *is any subset* A *of* X *containing an open set* O *containing* E: $E \subset O \subset A$.

Remark Every open set containing a point x is a neighborhood of x, but not every neighborhood of x is necessarily an open set.

1. Proposition *A set* O *is open if and only if it is a neighborhood of each of its points.*

Proof If O is open, then it follows from the last remark that O is a neighborhood of each of its points. Suppose now that O is a neighborhood of each of its points and let $x \in O$. Then there exists an open set E_x containing x and contained in O. Therefore $O = \bigcup_{x \in O} E_x$. Thus O, as a union of open sets, is open.

Remark It follows immediately from the definition of neighborhood that the topology of a space uniquely determines the neighborhoods of points. The above proposition shows that the converse is also true: If the neighborhoods of all the points of a topological space are known then the open sets, that is the topology of the space, is known. Therefore *two topologies on a set are identical iff they admit the same neighborhoods.*

Here are some properties of neighborhoods which, as the following Proposition 2 shows, can be taken together as the definition of a topological space.

V_1: Every point x has at least one neighborhood.

V_2: Every neighborhood of x contains x.

V_3: Every set containing a neighborhood of x is a neighborhood of x.

V_4: The intersection of two neighborhoods of x is a neighborhood of x.

V_5: If V is a neighborhood of x, there exists a neighborhood W of x such that $W \subset V$ and such that V is a neighborhood of each point of W.

2. Proposition *Let X be any nonempty set and suppose that with each point x in X there is associated a collection $\mathcal{V}(x)$ of subsets of X such that the properties V_1, V_2, V_3, V_4, and V_5 are satisfied. Then there is a topology $\mathcal{O}$ on X for which $\mathcal{V}(x)$ is the collection of neighborhoods of x, for x in X. Moreover the topology $\mathcal{O}$ is unique.*

The proof of this proposition is left to the

reader (see Prob. 5).

7.3 CLOSURE--INTERIOR--BOUNDARY

Definitions *Let X be a topological space, let E be a subset of X, and let x be a point of X.*

x is said to be a **point of closure** *of E iff every neighborhood of x contains a point of E.*

x is said to be an **accumulation point** *(also called* **cluster point** *or* **limit point**) *of E iff every neighborhood of x contains at least one point of E other than x (x itself may or may not belong to E). We denote by E' the set of accumulation points of a set E.*

x is said to be an **isolated point** *of E iff it belongs to E but is not an accumulation point of E. This means that there is a neighborhood of x which contains no point of E other than x. The set of points of X which are points of closure of the set E is called the* **closure** *of E, and it is denoted by $\overline{E}$.*

Obviously, for every E, $E \subset \overline{E}$. For any given set $E \subset X$ there is at least one closed set containing E, for example X itself. By property F_1, the intersection of all closed sets containing E must be a closed set; we call it the *smallest closed set* containing E.

3. Proposition *The smallest closed set S containing a set E is equal to the closure $\overline{E}$ of E.*

Proof Let $x \in S$. Then x must belong to $\overline{E}$. For, if it does not, then there is an open neighborhood V_x of x such that $V_x \cap E = \emptyset$. This implies that $S - V_x$ is a closed set containing E and contained in S, which contradicts the fact that S is the smallest closed set containing E. Therefore $S \subset \overline{E}$. Let now $y \in \overline{E}$. Then y must belong to S. But suppose not. Then y is an element of the open set $\complement S$, so that there is an open neighborhood V_y which contains no element of S. Since $E \subset S$ we also have $V_y \cap E = \emptyset$. But this contradicts the fact that $y \in \overline{E}$. Thus $\overline{E} \subset S$, and the proposition is proved.

4. Corollary *A set F is closed if and only if $F = \overline{F}$.*

Proof If F is closed, then it is equal to the smallest closed set containing F, that is $F = \overline{F}$. If $F = \overline{F}$, then since $\overline{F}$ is closed, F also is closed.

Notice that for any set S we always have $\overline{S} = \overline{\overline{S}}$.

5. Corollary *A set F is closed if and only if it contains all its accumulation points.*

Proof Suppose F closed. Then $F = \overline{F}$. By the definition of closure we have $\overline{F} = F \cup F'$, so that $F' \subset F$. Suppose $F' \subset F$. Then $F \cup F' \subset F \cup F = F$, or $\overline{F} \subset F$, which implies $\overline{F} = F$. Hence F is closed, since $\overline{F}$ is closed.

The following statements express some properties of the closure. The proofs are the same as those given in Chapter 2.

(1) $\overline{\emptyset} = \emptyset$; (2) $A \subset \overline{A}$; (3) $\overline{\overline{A}} = \overline{A}$;

(4) $\overline{A \cup B} = \overline{A} \cup \overline{B}$; (5) $\overline{A \cap B} \subset \overline{A} \cup \overline{B}$.

Property (4) can be extended to any finite number of sets but not to an infinite number because, in the second member, the arbitrary union of closed sets is not necessarily a closed set, while the first member, since it is a closure, is a closed set. For example consider the set $Q = \{r_n\}$ of all rational numbers on $\mathbb{R}$. Then

$\mathbb{R} = \overline{Q} \neq \bigcup_{n=1}^{\infty} \{r_n\} = Q$. We also notice that in (5) equality need not occur. For example, take on the real line A and B to be the sets of rationals and irrationals, respectively. Then $\overline{A} \cup \overline{B}$ equals the whole real line while $\overline{A \cap B} = \emptyset$.

It is worthwhile to mention here that properties (1), (2), (3), and (4) define a topology on X. More precisely let us call *closure operator* on a set X the operator that assigns to each subset A of X a subset $\overline{A}$ of X such that properties (1), (2), (3), and (4) are satisfied. We have the following theorems, the proof of which can be found in Kelley [16], page 43.

6. Theorem (Kuratowski) *Let* — *be a closure operator on a set* X*, let* $\mathcal{F}$ *be the collection of all subsets* A *of* X *for which* $\overline{A} = A$*, and let* $\mathcal{O}$ *be the family of complements of members of* $\mathcal{F}$*. Then* $\mathcal{O}$ *is a topology for* X*, and* $\overline{A}$ *is the closure in this topology of* A*, for each subset* A *of* X*.*

Definition *Let X be a topological space and $E \subset X$ a subset of X. Then the union of all open sets contained in E is said to be the* interior *of E. It is denoted by $\mathring{E}$, and it might be empty.*

An immediate consequence of this definition is that the interior of a set is an open set. Also a set E is open if and only if $E = \mathring{E}$. The interior of a set E is thus the largest open set in E.

Definition *Let X be a topological space and let E be a subset of X. Then the* boundary *$b(E)$ of E is defined to be the set of all points x such that each neighborhood of x intersects both E and $\complement E$.*

It follows from this definition that the points of the boundary of E are points of closure of E and $\complement E$. Hence $b(E) = \overline{E} \cap \overline{\complement E}$. Also it is clear that $b(E) = b(\complement E)$.

Definition *Let E be a subset of a topological space X. Then E is said to be* dense *in X iff $\overline{E} = X$.*

E is said to be nowhere dense *in X iff $\overline{E}$ has an empty interior.*

For example, on the real line, the set of rational numbers is dense while the set of integers is nowhere dense.

Here are some statements which derive directly from the definitions. Let X be a topological space.

(a) If E is dense in X and S is a set containing E, then S also is dense in X. Any subset of a nowhere dense set is nowhere dense.

(b) E is dense in X if and only if for every open set O we have $E \cap O \neq \emptyset$.

(c) E is nowhere dense in X if and only if $\overline{E}$ is nowhere dense in X.

(d) E is nowhere dense in X if and only if $\complement \overline{E}$ is dense in X, which is equivalent to saying that every nonempty open set O contains a nonempty open subset W such that $W \cap E = \emptyset$.

(e) If E and S are nowhere dense in X so is the set $E \cup S$.

We warn thc reader about the following fact: *If E is nowhere dense in X, then $\complement E$ is dense in X. But it is not true that if $\complement E$ is dense then E is nowhere dense.* It can happen that both E and $\complement E$ are dense in X. This is the case if E is the set of rational numbers.

7.4 CONTINUOUS FUNCTIONS

Definition *Let f be a mapping on a topological space X into a topological space Y. Then f is said to be* continuous *at a point $a \in X$ if and only if for every neighborhood V of $f(a)$, $f^{-1}[V]$ is a neighborhood of a. The mapping f is said to be continuous on the whole space X if and only if it is continuous at every point of X.*

The following theorem provides an equivalent definition of the continuity of f on X.

7. Theorem *Let X and Y be topological spaces and let f be a mapping on X into Y. Then*

f is continuous on X if and only if for every open set B in Y, $f^{-1}[B]$ *is an open set in X.*
Proof Suppose f is continuous on X. Since B is open in Y, it is a neighborhood of each of its points. By the definition of continuity at one point, $f^{-1}[B]$ is a neighborhood of each of its points, so that by Proposition 1, $f^{-1}[B]$ must be open. Suppose now that $f^{-1}[B]$ is open for every open set in Y, and let a be any point in X. Then for every neighborhood V of f(a) the set $f^{-1}[\mathring{V}]$ is an open (because $\mathring{V}$ is open) set containing a, therefore $f^{-1}[V]$ is a neighborhood of a. This proves that f is continuous at every point of X, so that f is continuous on X.

It is important to observe that *the continuity of* f *requires that the inverse images, and not the direct images of open sets under* f, *be open sets.*

In general the direct image of an open set by a continuous mapping is not open. For example the image of any nonempty open set under the constant function f(x) = 1 is not open.

Let X, Y, Z be topological spaces, f a mapping of X into Y, g a mapping of Y into Z, and h the mapping of X into Z defined by $h(x) = g(f(x))$ for all $x \in X$. Then h is the *composite mapping of* g *and* f and we write $h = g \circ f$.

8. Theorem *If f is continuous at* $a \in X$ *and g is continuous at the point f(a), then h is continuous at a.*

Proof Let V be a neighborhood of $h(a) = g(f(a))$.

Since g is continuous at $f(a)$, $g^{-1}[V]$ is a neighborhood of $f(a)$. Since f is continuous at a, we have that $f^{-1}[g^{-1}[V]] = h^{-1}[V]$ is a neighborhood of a. Thus h is continuous at a.

Definition *Let X and Y be topological spaces. A* homeomorphism *is a continuous one-to-one mapping f of X onto Y such that f^{-1} is also continuous.*

Two topological spaces are said to be *homeomorphic* iff there is a homeomorphism of one onto the other.

In Chapter 2 we discussed the general idea of isomorphism between two spaces. Homeomorphism is nothing else but an isomorphism of topological structures. If two topological spaces are homeomorphic every topological property which is true for one is also true for the other. Suppose now that a set X possesses one or several structures, one of which is a topological structure. Then a property of X is said to be *topological* iff it is possessed by every other topological space which is homeomorphic to X. In other words *a property is called topological iff it is preserved under every homeomorphism.* The property for example, that $\mathbb{R}$ contains a countable dense subset (the rationals) is a topological property. In general a property is topological iff it can be stated in terms of open sets or derived notions such as closed sets, accumulation points, dense sets, and the like.

7.5 BASES

It often happens that in order to exhibit a topological space $\langle X, \mathcal{O}\rangle$ it is not necessary to specify all the sets in the collection $\mathcal{O}$, but instead a certain subcollection of it. Such a subcollection will be a basis for the topology of X. We have the following definition.

Definition *Let* $\langle X, \mathcal{O}\rangle$ *be a topological space. A subcollection* $\mathcal{B} \subset \mathcal{O}$, *is said to be a* basis for the topology of X, *iff every open set can be written as a union of sets from* $\mathcal{B}$.

Observe that the collection $\mathcal{O}$ itself is a basis for the topology of X.

The collection of open intervals of real numbers is a basis for the usual topology of the real line. In the Euclidean plane with its usual metric topology, the collection of all open balls is a basis for its topology. Other bases for the same Euclidean plane are the collection of all open rectangles. In a general metric space $\langle X, d\rangle$ the collection of all open balls forms a basis for the metric topology of X.

Definition *Let* X *be a topological space. A collection* $\mathcal{B}_x$ *of open sets of* X *is said to be a* basis at a point $x \in X$ *if and only if for every open set* O *containing* x *there exists a set* $B \in \mathcal{B}_x$ *such that* $x \in B \subset O$.

We now observe that *a collection* $\mathcal{B}$ *of open sets in a topological space* $\langle X, \mathcal{O}\rangle$ *is a basis for the topology of* X *if and only if it contains a basis at each point of* X. To see this, assume first that $\mathcal{B}$ contains a basis at each point of X.

This means that for every open set $O \in \mathcal{O}$ and every $x \in O$ there is a $B_x \in \mathcal{B}$ such that $x \in B_x \subset O$. This implies that $O = \bigcup_{x \in O} B_x$ and $\mathcal{B}$ therefore is a basis. Next assume that $\mathcal{B}$ is a basis and let $O \in \mathcal{O}$. We have $O = \bigcup B_\lambda$ where $B_\lambda \in \mathcal{B}$, and if $x \in O$ then $x \in B_\lambda \subset O$ for some index λ. This proves that $\mathcal{B}$ contains a basis at each point of X.

9. Proposition *Let* X *be a set and* $\mathcal{B}$ *be a collection of subsets of* X. *Then* $\mathcal{B}$ *is a basis for some topology for* X *if and only if the following conditions are satisfied.*

(i) For each $x \in X$ *there is a* $B_x \in \mathcal{B}$ *such that* $x \in B_x$.

(ii) For every two sets B_1 *and* B_2 *in* $\mathcal{B}$, *if* $x \in B_1 \cap B_2$, *then there exists a set* $B_3 \in \mathcal{B}$ *such that* $x \in B_3 \subset B_1 \cap B_2$.

Proof Suppose $\mathcal{B}$ is a basis for some topology on X. Let $B_1, B_2 \in \mathcal{B}$ and $x \in B_1 \cap B_2$. Then since $B_1 \cap B_2$ is an open set, it is the union of sets in $\mathcal{B}$. Thus there is a $B_3 \in \mathcal{B}$ such that $x \in B_3 \subset B_1 \cap B_2$. Conversely let $\mathcal{B}$ be a collection of subsets of X, which satisfy conditions (i) and (ii). Define the collection $\mathcal{O}$ to consist of arbitrary unions of sets from $\mathcal{B}$. If we agree that the empty union of sets from $\mathcal{B}$ designates the empty set, we see that the empty set $\emptyset$ belongs to the collection $\mathcal{O}$. Also using condition (i) we have that $\bigcup_{x \in X} B_x = X \in \mathcal{O}$. Clearly arbitrary unions of sets in $\mathcal{O}$ belong to $\mathcal{O}$. We finally prove that $\mathcal{O}$ is closed under finite intersections. Let O_1, $O_2 \in \mathcal{O}$ and let $x \in O_1 \cap O_2$. Since O_1 and O_2 belong to $\mathcal{O}$, there are sets B_1 and B_2 in $\mathcal{B}$ such that $x \in B_1 \subset O_1$ and $x \in B_2 \subset O_2$. By (ii) there is a set $B_3 \in \mathcal{B}$ such that $x \in B_3 \subset B_1 \cap B_2 \subset O_1 \cap O_2$. The last relation shows that $O_1 \cap O_2$ can be written as the union of sets in $\mathcal{B}$, which proves that $\langle X, \mathcal{O} \rangle$ is a topological space, and $\mathcal{B}$ is a basis for its topology.

As we have seen earlier in this section, a topological space may admit more than one basis. In fact we observed that the set of all open triangles and the set of all open balls in the Euclidean plane, with its usual metric topology, form two different bases.

If the bases $\mathcal{B}_1$ and $\mathcal{B}_2$ determine the same topological space, then we say that $\mathcal{B}_1$ and $\mathcal{B}_2$ are *equivalent.*

10. Proposition *Let X be a set, and $\mathcal{B}_1$ and $\mathcal{B}_2$ two bases for topologies on X. Then $\mathcal{B}_1$ and $\mathcal{B}_2$ are equivalent if and only if for each $B_1 \in \mathcal{B}_1$ and each $x \in B_1$ there exists a $B_2 \in \mathcal{B}_2$ such that $x \in B_2 \subset B_1$, and for each $B_2 \in \mathcal{B}_2$ and*

each $y \in B_2$ *there exists a* $B_1 \in \mathcal{B}_1$ *such that* $y \in B_1 \subset B_2$.

The proof is left to the reader (see Prob. 11).

Definition *A topological space is said to satisfy the* first axiom of countability *iff there exists a countable basis at every point. It is said to satisfy the* second axiom of countability *iff there is a countable basis for its topology.*

It is clear that the second axiom of countability implies the first. Every metric space with its metric topology satisfies the first axiom of countability. For at every point x the collection of open balls $S_r(x)$, where r runs over the positive rational numbers, forms a countable basis at x. The Euclidean plane with its usual metric topology satisfies the second axiom of countability. The open balls $S_r(x)$ where x runs over all points with rational coordinates and r runs over the positive rational numbers, form a countable basis.

Let X be any arbitrary nonempty set and let $\mathfrak{F}$ be any nonempty collection of subsets of X. Obviously the collection $\{\mathcal{O}_\alpha\}$ of all possible topologies on X which contain $\mathfrak{F}$ as a subcollection (i.e., $\mathfrak{F} \subset \mathcal{O}_\alpha$ for every α) is not empty, since the discrete topology is one of them. The intersection of all the $\mathcal{O}_\alpha$ is a topology on X. In other words, for any given collection $\mathfrak{F}$ of subsets of X there is a topology containing $\mathfrak{F}$ which is the weakest topology containing $\mathfrak{F}$. This topology is uniquely determined by $\mathfrak{F}$. Another way to characterize this weakest topology is to take the collection of all arbitrary unions of finite intersections of sets in $\mathcal{F}$. To see this, we first remind the reader that here again we make the agreement that an empty union of subsets designates the empty set and an empty intersection the whole set X. To proceed with the proof we observe that the set of all finite intersections of sets from $\mathcal{F}$ satisfies conditions (i), (ii) of Proposition 9, and therefore it is a basis for a topology on X, which by construction is the weakest topology containing the collection $\mathcal{F}$.

Definition *A nonempty collection* $\mathcal{F}$ *of open subsets of a topological space* X *is said to be a* subbasis *for its topology if and only if each open set in* X *is the union of finite intersections of members of* $\mathcal{F}$. *In other words the collection of finite intersections of members of* $\mathcal{F}$ *forms a basis for the topology.*

7.6 WEAK TOPOLOGIES

Let X be an arbitrary nonempty set and let $\{X_\lambda, \mathcal{O}_\lambda\}_{\lambda \in \Lambda}$ be a collection of topological spaces. For each $\lambda \in \Lambda$ let f_λ be a mapping from X into X_λ. The problem is to determine the weakest topology on X such that for each $\lambda \in \Lambda$ the mapping

$$f_\lambda : X \to X_\lambda$$

is continuous. Such a topology exists and is unique. It is called the *weak topology generated by the collection* $\{f_\lambda\}_{\lambda\varepsilon\Lambda}$. To determine this topology we observe that since we want f to be continuous, we have by Theorem 7 that the set $f_\lambda^{-1}[O_\lambda]$ must be open for every open set O_λ of the space X_λ and each $\lambda \varepsilon \Lambda$. In other words the weak topology generated by $\{f_\lambda\}_{\lambda\varepsilon\Lambda}$ is nothing else but the topology we get if we take the collection $\{f_\lambda{}^{-1}[O_\lambda]\}$, where O_λ runs over all open sets of X_λ for each $\lambda \varepsilon \Lambda$, as a subbasis.

Remark One can easily prove (using the identity $f^{-1}\left[\bigcup_\alpha S_\alpha\right] = \bigcup_\alpha f^{-1}[S_\alpha]$) that instead of all open sets O_λ in X_λ it is sufficient to take as a subbasis the collection $\{f_\lambda{}^{-1}[B_\lambda]\}$ where B_λ runs over a basis of X_λ. The topology that we get is still the weak topology generated by $\{f_\lambda\}$.

Let us now consider an important particular case where all X_λ equal the real line (or the complex plane) with its usual topology. If O is any open set of real numbers, then the set $f^{-1}[O]$ can be written as the union of sets of the form

$$(*) \qquad \{x \varepsilon X: |f_\lambda(x) - f_\lambda(x_o)| < \varepsilon\}$$

with $\varepsilon > 0$ and $f(x_o) \varepsilon O$. This implies that all the sets of the form $(*)$ constitute a subbasis for the weak topology of X generated by $\{f_\lambda\}$. Hence all finite intersections of sets of the form

(*) is a basis for the weak topology generated by $\{f_\lambda\}$.

We now prove that instead of taking all finite intersections of the sets (*), it suffices to take all finite intersections of the form

$$V(x_o; f_1, f_2, \ldots, f_n, \varepsilon)$$

(**)

$$= \bigcap_{i=1}^{n} \{x \in X: |f_i(x) - f_i(x_o)| < \varepsilon\}$$

where x_o and ε are kept fixed, and $f_1, \ldots, f_n$ belong to the family $\langle f_\lambda \rangle$. To see this we make use of Proposition 9 by proving that the collection of sets of the form (**) satisfy conditions (i) and (ii) of this proposition. Since $x_o \in V(x_o; f_1, f_2, \ldots, f_n, \varepsilon)$, (i) is clearly satisfied. To prove that (ii) is satisfied let $V_1 = V(x_o; f_1, f_2, \ldots, f_n, \varepsilon_1)$ and $V_2 = (V_{y_o}; g_1, \ldots, g_m, \varepsilon_2)$ be any two sets of the form (**) and let $z \in V_1 \cap V_2$. Put

$$\max|f_i(z) - f_i(x_o)| = a < \varepsilon_1 \qquad (i = 1, 2, \ldots, n)$$

$$\max|g_j(z) - g_j(y_o)| = b < \varepsilon_2 \qquad (j = 1, 2, \ldots, m)$$

and let δ_1 and δ_2 be such that $0 < \delta_1 < \varepsilon_1 - a$, $0 < \delta_2 < \varepsilon_2 - b$. Then it is easily seen that

$$V(z;\ f_1,f_2,\ldots,f_n,\delta_1) \subset V_1$$

$$V(z;\ g_1,g_2,\ldots,g_m,\delta_2) \subset V_2$$

so that

$$V(x;\ f_1,f_2,\ldots,f_n,\delta_1) \cap V(z;\ g_1,g_2,\ldots,g_m,\delta_2)$$
$$\subset V_1 \cap V_2.$$

Also, if we take $\delta = \min(\delta_1,\delta_2)$ we get

$$V(z;\ f_1,f_2,\ldots,f_n,g_1,g_2,\ldots,g_m,\delta)$$
$$\subset V(z;\ f_1,f_2,\ldots,f_n,\delta_1) \cap V(z;\ g_1,g_2,\ldots,g_n,\delta_2).$$

By combining the last two relations we finally have

$$z \in V(z;\ f_1,f_2,\ldots,f_n,g_1,g_2,\ldots,g_m,\delta) \subset V_1 \cap V_2,$$

which proves that (ii) is satisfied. Thus the collection of sets of the form (**) constitute a basis for the weak topology generated by $\{f_\lambda\}$.

We now introduce the notion of a *subspace* of a topological space $\langle X, \mathcal{O}\rangle$. Let E be a nonempty subset of X. Among all topologies that can be defined on E consider those that make the identity mapping of E into X continuous. If f is this mapping, the inverse image $f^{-1}[O]$ of an open set O in X is the set $E \cap O$. Since F is supposed to be continuous, $E \cap O$ must be open in the topology sought for E. In other words in order that the identity mapping f be continuous it is necessary that the topology of E contains all sets of the form $E \cap O$ for all $O \in \mathcal{O}$. But on the other hand it is immediate that the collection of sets of the form $E \cap O$ with $O \in \mathcal{O}$, form a topology since they satisfy the axioms O_1, O_2, O_3 defining a topological space. The topology that the collection $\{E \cap O\}_{O\in\mathcal{O}}$ defines on E is the one we shall consider.

Definition *Let* $\langle X, \mathcal{O}\rangle$ *be a topological space and let* E *be a nonempty subset of* X. *Then* E *with the topology (sometimes called the* relative topology*) defined by the collection* $\{E \cap O\}_{O\in\mathcal{O}}$ *is said to be a* subspace *of* X. *We also say that the topology of* E *is* induced *by the topology of* X. *The identity mapping of the subspace* E *into* X *is then continuous.*

If E is a subspace of $\langle X, \mathcal{O}\rangle$ then it follows from the above definition that a closed set of E is of the form $E - E \cap O = E \cap \complement O$ where $O \in \mathcal{O}$. This means that any closed set in the subspace E is the intersection of E and a closed set in X. Observe that *a set which is open* (closed) *in* E *is not necessarily open* (closed) *in* X. For example, consider the real line $\mathbb{R}$ as a subspace of the complex plane. Then an open interval is an open set of $\mathbb{R}$, but it is not an open set of the plane. However it is easy to see that every open (closed) subset of E is open (closed) in X if and only if E is open (closed) in X.

Example

Let $\langle X, \prec \rangle$ be a totally ordered set with its order topology (see sec. 7.2, Ex. 3). Let A be any interval in X with end points a and b. Then A is also a totally ordered set by $\prec$, so that we can consider A as a topological space with its order topology. This order topology on A is identical with the topology on A induced by the order topology of X, that is if A is considered as a subspace of X. For, every open interval in A is of the form (c,d) or (a,c) or (c,b) with c and d points of A. But such an interval is the intersection of A with an open interval of X. On the other hand the union of such sets (which by definition is an open set in the order topology of A) is the intersection of A with an open set in X. This proves that both topologies on A coincide.

We shall now use the notion of weak topology generated by a collection of functions to define a topology on the Cartesian product $X = \prod_{\lambda \varepsilon \Lambda} X_\lambda$ of an arbitrary family $\langle X_\lambda \rangle_{\lambda \varepsilon \Lambda}$ of topological spaces. We recall that the Cartesian product X consists of all functions x defined on the index set Λ with values in $\bigcup_{\lambda \varepsilon \Lambda} X_\lambda$, and such that $x(\lambda) \varepsilon X_\lambda$. If Λ is a finite set with n elements then the Cartesian product $\prod_{i=1}^{n} X_i$ consists of all sequences $\langle x_i \rangle_{i=1}^{n}$ where $x_i \varepsilon X_i$.

For each $\lambda \varepsilon \Lambda$ consider the mapping

$$p_\lambda : X \to X_\lambda$$

defined for each $x \in X$ by

$$p_\lambda(x) = x(\lambda).$$

The mapping p_λ is called *the projection of* X *onto* X_λ, and $x(\lambda)$ is the λ-th coordinate of the point x. We also write x_λ instead of $x(\lambda)$.

The weak topology generated by the family $\langle p_\lambda \rangle_{\lambda\varepsilon\Lambda}$ of all projections is the topology that we shall take for X. It is called the *product topology* or the *Tychonoff topology*. A subbasis for the product topology in X is the collection $\{p_\lambda^{-1}[O_\lambda]\}_{\lambda\varepsilon\Lambda}$ where O_λ runs over the open sets of X_λ. For a particular $\lambda_o \varepsilon \Lambda$, the set $p_{\lambda_o}^{-1}\left[O_{\lambda_o}\right]$ is the set of all x in X such that $p_{\lambda_o}(x) = x(\lambda_o) \varepsilon O_{\lambda_o}$. In other words $p_{\lambda_o}^{-1}\left[O_{\lambda_o}\right]$ equals the Cartesian product $\prod_{\lambda\varepsilon\Lambda} U_\lambda$ where $U_\lambda = X_\lambda$ for $\lambda \neq \lambda_o$ and $U_{\lambda_o} = O_{\lambda_o}$. Now a finite intersection of sets of this type gives a set of the form $\prod_{\lambda\varepsilon\Lambda} V_\lambda$ where all but a finite number of the V_λ equal X_λ, and the others are open sets in the respective spaces.

Therefore the weak topology generated by $\{p_\lambda\}_\lambda$, that is the product topology, is determined by taking all sets of the form $\prod_{\lambda \varepsilon \Lambda} V_\lambda$ as a basis.

In the particular case in which Λ contains only two elements we only have two topological spaces say X_1 and X_2, and a basis for the product topology on $X_1 \times X_2$ is, according to the above discussion, the collection of all sets of the form $O_1 \times O_2$ where O_1 is an open set in X_1 and O_2 an open set in X_2.

7.7 SEPARATION

Definition *Let* $\langle x_n \rangle$ *be a sequence of points in a topological space* X. *Then* $\langle x_n \rangle$ *is said to converge to a point* $x_o \varepsilon X$ *iff for every neighborhood* V *of* x_o *there is an integer* N *such that for every* $n \geq N$ *we have* $x_n \varepsilon V$. *We also say that* x_o *is the* limit *of* $\langle x_n \rangle$.

The reader recalls that a sequence of real or of complex numbers, or more generally a sequence of points in a metric space, has at most one limit. Unfortunately this is not true for an arbitrary topological space. As an example consider any non-empty set X with the trivial topology. Then every point of X is a limit of every sequence in X. However, there are topological spaces in which any sequence converges to at most one point. These spaces are the more useful and they satisfy some additional conditions which provide, so to speak, more "separation" among their points.

Definitions *Let* X *be a topological space. Then*

T_0: X *is said to be a* T_0*-space iff given any two distinct points of* X, *there is an open set which contains one of them but not the other.*

If X contains more than one point and has the trivial topology then X is not a T_0-space.

T_1: X *is said to be a* T_1*-space iff given any two distinct points* x *and* y *of* X, *there is an open set containing* x *but not* y, *and an open set containing* y *but not* x.

Example

Let $X = \{x \in \mathbb{R}: 0 \leq x < 1\}$ where every open set is by definition empty or of the form $[0,a)$ with $0 \leq a \leq 1$. Then X is a T_0-space but not a T_1-space. (Prove this!)

T_2: X *is said to be a* T_2*-space or a* Hausdorff space *iff given any two distinct points* x *and* y *of* X *there are two disjoint open sets containing* x *and* y *respectively.*

An infinite topological space with the cofinite topology (sec. 7.2, Ex. 2) is a T_1-space but not a T_2-space.

It is now the proper time to ask whether or not a topological space X with the topology generated by a collection $\{f_\lambda\}_{\lambda\in\Lambda}$ of real-valued (or complex-valued) functions is a Hausdorff space. The answer is yes, if we make the additional assumption that the collection $\{f_\lambda\}_{\lambda\varepsilon\Lambda}$ *separates points in* X. By this we mean that for any two distinct points x_0 and y_0 in X there is an $f \varepsilon \{f_\lambda\}_{\lambda\varepsilon\Lambda}$ such that $f(x_0) \neq f(y_0)$. In fact, if such an f exists, then choose ε so that $0 < \varepsilon < |f(x_0) - f(y_0)|$ and consider the open sets $V(x_0; f, \varepsilon/2)$ and $V(y_0; f, \varepsilon/2)$ (see sec. 7.6). We obviously have $V(x_0; f, \varepsilon/2) \cap V(y_0; f, \varepsilon/2) = \emptyset$. This proves that X with the weak topology generated by $\{f_\lambda\}_{\lambda\varepsilon\Lambda}$ is a Hausdorff space provided that the family $\{f_\lambda\}$ separates points in X.

It is also easy to see that every metric space is a Hausdorff space.

One can prove in exactly the same way as in the case of the real line, that *in a Hausdorff space every sequence of points has at most one limit.*

11. Proposition *Let* $\langle X_\lambda \rangle_{\lambda\varepsilon\Lambda}$ *be a family of Hausdorff spaces. Then the product space* $X = \prod_{\lambda\varepsilon\Lambda} X_\lambda$ *is also a Hausdorff space.*

Proof Let x and y be two distinct points in X. Since $x \neq y$, we must have $x(\lambda) \neq y(\lambda)$ for some $\lambda \varepsilon \Lambda$. Since X_λ is Hausdorff there are disjoint open sets V and U containing $x(\lambda)$ and $y(\lambda)$ respectively. This implies that $p_\lambda^{-1}[V]$ and $p_\lambda^{-1}[U]$ (p_λ is the projection of X onto X_λ) are disjoint open sets containing x and y respectively in the product X. Hence X is Hausdorff.

12. Proposition *A topological space X is a T_1-space if and only if every set consisting of only one point is closed.*

Proof Suppose that X is a T_1-space and let x be any point in X. If y is any point in $X - \{x\}$, then there is an open set O containing y but not x. This means that no point of the set $X - \{x\}$ belongs to the closure of the set $\{x\}$. Therefore $\overline{\{x\}} = \{x\}$, so that $\{x\}$ is closed.

Next suppose that for every $x \varepsilon X$, $\{x\}$ is a closed set. Let x and y be two distinct points of X. Then $X - \{x\}$ and $X - \{y\}$ are open sets for which we have $y \varepsilon X - \{x\}$ but $x \not\varepsilon X - \{x\}$, and $x \varepsilon X - \{y\}$ but $y \not\varepsilon X - \{y\}$. This proves that X is a T_1-space.

Definition *A topological space is said to be* regular *or a T_3-space, iff it is a T_1-space and iff for any closed set F and any point x not in F there are disjoint open sets O_1 and O_2 such that $x \varepsilon O_1$ and $F \subset O_2$.*

Definition *A topological space is said to be* normal *or a T_4-space if and only if it is T_1 and if and only if given two disjoint closed sets F_1 and F_2 there are disjoint open sets O_1 and O_2 such that $F_1 \subset O_1$ and $F_2 \subset O_2$.*

The T-spaces form a hierarchy. By this we mean that a T_4-space is also a T_3-space. A T_3-space is also a T_2-space and so on.

(*Note:* The reader is warned that not all authors

define a regular space to be synonymous with a T_3-space nor a normal space to be synonymous with a T_4-space. Sometimes a regular space or a normal space is defined in the same way as above except for the condition T_1 which is not necessarily assumed to hold in these spaces.)

We close this section with the following two important theorems concerning normal topological spaces. Proofs of these theorems can be found in Simmons [25].

13. Theorem (Urysohn's Lemma) *Let X be a normal topological space and A and B two disjoint closed sets of X. Then there is a continuous real-valued function f defined on X such that $0 \leq f \leq 1$ on X with $f(x) = 0$ for all $x \in A$ and $f(x) = 1$ for all x in B.*

14. Theorem (Urysohn's Metrization Theorem) *Every normal topological space satisfying the second axiom of countability is metrizable.*

7.8 COMPACTNESS

The notion of a compact set is suggested by the Heine-Borel-Lebesgue theorem established in Chapter 2. The theorem asserts that every open covering of a closed and bounded interval (or set) of real numbers has a finite subcovering. The consequences of this theorem are very profound and important and they lie in the fact that one is allowed to replace certain global studies with a local study, that is, a study reduced to each of the members of the finite open subcovering. As in the case of the real line, an *open covering* of a subset A of a topological space X is a collection of open sets such that their union contains A.

Definition *A topological space X is said to be* compact *if and only if every open covering* $\mathcal{U}$ *of X has a finite subcovering, that is, there is a finite number* $O_1, O_2, \ldots, O_n$ *of open sets of the collection* $\mathcal{U}$ *such that* $O_1 \cup O_2 \cup \ldots \cup O_n = X$.

A subset K of X is said to be *compact* iff K considered as a subspace of X (that is, with the relative topology induced by the topology of X) is compact.

(*Note:* Some authors, especially French, in order to exclude spaces of minor interest, such as those with the trivial topology, reserve the term "compact" for compact Hausdorff spaces.)

The following proposition characterizes compact sets in terms of closed sets.

Definition *A collection* $\mathcal{F}$ *of sets has the* finite intersection property *if and only if every finite subcollection of* $\mathcal{F}$ *has nonempty intersection.*

15. Proposition *A topological space X is* compact *if and only if every collection of closed sets in X, which has the finite intersection property, has nonempty intersection.*

The proof follows easily from De Morgan's laws (see Chap. 2) and it is left to the reader.

16. Corollary *A topological space X is compact if and only if for every collection* $\{E_\lambda\}_{\lambda \in \Lambda}$ *that has the finite intersection property we have* $\bigcap_{\lambda \in \Lambda} \overline{E}_\Lambda \neq \emptyset$.

Proof Suppose X compact and let $\{E_\lambda\}_{\lambda\varepsilon\Lambda}$ have the finite intersection property. Then for any finite subset J of Λ we have

$$\bigcap_{\lambda\varepsilon J} E_\lambda \subset \overline{\bigcap_{\lambda\varepsilon J} E_\lambda} \subset \bigcap_{\lambda\varepsilon J} \overline{E}_\lambda .$$

Thus the collection $\{\overline{E}_\lambda\}_{\lambda\varepsilon\Lambda}$ has also the finite intersection property and since X is assumed to be compact, we have by Proposition 15 that $\bigcap_{\lambda\varepsilon\Lambda} \overline{E}_\lambda \neq \emptyset$.

If X has the property stated in the corollary then clearly X is compact.

The reader can easily observe that Theorem 8 and Proposition 11 of Chapter 2 imply that *a set of real numbers is compact if and only if it is bounded and closed.*

17. Proposition *Let K be a subspace of a topological space X. Then K is compact if and only if every collection of open sets in X which covers K contains a finite subcollection which covers K.*

Proof Suppose K compact and let $\mathcal{U}$ be a collection of open sets in X which covers K. Then the sets $\{K \cap O\}_{O\varepsilon\mathcal{U}}$, since each $K \cap O$ is open in the subspace K, form an open covering of the compact subspace K. Hence there is a finite number of sets $\{K \cap O_i\}_{i=1}^n$, with $O_i \;\varepsilon\; \mathcal{U}$ $(i = 1,2,\ldots,n)$ such that $\bigcup_{i=1}^n (K \cap O_i) \supset K$. This implies that the collection $\{O_i\}_{i=1}^n$ covers K.

Next suppose that every collection of open sets in X which covers K contains a finite subcollection which covers K, and let $\mathcal{A}$ be a collection of open sets in K which covers K. Every set in $\mathcal{A}$ is of the form $K \cap O$ where O is open in X. The collection of all open sets O in X such that $K \cap O \in \mathcal{A}$, clearly covers K. By assumption there is a finite subcollection $\{O_i\}_{i=1}^m$ with $K \cap O_i \in \mathcal{A}$, $(i = 1,2,\ldots,m)$ such that $\bigcup_{i=1}^m O_i \supset K$. But

$$\bigcup_{i=1}^m (K \cap O_i) = K \cap \bigcup_{i=1}^m O_i = K,$$

which implies that the finite subcollection $\{K \cap O_i\}_{i=1}^m$ of $\mathcal{A}$ covers K. This proves that K is compact.

18. Proposition *A compact subspace of a Hausdorff space is closed.*

Proof Let K be a compact subspace of a Hausdorff space X. To prove that K is closed it suffices to prove that $\complement K$ is open. Let a be a fixed but arbitrary point of $\complement K$. Since X is Hausdorff, for every $x \in K$ there are open disjoint neighborhoods in X, V_x and W_x of a and x respectively. The collection $\{W_x\}_{x \in K}$ is an open covering of K, so that by Proposition 17, there is a finite subcollection $\{W_{x_i}\}_{i=1}^n$ of the open sets W_x, which covers K. Put $V = \bigcap_{i=1}^n V_{x_i}$. Clearly V is open and $V \cap W_{y_i} = \emptyset$ for $i = 1,2,\ldots,n$. This implies that $V \cap K = \emptyset$, which means that $V \subset \complement K$. Since a was an arbitrary point of $\complement K$, it follows that $\complement K$ is open.

Observe that the hypothesis that X is Hausdorff is essential. For if X is any set containing more than one point, and it has the trivial topology, then for every $x \in X$ the set $\{x\}$ is a compact subspace of X, but $\{x\}$ is not closed.

19. Proposition *In a compact topological space X, every closed subset is a compact subspace.*

Proof We use Proposition 15. Let K be a closed set in X and let $\{F_\lambda\}_{\lambda\in\Lambda}$ be a collection of closed sets in the subspace K, having the finite intersection property. Since K is closed in X the F_λ are also closed sets in X. Since X is compact it follows that $\bigcap_{\lambda\in\Lambda} F_\lambda \neq \emptyset$, which implies that K is compact.

20. Proposition *Let X and Y be two topological spaces and let f be a continuous mapping of X into Y. Then if X is compact, f[X] is a compact subspace of Y.*

Proof Let $\{O_\lambda\}_{\lambda\in\Lambda}$ be any covering of f[X] by open sets in the subspace f[X]. Then since f is continuous, the collection $\{f^{-1}[O_\lambda]\}_{\lambda\in\Lambda}$ is an open covering of X. Since X is compact there exists a finite covering $\{f^{-1}[O_i]\}_{i=1}^n$ with

$O_i \in \{O_\lambda\}_{\lambda\in\Lambda}$ $(i = 1,2,\ldots,n)$. We have for each i $(i = 1,2,\ldots,n)$ $f[f^{-1}[O_i]] = O_i$. Hence the collection $\{O_i\}_{i=1}^n$ covers $f[X]$. Therefore $f[X]$ is compact.

21. Corollary *Let* X *be a compact topological space and* Y *a Hausdorff space. Then any continuous one-to-one mapping* f *of* X *onto* Y *is a homeomorphism.*

Proof Since f is one-to-one f^{-1} exists. To prove the corollary it suffices to show that f^{-1} is continuous, or equivalently that for every closed set F in X, $f[F]$ is closed in Y. Since X is compact, F as a closed subset of X is compact. Thus $f[F]$ is a compact subspace of Y, and hence also closed by Proposition 18.

Let $\langle x_n \rangle$ be a sequence in a topological space X. A point $a \in X$ is said to be a *cluster point* of $\langle x_n \rangle$ iff for each open set O containing a, and for each positive integer N, there is an $n \geq N$ with $x_n \in O$. In other words, every neighborhood of a contains an infinite number of terms of the sequence $\langle x_n \rangle$.

The reader should carefully observe that a cluster point of a sequence $\langle x_n \rangle$ is not in general an accumulation point of the range $\{x_n\}$ of $\langle x_n \rangle$. A point a can be a cluster point of $\langle x_n \rangle$ either because it is an accumulation point of the set $\{x_n\}$ or because it coincides with x_n for infinitely many n.

A topological space X is said to be *countably compact* if and only if every countable covering of X has a finite subcovering.

22. Proposition *A topological space X is countably compact if and only if every sequence $\langle x_n \rangle_{n=1}^{\infty}$ in X has at least one cluster point.*

Proof Suppose X countably compact, and let $A_n = \{x_i\}_{i=n}^{\infty}$ be the range of the sequence $\langle x_i \rangle_{i=n}^{\infty}$. The collection $\{\bar{A}_n\}_{n=1}^{\infty}$, as a nested collection of closed sets, has the finite intersection property. We prove that $\bigcap_{n=1}^{\infty} \bar{A}_n \neq \emptyset$. To see this suppose on the contrary that $\bigcap_{n=1}^{\infty} \bar{A}_n = \emptyset$, or equivalently $\bigcup_{n=1}^{\infty} \complement \bar{A}_n = X$. Since X is countably compact, for a finite number of the open sets $\complement \bar{A}_n$ say $\complement \bar{A}_{n_1}$, $\complement \bar{A}_{n_2}, \ldots, \complement \bar{A}_{n_k}$, we have $\complement \bar{A}_{n_1} \cup \complement \bar{A}_{n_2} \cup \ldots \complement \bar{A}_{n_k} = X$ or equivalently $\bar{A}_{n_1} \cap \bar{A}_{n_2} \cap \ldots \bar{A}_{n_k} = \emptyset$. But this last equality contradicts the fact that the $\langle \bar{A}_n \rangle$ have the finite intersection property. Therefore $\bigcap_{n=1}^{\infty} \bar{A}_n \neq \emptyset$. Let $a \in \bigcap_{n=i}^{\infty} \bar{A}_n$, then a is a cluster point of $\langle x_n \rangle$. For if O is any open set containing a and N any positive integer, we have $a \in \bar{A}_N$, so that there is a positive integer $n \geq N$ with $x_n \in O$. Next suppose that every sequence in X has a cluster point and let $\{O_i\}_{i=1}^{\infty}$ be any countable open covering of X. We want to prove that $\{O_i\}$ contains a finite subcovering. Suppose not. Then for any finite number $O_{n_1}, O_{n_2}, \ldots, O_{n_k}$ of sets in $\{O_i\}$ we have

$\complement(O_{n_1} \cup O_{n_2} \cup \ldots \cup O_{n_k}) \neq 0$ or

$\complement O_{n_1} \cap \complement O_{n_2} \cap \ldots \cap \complement O_{n_k} \neq \emptyset$. In other words the collection $\{\complement O_i\}$ of closed sets has the finite intersection property. Put $B_n = \bigcap_{k=i}^{n} \complement O_k$. For each n, B_n is a nonempty closed set, so that for each n we can choose a point $b_n \in B_n$. The sequence $\langle b_n \rangle$ has by assumption a cluster point, say b. For any set $\complement O_i$ of the collection $\{\complement O_i\}$, b_n belongs to $\complement O_i$ for all $n \geq i$, which implies that the cluster point b must belong to $\complement O_i$. Thus, since b belongs to every $\complement O_i$, we have that

$b \in \bigcap_{i=1}^{\infty} \complement O_i$. But this means that

$\complement\left(\bigcap_{i=1}^{\infty} \complement O_i\right) = \bigcup_{i=1}^{\infty} O_i \neq X$ which contradicts the assumption that $\{O_i\}$ is an open covering of X. The proposition is proved.

Clearly, if X is compact then X is countably compact. On the other hand, if X is countably compact and satisfies the second axiom of countability then X is compact. For it is easy to see that in every space that satisfies the second axiom of countability every open covering has a countable subcovering.

In dealing with sets of real numbers the reader is familiar with the fact that if a real number x_o is an accumulation point of a set S then some sequence $\langle x_n \rangle$ in S converges to x_o. The following example shows that this is not true in general.

Example

Let X = [0,1]. A set O in X is defined to be open iff it is empty or $\complement O$ is countable. With this definition of open set X is clearly a topological space.

Consider the subset [(1/2),1] of X. Then one easily proves that the point 1/4 for instance is an accumulation point of [(1/2),1] but no sequence of points in [(1/2),1] converges to 1/4.

Definition *A space X is said to be* sequentially compact *if and only if every infinite sequence from X has a convergent subsequence.*

It follows from Proposition 22 that *a sequentially compact space is countably compact.* On the other hand one can prove (see Prob. 37), that *a countably compact space satisfying the first countability axiom is sequentially compact.*

In this book we make use only of the notion of compactness, the other two notions of countable and sequential compactness being cited merely for reference purposes.

The next lemma is needed for the proof of Tychonoff's theorem, one of the most important theorems in topology, with many important applications in analysis especially in the theory of function spaces.

23. Lemma *Let* X *be any nonempty set and let* $\mathcal{A}$ *be a collection of subsets of* X *which has the finite intersection property. Then there is a collection* $\mathcal{F}$ *of subsets of* X *containing* $\mathcal{A}$ *such that* $\mathcal{F}$ *has the finite intersection property and such that no collection properly containing* $\mathcal{F}$ *has the finite intersection property. (Such collection* $\mathcal{F}$ *is said to be* maximal *with respect to the finite intersection property.)*

Proof Let $\mathcal{C}$ be the collection of all collections containing $\mathcal{A}$ and having the finite intersection property. Then $\mathcal{C}$ is partially ordered by inclusion so that by the Hausdorff maximal principle there is a maximal linearly ordered subset of $\mathcal{C}$ say $\mathcal{B}$. The collection $\mathcal{F}$ sought by the lemma is the union of all collections in $\mathcal{B}$. To see this let $E_1, E_2, \ldots, E_n$ be in $\mathcal{F}$. Then each E_i $(i = 1,2,\ldots,n)$ belongs to some $\mathcal{H}_i \in \mathcal{B}$. The set $\{\mathcal{H}_1, \mathcal{H}_2, \ldots, \mathcal{H}_n\}$ as a finite subset of the linearly ordered set $\mathcal{B}$, has a maximum element say $\mathcal{H}_p$ $(1 \leq p \leq n)$, that is $\mathcal{H}_i \subset \mathcal{H}_p$ for $i = 1,2,\ldots,n$. This implies that all E_i belong to $\mathcal{H}_p$. Now $\mathcal{H}_p$ as an element of $\mathcal{C}$ has the finite intersection property so that

$\bigcap_{i=1}^{n} E_i \neq \emptyset$, which proves that $\mathcal{F}$ has the finite intersection property. It remains to prove that if $\mathcal{F}_1$ is another collection containing $\mathcal{F}$ properly, then $\mathcal{F}_1$ cannot have the finite intersection property. To show this suppose that $\mathcal{F}_1$ has the finite intersection property. Since $\mathcal{F} \subsetneq \mathcal{F}_1$ we have $\mathcal{H} \subset \mathcal{F}_1$ for every $\mathcal{H} \in \mathcal{B}$. This implies that $\mathcal{F}_1$ must belong to $\mathcal{B}$. For if not, the set $\mathcal{B} \cup \{\mathcal{F}_1\}$ would be a linearly ordered subset of $\mathcal{C}$ containing $\mathcal{B}$ which would contradict the maximality of $\mathcal{B}$. Thus $\mathcal{F}_1 \in \mathcal{B}$. But since $\mathcal{F}$ is the union of all collections in $\mathcal{B}$ and since $\mathcal{F}_1$ is in $\mathcal{B}$ we must have $\mathcal{F}_1 \subset \mathcal{F}$ which contradicts $\mathcal{F} \subsetneq \mathcal{F}_1$. This proves the lemma.

24. Theorem (Tychonoff) *Let* $\langle X_\lambda \rangle_{\lambda \in \Lambda}$ *be an arbitrary family of compact topological spaces. Then the Cartesian product* $X = \prod_{\lambda \in \Lambda} X_\lambda$ *is compact with respect to the product topology (see sec. 7.6).*

Proof By Corollary 16, it suffices to prove that for every collection $\{E_\alpha\}_{\alpha \in \Gamma}$ of sets in X, that has the finite intersection property, we have $\bigcap_{\alpha \in \Gamma} \overline{E}_\alpha \neq \emptyset$. By Lemma 23, there is a collection $\mathcal{F}$ of subsets of X containing $\{E_\alpha\}_{\alpha \in \Gamma}$, having the finite intersection property and such that $\mathcal{F}$ is maximal with respect to the finite intersection property. In order to simplify the writing we may assume that $\mathcal{F} = \{E_\alpha\}_{\alpha\varepsilon\Gamma}$. For it is clear that if we prove that the intersection of all sets $\mathcal{F}$ is nonempty then a fortiori we shall have

$\bigcap_{\alpha\varepsilon\Gamma}\overline{E}_\alpha \neq \emptyset$, since $\mathcal{F}$ contains the collection

$\{E_\alpha\}_{\alpha\varepsilon\Gamma}$. Thus we assume that $\{E_\alpha\}_{\alpha\varepsilon\Gamma}$ is maximal, with respect to the finite intersection property.

Let λ be any element in Λ, and consider the collection $\{P_\lambda[E_\alpha]\}_{\alpha\varepsilon\Gamma}$ of the projections of the sets E_α onto the space X_λ. If J is any finite subset of Γ we have $\bigcap_{\alpha\varepsilon J} P_\lambda[E_\alpha] \neq \emptyset$. For

suppose on the contrary that $\bigcap_{\alpha\varepsilon J} P_\lambda[E_\alpha] = \emptyset$.

Then since $P_\lambda\left(\bigcap_{\alpha\varepsilon J} E_\alpha\right) \subset \bigcap_{\alpha\varepsilon J} P_\lambda[E_\alpha]$ we also have

$P_\lambda\left(\bigcap_{\alpha\varepsilon J} E_\alpha\right) = \emptyset$, which implies $\bigcap_{\alpha\varepsilon J} E_\alpha = \emptyset$.

But this last equality contradicts the fact that $\{E_\alpha\}_{\alpha\varepsilon\Gamma}$ has the finite intersection property.

Thus $\bigcap_{\alpha\varepsilon J} P_\lambda[E_\alpha] \neq \emptyset$, so that $\{P_\lambda[E_\alpha]\}_{\alpha\varepsilon\Gamma}$ has the finite intersection property. Since X_λ is compact we have by Corollary 16 that

$\bigcap_{\alpha\varepsilon\Gamma} \overline{P_\lambda[E_\alpha]} \neq \emptyset$.

Let $x_\lambda \varepsilon \bigcap_{\alpha\varepsilon\Gamma} \overline{P_\lambda[E_\alpha]}$ and let x be that point of X whose λ-th coordinate is x_λ. We want to prove that $x \in \bigcap_{\alpha\in\Gamma} \overline{E}_\alpha$. To do this it suffices to show that for each $\alpha \in \Gamma$, every open set containing x also contains an element of E_α, or equivalently, every basis open set at x contains an element of E_α.

We now recall (see sec. 7.6) that a subbasis for the product topology of X consists of all sets of the form

$$V_\lambda(x) = \{y \in X: y \in V(x_\lambda)\}$$

where $V(x_\lambda)$ is an open set in X_λ containing x_λ. Hence a basis open set V(x) at x is a finite intersection of sets of the form $V_\lambda(x)$, that is $V(x) = \bigcap_{\lambda\in I} V_\lambda(x)$ where I is a finite subset of Λ. But we already know that $x_\lambda \in \bigcap_{\alpha\in\Gamma} \overline{P_\lambda[E_\alpha]}$, and therefore, $V(x_\lambda) \cap P_\lambda[E_\alpha] \neq \emptyset$ for all $\alpha \in \Gamma$ and each $\lambda \in I$. It follows that

$$(*) \qquad V_\lambda(x) \cap E_\alpha \neq \emptyset$$

for all $\alpha \in \Gamma$ and each $\lambda \in I$.

Next we observe that since $\{E_\alpha\}_{\alpha\in\Gamma}$ is maximal with respect to the finite intersection property, the intersection of any finite number of sets in this collection is again in this collection.

For if $\mathcal{H}$ is the collection of all sets which are finite intersections of sets in $\{E_\alpha\}_{\alpha\varepsilon\Gamma}$ then $\mathcal{H}$ is a collection having the finite intersection property and containing $\{E_\alpha\}_{\alpha\varepsilon\Gamma}$. Since $\{E_\alpha\}_{\alpha\varepsilon\Gamma}$ is maximal we have $\mathcal{H} = \{E_\alpha\}_{\alpha\varepsilon\Gamma}$. Therefore the relation (*) implies that each $V_\lambda(x)$ has a non-empty intersection with any finite intersection of sets from the collection $\{E_\alpha\}_{\alpha\varepsilon\Gamma}$. But again by the maximality of $\{E_\alpha\}_{\alpha\varepsilon\Gamma}$ we have that $V_\lambda(x) \;\varepsilon\; \{E_\alpha\}_{\alpha\varepsilon\Gamma}$ for all $\lambda \;\varepsilon\; I$. Thus

$$V(x) = \bigcap_{\lambda\varepsilon I} V_\lambda(x) \;\varepsilon\; \{E_\alpha\}_{\alpha\varepsilon\Gamma}.$$

Finally, since $\{E_\alpha\}_{\alpha\varepsilon\Gamma}$ has the finite intersection property and $V(x)$ is a member of the collection, we have $V(x) \cap E_\alpha \neq \emptyset$ for all $\alpha \;\varepsilon\; \Gamma$. This proves the theorem.

Let us consider the space $\mathbb{R}^n$, that is the Cartesian product of n spaces identical with the real line $\mathbb{R}$, the topology on $\mathbb{R}^n$ being the product topology. A subset K of $\mathbb{R}^n$ is said to be *bounded* iff there are n bounded intervals $I_1, I_2, \ldots, I_n$ of $\mathbb{R}$ such that $K \subset \prod_{i=1}^{n} I_i$. The set $\prod_{i=1}^{n} I_i$ is sometimes called an *interval* of $\mathbb{R}^n$. The reader will recall that a set of real numbers is compact if and only if it is bounded and closed. The following proposition asserts that the same holds in $\mathbb{R}^n$ with the product topology.

25. Corollary *A subset* K *of* $\mathbb{R}^n$ *is compact if and only if it is bounded and closed.*

Proof Let K be a compact subset of $\mathbb{R}^n$. Since $\mathbb{R}^n$ is Hausdorff, K is closed by Proposition 18. Also, since projection mappings are continuous we have by Proposition 20 that the projection of K onto each factor $\mathbb{R}$ is compact and therefore contained in a bounded interval of $\mathbb{R}$. Thus K is contained in a bounded interval of $\mathbb{R}^n$ so that K is bounded.

Next let K be a closed and bounded set in $\mathbb{R}^n$. Then there is a finite number $I_1, I_2, \ldots, I_n$ of compact intervals of $\mathbb{R}$, such that $K \subset \prod_{i=1}^{n} I_i$. But $\prod_{i=1}^{n} I_i$ is compact by Theorem 24, hence so is K as a closed subset of a compact set.

7.9 LOCALLY COMPACT TOPOLOGICAL SPACES--COMPACTIFICATION

A topological space X is said to be *locally compact* iff each point $x \in X$ has at least one compact neighborhood.

Clearly every compact space is locally compact. The real line with its usual metric topology is locally compact but not compact.

The subspace Q of ℝ, where Q is the set of rational numbers is neither compact nor locally compact (see Prob. 39). Proposition 20 does not have an equivalent statement for locally compact spaces. In other words *it is not true that the continuous image of a locally compact space is locally compact.* For let N be the set of all integers with the discrete topology and Q the set of all rationals considered as a subspace of ℝ. Since N and Q are both denumerable there is a one-to-one correspondence f, mapping N onto Q. Then f is continuous, N is locally compact, but as we noticed before, $f[N] = Q$ is not locally compact.

Given a topological space X which is not compact, we can construct by adjoining extra points to X another topological space X' which is compact and such that X is a dense subspace of X' (or more precisely X is homeomorphic to a dense subspace of X'). Such a space X' is said to be a *compactification* of X. The compactification of a space is a convenience, for it allows the use of the standard compactness arguments and many proofs are considerably simplified. The extra points adjoined to obtain the compactification of a space X are called the *points at infinity* of X. The following two examples show that *a space can be compactified in more than one way.*

Example 1

Let ℝ be the real line with its usual metric topology. We shall compactify ℝ (which is not compact) by adjoining only one point at infinity. Consider the complex plane $\mathbb{R}^2$, and let f be a mapping of $\mathbb{R}^2 - \{0\}$ into itself, defined by $f(z) = z/|z|^2$

for each complex number $z = x + iy$ in $\mathbb{R}^2 - \{0\}$. It is easily seen that f is a homeomorphism of $\mathbb{R}^2 - \{0\}$ onto itself. The image of the straight line $y = 1$ is the set $S - \{0\}$ where S is the circumference of the circle centered at the point $(0,(1/2))$ with radius $1/2$. Now the mapping g defined for each $x \varepsilon \mathbb{R}$ by $g(x) = (x,1)$ is a homeomorphism between $\mathbb{R}$ and the line $y = 1$. Thus the composite mapping $f \circ g$ of $\mathbb{R}$ onto $S - \{0\}$ is a homeomorphism. Let us now identify the two homeomorphic spaces $\mathbb{R}$ and $S - \{0\}$. Then clearly the circumference S is a compactification of $\mathbb{R}$. Its unique point at infinity is the point 0.

Example 2

Consider the extended real line $\overline{\mathbb{R}}$ and the interval $[-1,1]$, both considered as topological spaces with their order topology (see sec. 7.2, Ex. 3). The mapping $f: \overline{\mathbb{R}} \to [-1,1]$ defined by

$$f(-\infty) = -1$$

$$f(+\infty) = 1,$$

$$f(x) = \frac{x}{1+|x|} \qquad \text{for } x \varepsilon \mathbb{R}$$

is an increasing mapping and it is easily seen to be a homeomorphism of $\overline{\mathbb{R}}$ onto $[-1,1]$. But in the example given in section 7.6 we proved that the order topology on $[-1,1]$ is identical with the one induced by the usual metric topology of $\mathbb{R}$. Hence $[-1,1]$ is compact and the same is true for $\overline{\mathbb{R}}$. Furthermore, the topology induced by $\overline{\mathbb{R}}$ on its subspace $\mathbb{R}$ is identical with the usual metric topology on $\mathbb{R}$. This proves that $\overline{\mathbb{R}}$ is a compactification of $\mathbb{R}$. There are two points at infinity of $\mathbb{R}$, namely $-\infty$ and $+\infty$.

Among the various ways of compactifying a topological space we shall consider only the one familiar in analysis, suggested by Example 1 above, and which is made by adjoining a single point at infinity. It is called the *one-point compactification* of a topological space, and it was formulated by the Russian mathematician B. Alexandroff. It is as follows:

Let X be a topological space and let ω be any object which is not an element of X. The one-point compactification of X is the topological space $X^* = X \cup \{\omega\}$ where a subset $O^* \subset X^*$ is defined to be open if O^* is an open set of the space X (in this case $\omega \notin O^*$) or if $X^* - O^*$ is a closed compact subset of X (in this case $\omega \in O^*$). It can be proved that with this definition of open set, X^* becomes a compact topological space and that X is a subspace of X^*. Moreover the space X^* is Hausdorff if and only if X is locally compact and Hausdorff.

If X is already compact, then the one-point compactification is still possible. In this case the point at infinity ω is an isolated point of X^*. On the other hand if ω is an isolated point of X^*, then X is closed in X^* and therefore is compact. All proofs of the above statements concerning the compactification of a space can be found in [16] page 150.

7.10 THE STONE-WEIERSTRASS THEOREM

The problem we are concerned with in this section can be outlined as follows:

Let X be a compact Hausdorff space and let $\mathcal{C}(X,\mathbb{R})$ be the set of all real-valued continuous functions on X. We shall prove that every collection of functions in $\mathcal{C}(X,\mathbb{R})$ which is sufficiently "rich" and which is closed under certain operations can be used to "approximate" every function in $\mathcal{C}(X,\mathbb{R})$. Some preliminaries are needed.

A subset A of $\mathcal{C}(X,\mathbb{R})$ is said to be an *algebra* iff for every pair f and g in A, and for all real numbers a and b, we have $af + bg \in A$ and $fg \in A$, where the sum $f + g$, the products af, and fg are defined in the usual manner, that is:

$$(f + g)(x) = f(x) + g(x)$$

$$(af)(x) = af(x)$$

$$(fg)(x) = f(x)g(x).$$

In other words A is a real linear space closed under the multiplication between elements in A. $\mathcal{C}(X,\mathbb{R})$ itself is of course an algebra of functions. Since X is compact, $f[X]$ is a compact set of real numbers for every $f \in \mathcal{C}(X,\mathbb{R})$. If we define the norm of f by $\|f\| = \max_{x\in X} f(x)$, since $\|f\|<+\infty$, $\mathcal{C}(X,\mathbb{R})$ becomes a *normed algebra*. For f and g in $\mathcal{C}(X,\mathbb{R})$ if we set $d(f,g) = \|f - g\|$, then $\mathcal{C}(X,\mathbb{R})$ becomes a metric space. The corresponding metric topology on $\mathcal{C}(X,\mathbb{R})$ is also called the *topology of uniform convergence* (sometimes *uniform topology*). This terminology is due to the fact that a sequence $\langle f_n \rangle$ of real-valued functions on X converges to a function f if and only if, in the space $\mathcal{C}(X,\mathbb{R})$ with the topology of uniform convergence, the sequence of points $\langle f_n \rangle$ converges to the point f. For $\langle f_n \rangle$ converges to f if and only if for every $\varepsilon > 0$ and for n sufficiently large, we have $d(f(x),f_n(x)) < \varepsilon$.

A subset A of $\mathcal{C}(X,\mathbb{R})$ is said to be a *lattice* iff for all f and g in A the function $f \wedge g$ defined by $(f \wedge g)(x) = \min[f(x),g(x)]$ and the function $f \vee g$ defined by $(f \vee g)(x) = \max[f(x),g(x)]$, both belong to A.

The space $\mathcal{C}(X,\mathbb{R})$ is itself a lattice since it is easily seen that if f and g are continuous then $f \vee g$ and $f \wedge g$ are also continuous.

If A is an algebra in $\mathcal{C}(X,\mathbb{R})$, then it can be shown (see Prob. 45) that its closure $\bar{A}$ is also an algebra.

26. Proposition *Let X be a compact Hausdorff space, let A be a lattice in $\mathcal{C}(X,\mathbb{R})$, and let $f \in \mathcal{C}(X,\mathbb{R})$. Then $f \in \bar{A}$ if and only if on every two point set $\{x,y\}$ where x and y are points of X, f is a uniform limit of elements of A (i.e., for every two elements x and y and for every $\varepsilon > 0$, there is a function $g_{x,y} \in A$ such that*

$$|f(x) - g_{x,y}(x)| < \varepsilon, \ |f(y) - g_{x,y}(y)| < \varepsilon).$$

Proof Let $f \in \mathcal{C}(X, \mathbb{R})$ and suppose that on every two-point set $\{x,y\}$ f is a uniform limit of elements of A. Then for every $\varepsilon > 0$ there is a $g_{x,y} \in A$ such that

$$(1) \qquad |f(x) - g_{x,y}(x)| < \varepsilon$$

$$(2) \qquad |f(y) - g_{x,y}(y)| < \varepsilon.$$

Set $E_{x,y} = \{z \in X\colon g_{x,y}(z) < f(z) + \varepsilon\}$. Then $E_{x,y}$ being the inverse image of the open set $(\varepsilon, +\infty)$ by the continuous function $g_{x,y} - f$ is open and it contains y because of (2). If x is fixed, the collection $\{E_{x,y}\}_{y \in X}$ is then an open covering of the compact space X. Thus there is a finite subcovering $\{E_{x,y_i}\}$, where i runs over a finite subset of positive integers. Put $g_x = \inf_i (g_{x,y_i})$. If z is any point of X, then z belongs to E_{x,y_i} for some i, so that

$$g_x(z) \leq g_{x,y_i}(z) < f(z) + \varepsilon.$$

Next put $E_x = \{z \in X\colon g_x(z) > f(z) - \varepsilon\}$. Then the continuity of $g_x - f$ implies that E_x is open, and because of (1), E_x contains x. Thus $\{E_x\}_{x \in X}$ is again an open covering of the compact space X, so that there is a finite subcovering

$\{E_{x_j}\}$, where j runs over a finite subset of positive integers.

Next put $g = \sup_j(g_j)$. Then since A is a lattice, and the g_j are finite in number, we have that $g \in A$, $g < f + \varepsilon$, and $g > f - \varepsilon$ on X. This implies that in every neighborhood of f there is an element of A. Therefore $f \in A$.

On the other hand if $f \in \overline{A}$, then obviously f is a uniform limit of elements of A on any subset of X, and in particular on any pair $\{x,y\}$. The proposition is proved.

27. Proposition *Let X be a compact Hausdorff space, let A be an algebra in $\mathcal{C}(X,\mathbb{R})$, and assume that A is a closed set. Then if f belongs to A, so does $|f|$.*

Proof If $f = 0$ then there is nothing to prove. So assume that $f \neq 0$, and suppose without loss of generality that $\|f\| \leq 1$. If $x \in X$ put $f(x) = t$. Then $-1 \leq t \leq 1$. Let ε be an arbitrary positive number and define $z(t) = (t^2 - 1)/(1 + \varepsilon^2)$. Clearly we have

$$t^2 + \varepsilon^2 = (1 + \varepsilon^2)(1 + z) \quad \text{and } |z| \leq (1 + \varepsilon^2)^{-1} < 1.$$

Hence the Taylor series of $\sqrt{1 + z}$ converges uniformly to $\sqrt{1 + z}$ for z as restricted above. Therefore there exists a polynomial $p(t)$ such that

$$(1) \qquad |(t^2 + \varepsilon^2)^{1/2} - p(t)| \leq \varepsilon$$

$$\text{for all } t \in [-1,1].$$

For $t = 0$, we get from (1)

$$|p(0)| \leq 2\varepsilon. \tag{2}$$

Also one easily verifies that for every $\varepsilon > 0$

$$0 \leq |(t^2+\varepsilon^2)^{\frac{1}{2}} - |t|| \leq \varepsilon. \tag{3}$$

If we put $P(t) = p(t) - p(0)$, we get using (1), (2) and (3)

$$|P(t) - |t|| = |p(t) - p(0) - |t||$$

$$\leq |p(t) - (t^2+\varepsilon^2)^{\frac{1}{2}}| + |(t^2+\varepsilon^2)^{\frac{1}{2}} - |t|| + |p(0)|$$

$$\leq \varepsilon + \varepsilon + 2\varepsilon = 4\varepsilon.$$

Replacing t by $f(x)$ we obtain

$$|P(f(x)) - |f(x)|| \leq 4\varepsilon \quad \text{for all } x \in X.$$

Now the polynomial P has a zero constant term. This implies that $P(f)$ is an element of A, so that the last inequality can be written as

$$\|P(f) - |f|\| \leq 4\varepsilon.$$

This proves that $|f|$ is a point of closure of the closed set A, so that $|f| \in A$.

28. Proposition *If* A *is as in Proposition 27 then* A *is a lattice.*

Proof By Proposition 27 if f and g are in A

then $|f + g|$ and $|f - g|$ are also in A. On the other hand one easily sees that

$$f \vee g = \tfrac{1}{2}[(f + g) + |f - g|];$$

$$f \wedge g = \tfrac{1}{2}[(f + g) - |f - g|].$$

Therefore if f and g are in A, $f \vee g$ and $f \wedge g$ are also in A. Hence A is a lattice.

29. Theorem (Stone-Weierstrass) *Let X be a compact Hausdorff space, and let A be an algebra of functions in* $\mathcal{C}(X,\mathbb{R})$ *such that (i) A separates the points of X (see sec. 7.7), (ii) for every* $x \in X$ *there exists an* $f \in A$ *such that* $f(x) \neq 0$. *Then* $\overline{A} = \mathcal{C}(X,\mathbb{R})$.

Proof First we shall show that for all x and y in X, with $x \neq y$ there exists a g in A such that we have $g(x) \neq g(y)$ and $g(x)g(y) \neq 0$. Indeed, by hypothesis, there exists an $h \in A$ such that $h(x) \neq h(y)$. If $h(x)h(y) \neq 0$ then we are done, for we can take $g = h$. If $h(x)h(y) = 0$ then either $h(x) = 0$ or $h(y) = 0$ (but not both). Suppose for definiteness $h(x) = 0$. By hypothesis there exists an $h_1 \in A$ such that $h_1(x) \neq 0$. In this case take for g the function $g = h + \varepsilon h_1$ with $\varepsilon \neq 0$ and such that $g(x) \neq g(y)$ and $g(y) \neq 0$. We have $g(x) \neq g(y)$ and $g(x)g(y) \neq 0$.

Next let f be any function in $\mathcal{C}(X,\mathbb{R})$ and x and y any two distinct points in X. Put $f(x) = a$ and $f(y) = b$. Consider the function k defined by $k = c_1 g + c_2 g^2$, where g is in A, and is such that $g(x) \neq g(y)$ and $g(x)g(y) \neq 0$, and c_1, c_2 are scalars satisfying the system of equations

$$a = c_1 g(x) + c_2 g^2(x);$$

$$b = c_1 g(y) + c_2 g^2(y).$$

The existence of the scalars c_1 and c_2 is guaranteed by the fact that the determinant of this system does not vanish. We also observe that k belongs to A.

As we have seen earlier, since A is an algebra, $\bar{A}$ also is an algebra. Then it follows from Proposition 28 that $\bar{A}$ is a lattice. Finally, since for arbitrary scalars a and b there is a function k in A such that $k(x) = a$ and $k(y) = b$, Proposition 26 asserts that $\bar{A} = \mathcal{C}(X, \mathbb{R})$. The theorem is proved.

Remark Observe that in Theorem 29, if the algebra A contains the constant functions then condition (ii) is satisfied.

30. Corollary *Let X be a compact set in $\mathbb{R}^n$ and let f be a continuous real-valued function on X. Then for any given $\varepsilon > 0$ there is a polynomial P in n variables (the coordinates) with real coefficients such that $|f(x) - P(x)| < \varepsilon$ for all x in X.*

Proof Let us call *coordinate function* a function on $\mathbb{R}^n$ into $\mathbb{R}$ defined by $x \to x_i$ where $x = (x_1, \ldots, x_n)$, $1 \le i \le n$. Then the set of all polynomials in the coordinate functions forms an algebra A on X. Clearly A contains the constants, so that by the above remark, condition (ii) of Theorem 29 is satisfied. Furthermore A separates the points of X, for if $x = \langle x_1,\ldots,x_n\rangle$, $y = \langle y_1,\ldots,y_n\rangle$ are two distinct points of X, then $x_i \neq y_i$ for at least one i. This implies that at least one of the coordinate functions takes different values on x and y. Thus Theorem 29 applies and the corollary follows.

(*Note:* For $n = 1$ and $X = [a,b]$ we get from Corollary 30, the classical *Weierstrass approximation theorem.*)

7.11 *CONNECTIVITY*

Let us first consider subsets of the complex plane or of the real line. Intuitively speaking, a subset is *connected* if it is made of only one "solid piece." Also, if a subset X is the union of two nonempty sets A_1 and A_2 and if there are disjoint closed sets B_1 and B_2 such that $A_1 \subset B_1$ and $A_2 \subset B_2$ then it is natural to say that X is not connected. These intuitive remarks lead to the following precise notion of connectedness.

Definition *A topological space* X *is said to be* connected *if and only if it is not the union of two nonempty closed sets. A subset* S *of* X *is said to be connected iff* S, *considered as a subspace of* X, *is connected.*

Here are two propositions which can also be taken as equivalent definitions to the one given above.

X *is connected if and only if there are not two open disjoint nonempty sets* O_1 *and* O_2 *such that* $X = O_1 \cup O_2$.

X *is connected if and only if the only subsets of* X *which are both open and closed are* X *and* $\emptyset$.

To prove the first of these propositions observe that each of the sets O_1 and O_2 is the complement of the other, so that they are closed sets as well as open sets. The second proposition follows from the first and the definition of connectedness.

31. Proposition *Let* $\mathcal{C}$ *be any collection of connected subsets of a topological space* X, *such that* $\bigcap_{A \in \mathcal{C}} A \neq \emptyset$. *Then the set* $S = \bigcup_{A \in \mathcal{C}} A$ *(the union of all sets in* $\mathcal{C}$*) is connected.*

Proof Assume S is not connected and let O_1 and O_2 be two nonempty disjoint open subsets of the subspace S such that $S = O_1 \cup O_2$. For every $A \in \mathcal{C}$ the sets $A \cap O_1$ and $A \cap O_2$ are open sets in the subspace A. Since A is connected, either $A \cap O_1$ or $A \cap O_2$ is empty. Thus A is contained in O_1 or in O_2. Since the sets of the collection $\mathcal{C}$ have at least one element a in common, a must belong to O_1 or O_2. Suppose $a \in O_1$. This implies that

O_1 contains all the sets A of $\mathcal{C}$, so that $O_2 = \emptyset$. But this contradicts the assumption that O_2 is not empty. Hence S is connected.

32. Proposition *Let f be a continuous mapping of a connected topological space X onto a topological space Y. Then Y is connected.*

Proof Let E be a subset of Y that is both open and closed. Since f is continuous $f^{-1}[E]$ is both open and closed in X. Since X is connected we have either $f^{-1}[E] = X$ or $f^{-1}[X] = \emptyset$. But $f[f^{-1}[E]] = E$ and the mapping f is onto so that either $E = Y$ or $E = \emptyset$. Hence Y is connected.

33. Proposition *Let A be any connected set of a topological space X and B any other subset of X such that $A \subset B \subset \bar{A}$. Then B is connected.*

Proof Let O_1 and O_2 be two open nonempty disjoint sets of the subspace B. Then $A \cap O_1$ and $A \cap O_2$ are open nonempty disjoint subsets of the subspace A. Since A is connected either $A \cap O_1$ or $A \cap O_2$ is empty. Assume $A \cap O_1 = \emptyset$. Since $B \subset \bar{A}$ and $A \subset B$, A is dense in B. This implies that $O_1 = \emptyset$, for otherwise $A \cap O_1$ would not be empty. Hence B is connected.

By taking $\bar{A} = B$ we see that *the closure of every connected set is connected.*

A metric space $\langle X,d \rangle$ is said to be *well-linked* iff for every two points a and b in X and for every $\varepsilon > 0$ there is a finite sequence $\langle a_1, a_2, \ldots, a_n \rangle$ with $a_1 = a$ and $a_n = b$ such that

$d(a_i,a_{i+1}) \leq \varepsilon$ for $i = 1,2,\ldots,n-1$.

Whenever such a sequence exists we say that a and b can be joined by an ε-chain.

34. Theorem *Let $\langle X,d\rangle$ be a metric space and consider X as a topological space with its metric topology. Then*

(a) If X is connected then X is well-linked.

(b) If X is compact then X is connected if and only if it is well-linked.

Proof of (a) For $a \varepsilon X$ and $\varepsilon > 0$, let $A(a,\varepsilon)$ be the set of all points that can be joined to a with an ε-chain. We want to prove that $X = A(a,\varepsilon)$. Clearly $A(a,\varepsilon) \neq \emptyset$ since it contains a. We prove that $A(a,\varepsilon)$ is open.

Let $x \varepsilon A(a,\varepsilon)$ and let $S_\varepsilon(x)$ be the open ball centered at x with radius ε. There is a sequence $\langle a_i\rangle_{i=1}^n$ with $a_1 = a$, $a_n = x$ such that $d(a_i,a_{i+1}) \leq \varepsilon (i = 1,2,\ldots,n-1)$. If $y \varepsilon S_\varepsilon(x)$ then $d(x,y) < \varepsilon$. Thus a and y can be joined by the ε-chain $\langle y,a_1,a_2,\ldots,a_n\rangle$. Hence $S_\varepsilon(x) \subset A(a,\varepsilon)$ so that $A(a,\varepsilon)$ is open. $A(a,\varepsilon)$ is also closed. To see this let x be an accumulation point of $A(a,\varepsilon)$. Then there is a point y of $A(a,\varepsilon)$ in the open ball $S_\varepsilon(x)$. We have $d(x,y) < \varepsilon$ so that x and a can be joined by an ε-chain which consists of the union of the ε-chain joining y and a, and the set $\{x\}$. Thus $x \varepsilon A(a,\varepsilon)$ so that $A(a,\varepsilon)$ is closed. Since X is connected and $A(a,\varepsilon)$ is both open and closed and nonempty, we have $A(a,\varepsilon) = X$, which proves that X is well-linked.

Proof of (b) If X is connected and compact, then by (a) X is well-linked (even if X is not compact). Next let X be compact and well-linked and suppose that X is not connected. Then there exist two nonempty disjoint closed sets F_1 and F_2 such that $X = F_1 \cup F_2$. The F_1 and F_2 as closed subsets of a compact set are also compact; since they are disjoint their distance must be positive. That is

$$d(F_1,F_2) = \inf\{d(x_1,x_2): x_1 \in F_1, x_2 \in F_2\}$$
$$= \delta > 0.$$

This shows that no point in F_1 can be joined to a point in F_2 by an ε-chain with $\varepsilon < \delta/2$. For if $\langle a_1,a_2,\ldots,a_n\rangle$ is such a chain, with $a_1 \in F_1$ and $a_n \in F_2$, let i_0 be the smallest index such that $a_{i_0} \in F_2$. Then clearly $i_0 > 1$ and $a_{i_0-1} \in F_1$. We have $d(a_{i_0-1}, a_{i_0}) < \delta/2$ which contradicts the fact that $d(F_1,F_2) = \delta$. This proves that X is connected.

35. Corollary *A subset of the real line is connected if and only if it is an interval (open, closed, or half-open).*

Proof Let $[a,b] \subset \mathbb{R}$ be a compact interval. Clearly $[a,b]$ is well-linked, so that by Theorem 29 $[a,b]$ is compact. Next let I be any interval in R and let $a \in I$. Then I is the union of closed intervals each of which is of the form $[a,y]$ all of which have the point a in common so that by Proposition 31 I is connected.

PROBLEMS

1. Show that if $\{\sigma_\alpha\}_{\alpha\in\Lambda}$ is a collection of topologies on a set X then $\bigcap_{\alpha\in\Lambda} \sigma_\alpha$ is also a topology on X.
2. If X is a metric space and if σ is the collection of all sets $E \subset X$ which are arbitrary unions of open balls, show that σ is a topology on X.
3. Let X be an arbitrary infinite set. Define a subset $O \subset X$ to be open if $O = \emptyset$ or if $\complement O$ is finite. Show that the so defined collection of open sets is a topology on X (called the *cofinite topology*).
4. Prove the statements F_1, F_2 and F_3 given in section 7.2.
5. Prove Proposition 2.
   (Hint: Let σ be the collection of all subsets A of X such that for every $x \in A$ we have $A \in \mathcal{V}(x)$. Show that σ is a topology on X. Clearly every neighborhood of x belongs to $\mathcal{V}(x)$. Conversely let $V \in \mathcal{V}(y)$, and let U be the set of all points y such that $V \in \mathcal{V}(y)$. Show that $x \in U$, $U \subset V$ and $U \in \sigma$.)

6. Show that a set E is open iff $E = \overset{\circ}{E}$.
7. Prove Propositions (a) through (e) stated immediately after the definition of a dense set (see p. 325).
8. Let X and Y be topological spaces. Show that a mapping $f: X \to Y$ is continuous iff for every closed set F in Y $f^{-1}[F]$ is a closed set in X.
9. Let X be the real line with its usual metric topology, and let Y be the real line with the discrete topology. Show that the identity mapping $x \to x$ is not a homeomorphism.
10. Let $X = [0,1) \cup \{2\}$, $Y = [0,1]$, be topological spaces with the induced topology by the metric topology of the real line. Let $f: X \to Y$ be defined by

$$f(x) = x \quad \text{if} \quad x \in [0,1) \quad \text{and} \quad f(2) = 1.$$

Show that f is not a homeomorphism.
11. Prove Proposition 10.
12. Show that the Euclidean plane with its usual metric topology satisfies the second axiom of countability.
13. Show that every metric space with its metric topology satisfies the first axiom of countability.
14. Let X be a topological space which satisfies the second axiom of countability, and let $\{G_\alpha\}$ be an open covering of X. Show that a countable number of G_α covers X.
15. Give an example of a topological space which

satisfies the first axiom of countability but not the second.
(Hint: Consider an uncountable discrete topological space.)

16. Let X be a nonempty set and let $\mathcal{F}$ be any collection of subsets of X. Show that there is a weakest topology $\mathcal{O}$ which contains $\mathcal{F}$.
(Hint: Fill in the details in the proof given in the text.)

17. Let $f: X \to Y$ be a mapping from a topological space X into a topological space Y. Let S be a subbasis for the topology of Y, and let $\mathcal{B}$ be the basis generated by S. Let $\mathcal{B}_1$, be a basis for the topology of X.
(a) Show that f is continuous iff the inverse image of each set in $\mathcal{B}$ is open.
(b) Show that the inverse image of each set in $\mathcal{B}$ is open iff the inverse image of each set in S is open.
(c) Show that for each open set O in X, $f[O]$ is open iff the image of each set in $\mathcal{B}_1$ is open.
The mapping $f: X \to Y$, where X and Y are topological spaces is said to be *open* if for every open set O in X, $f[O]$ is open in Y.

18. Let X be a topological space and E a subspace of X. Show that every open (respectively closed) set in E is an open (respectively closed) set in X, iff E is an open (respectively closed) set in X.

19. Let X and Z be topological spaces, and Y be a subspace of Z. Let $f: X \to Y$ be a mapping of X into Y. Show that f is con-

tinuous at a point $x_o \in X$ iff the mapping f, regarded as a mapping from X into Z, is continuous at x_o.

20. Let $\{f_\alpha\}$ be a collection of real-valued functions defined on the real line $\mathbb{R}$. Give a complete description of the weak topology generated on $\mathbb{R}$ by the collection $\{f_\alpha\}$, in each of the following cases.

    (a) $\{f_\alpha\}$ consists of all constant functions.

    (b) $\{f_\alpha\}$ consists of a single function f defined by $f(x) = 0$ if $x \leq 0$ and $f(x) = 1$ if $x > 0$.

    (c) $\{f_\alpha\}$ consists of a single function f, defined by $f(x) = -1$ if $x < 0$, $f(0) = 0$ and $f(x) = 1$ if $x > 0$.

    (d) $\{f_\alpha\}$ consists of a single function f, defined by $f(x) = x$ for all x.

    (e) $\{f_\alpha\}$ consists of all bounded functions which are continuous with respect to the metric topology of $\mathbb{R}$.

    (f) $\{f_\alpha\}$ consists of all functions which are continuous with respect to the metric topology of $\mathbb{R}$.

21. Show that every sequence of points in a Hausdorff space has at most one limit.

22. Show that the weak topology generated by a collection of functions $\{f_\lambda\}_{\lambda\in\Lambda}$ provides a Hausdorff space iff the collection $\{f_\lambda\}_{\lambda\in\Lambda}$ separates points.

23. Let X be a Hausdorff space. Show that the set

$$\Delta = \{(x,x): x \in X\}$$

is a closed set in the product topology of $X \times X$.

24. Let f and g be continuous mappings of a topological space X into a Hausdorff space Y. Show that the set $\{x \in X: f(x) = g(x)\}$ is closed.
(Hint: Consider the mapping $h: X \to X \times X$ defined by $h(x) = \langle f(x), g(x) \rangle$. Show that h is continuous. Use Problem 23.)

25. (a) Let $f: X \to Y$ be a continuous mapping of a topological space X into a Hausdorff space Y. Show that the set $\{\langle x, f(x) \rangle: x \in X\}$ is closed in $X \times Y$.
(b) Show that the converse of (a) is false.
(Hint: $f: \mathbb{R} \to \mathbb{R}$ with $f(x) = 1/x$ if $x \neq 0$ and $f(0) = 0$.)

26. Let f be a continuous mapping of a topological space X into a topological space Y. Show that the subspace $\{\langle x, f(x) \rangle: x \in X\}$ of $X \times Y$ and the space X are homeomorphic.

27. Show that a Hausdorff space is normal iff for every closed set F, and an open set O containing F there is an open set G such that $F \subset G$ and $\overline{G} \subset O$.

28. Let X be a T_1 space with the property that if x_0 is any point in X and F any closed subspace of X which does not contain x_0,

then there exists a real-valued continuous function f on X such that $0 \leq f(x) \leq 1$ for all $x \in X$, and such that $f(x) = 0$ and $f[F] = \{1\}$. Then we say that X is a *completely regular space.*

(a) Show that every normal space is completely regular.

(b) Let X be a T_1-space, such that for each $x_0 \in X$ and F a closed set not containing x_0, there is a continuous function f on X such that $f[F] = \{0\}$, $f(x_0) = 1$ and $0 \leq f(x) \leq 1$ for all $x \in X$. Show that X is a completely regular space.

(c) Show that a completely regular space is regular.

The completely regular spaces are sometimes called $T_{3\frac{1}{2}}$ spaces.

29. Let X be a completely regular space, and let $\mathcal{C}(X,\mathbb{R})$ be the set of all bounded continuous real-valued functions defined on X. Show that the weak topology generated by $\mathcal{C}(X,\mathbb{R})$ is the given topology of the space X.

30. Prove Proposition 15.

31. Prove that a compact Hausdorff space is regular.

32. Prove that a compact Hausdorff space is normal.

33. Let $f: X \to \mathbb{R}$ be a continuous function on a compact topological space X. Show that f is bounded and attains its supremum and infimum on X.

34. Let X be a compact topological space and let $f_n: X \to \mathbb{R}$ $(n = 1,2,\ldots)$ be a monotone sequence

of real-valued continuous functions (i.e., $f_n \leq f_{n+1}$ for all n, or $f_{n+1} \leq f_n$ for all n) which converges pointwise to a continuous real-valued function f defined on X. Show that $\langle f_n \rangle$ converges to f uniformly. (This is known as Dini's theorem.)

35. Let X be a compact topological space. We know that every closed subspace of X is compact. Give an example to show that a compact subspace of X need not be closed.

36. Let X be a compact Hausdorff space.
(a) Show that every infinite subset of X has an accumulation point in X.
(b) Show that every subset of X which does not have an accumulation point in X is finite.
(*Note: A point* a *is an accumulation point of a set* S *iff every open set containing* a *contains an infinite number of points of* S.)

37. Show that a countably compact space satisfying the first countability axiom is sequentially compact.

38. The sphere S_{n-1} in $\mathbb{R}^n$ is defined to be the set consisting of all points $x = \langle x_1, \ldots, x_n \rangle$ such that $\sum_{i=1}^{n} x_i^{\,2} = 1$. The Cartesian product $(S_1)^p$ is called the p-dimensional *torus*. Show that S_{n-1} and $(S_1)^p$ are both compact.

39. Show that the subspace Q of $\mathbb{R}$, where Q is the set of rational numbers with the induced metric topology of $\mathbb{R}$, is neither compact nor locally compact.

(Hint: Let V be a compact neighborhood of 0 in Q. Then V contains a neighborhood of the form $V_1 = Q \cap [-a,a]$. But then V_1 as a closed subset of V would be compact. Show that this is false by considering the decreasing sequence $V \cap [x - (1/n), x + (1/n)]$ for x irrational in $[-a,a]$.)

40. Let $\langle X_\lambda \rangle_{\lambda \varepsilon \Lambda}$ be a family of topological spaces such that the Cartesian product

$$X = \prod_{\lambda \varepsilon \Lambda} X_\lambda$$

is compact in the product topology. Show that X_λ is compact for each $\lambda \varepsilon \Lambda$. Compare with Theorem 24.

41. Prove that if X is a Hausdorff space, and if E_1 and E_2 are disjoint compact sets in X, then there exist open sets O_1 and O_2 such that $O_1 \supset E_1$, $O_2 \supset E_2$ and $O_1 \cap O_2 = \emptyset$.

42. Let X be a locally compact topological space but not compact. Let f be a continuous real-valued function on X such that for every $A > 0$ there is a compact set $K \subset X$ such that $f(x) > A$ for $x \not\varepsilon K$. Show that there is a point $a \varepsilon X$ such that $f(a) \leq f(x)$ for all $x \varepsilon X$.

(Hint: Let x_0 be any point in X, and let $A > f(x_0)$. Then there is a compact set K such that $f(x) > A$ for $x \not\varepsilon K$. The restriction of f to K is continuous, hence f has an infimum m which is attained at some point $a \in K$.)

43. Let P be an arbitrary polynomial, with complex coefficients, of the complex variable z. The function $z \to |P(z)|$ is continuous on the complex plane. Use Problem 42 to prove that $|P(z)|$ attains its infimum at some point.

44. Let X be a compact Hausdorff space. Show that $\mathcal{C}(X,\mathbb{R})$ is a lattice.

45. Let X be as in Problem 44. Show that if A is an algebra in $\mathcal{C}(X,\mathbb{R})$ then $\overline{A}$ is also an algebra.

46. (a) Let X be a compact Hausdorff space, $\mathbb{C}$ the field of complex numbers and $\mathcal{C}(X,\mathbb{C})$ the set of all complex-valued continuous functions on X. The space $\mathcal{C}(X,\mathbb{C})$ is a normed algebra with respect to the same operations and norm as these are defined in $\mathcal{C}(X,\mathbb{R})$. Let A be a subalgebra of $\mathcal{C}(X,\mathbb{C})$, such that:

    1) A separates points of X.

    2) For every $x \in X$ there exists an $f \in A$ such that $f(x) \neq 0$.

    3) For every $f \in A$ we also have $\overline{f} \in A$ (where $\overline{f}$ denotes the complex conjugate of f).

    Show that

$$\overline{A} = \mathcal{C}(X,\mathbb{C}).$$

(Hint: The subalgebra B over the real numbers consisting of the real-valued functions in A satisfies conditions (i) and (ii) of Theorem 29. To see this suppose that f separates the points x and y. Then $R_e(f)$ or $R_e(if)$ (where $R_e f$ denotes the real part of f) also separates x and y. Similarly if $f(x) \neq 0$, then either $R_e f(x) \neq 0$ or $R_e(if)(x) \neq 0$. Therefore $\overline{B} = \mathcal{C}(X,\mathbb{R})$ and $\overline{B + iB} = \mathcal{C}(X,\mathbb{C})$.)

(b) Show that if condition 3) in (a) is not satisfied then the conclusion $\overline{A} = \mathcal{C}(X,\mathbb{C})$ is in general false.

(Hint: Let X be the unit disc on $\mathbb{C}$ and let A be the set of all polynomials in the complex variable z. Then A satisfies 1) and 2) but not 3). Show that $\overline{A} \neq \mathcal{C}(X,\mathbb{C})$.)

47. Show that every continuous complex-valued function on the real line $\mathbb{R}$ which is periodic with period 2π, is the uniform limit of trigonometric polynomials $\sum_{p=-n}^{n} a_p e^{ipt}$.

(Hint: Let $X = \{z \in \mathbb{C}: |z| = 1\}$ be the unit circle of $\mathbb{C}$. The function $z \to z$ separates the points of X and does not vanish on X. It follows that the algebra generated by z and $\overline{z}$ is dense in $\mathcal{C}(X,\mathbb{C})$. Consider the mapping $\phi : \mathbb{R} \to X$ defined by $\phi(t) = e^{it}$. For every $f \in \mathcal{C}(X,\mathbb{C})$, the composite mapping $f \circ \phi$ is continuous on $\mathbb{R}$ periodic with period 2π, and it is also known that every continuous periodic function with period 2π on $\mathbb{R}$ is of this form. Since f is the uniform limit of polynomials in z and $\overline{z}$, it follows that $f \circ \phi$ is the uniform limit of polynomials in

e^{it} and e^{-it}.)

48. Show that a topological space X is connected iff there are no two open disjoint sets O_1 and O_2 such that $X = O_1 \cup O_2$.

49. Show that a topological space X is connected iff the only subsets of X which are both open and closed are X and $\emptyset$.

50. A subset E of a topological space X is said to be connected iff E, considered as a subspace of X, is connected. Show that the closure of every connected set is connected.

51. Let E and S be subsets of a topological space X such that S is connected and intersects E and $\complement E$ in at least one point each (that is $S \cap E \neq \emptyset$, $S \cap \complement E \neq \emptyset$). Then S contains at least one point of the boundary of E.

52. Let X be a connected topological space and let f be a continuous real-valued function on X. Suppose a and b are two points of f[X] such that $a < b$. Then for any c such that $a < c < b$ there exists an $x \in X$ such that $f(x) = c$.

53. Give an example to show that a well-linked metric space is not necessarily connected.

54. Show that if a connected topological space X has a nonconstant continuous real valued function defined on it, then X is an uncountable set.

55. Show that a topological space X is not connected iff there exists a continuous mapping of X onto the discrete two-point space {0,1}.

56. Let X, $\mathcal{C}$ and S be as in Proposition 31. Show that if every pair A and B of $\mathcal{C}$ has a nonempty intersection, then S is connected.

CHAPTER 8

Banach Spaces

8.1 VECTOR SPACES

The structure of *vector space* (*linear space*) is one of the most important in mathematical analysis. Historically vector spaces were used as a convenient means of expressing properties of systems of linear equations. Later on, it was realized that the structure of vector space occurs in every branch of analysis, as for example in the theory of linear differential equations, integration, the theory of integral equations and others, and that the use of a kind of geometric language could simplify and make the various questions intuitively clearer. At the present time vector spaces have become a quite fundamental tool in mathematics.

Definition *A set* X *(whose elements will be called* vectors*) is called a* vector space *over the reals (or a field* F*) if and only if we have a function* $+$ *on* $X \times X$ *into* X *and a function* $\cdot$ *on* $\mathbb{R} \times X$ *into* X *such that*

1) $x + y = y + x$; $x \in X$, $y \in X$.

2) $(x + y) + z = x + (y + z)$; $x \in X$, $y \in X$, $z \in X$.

3) There is a vector θ *in* X *such that* $x + \theta = x$ *for all* x *in* X.

4) $\lambda\cdot(x + y) = \lambda\cdot x + \lambda\cdot y;\quad \lambda \in \mathbb{R},\ x,y \in X.$

5) $(\lambda + \mu)\cdot x = \lambda\cdot x + \mu\cdot x;\quad \lambda,\mu \in \mathbb{R},\ x \in X.$

6) $\lambda\cdot(\mu\cdot x) = (\lambda\cdot\mu)\cdot x;\quad \lambda,\mu \in \mathbb{R},\ x \in X.$

7) $0\cdot x = \theta,\quad 1\cdot x = x.$

We notice that X is a commutative group with respect to + (called addition). This implies that the element θ defined in *3)* is unique. Also the element $(-1)\cdot x$ written $-x$ and called the *negative of* x is unique. We call the function $\cdot$, multiplication by scalars and we interchangeably write λx for $\lambda\cdot x$. The conditions *4)*, *5)*, *6)*, and *7)* of the definition merely indicate that there is an addition and a multiplication on $\mathbb{R}$. Therefore, it seems that we could also define a vector space over a ring A which would play the role that $\mathbb{R}$ plays in the preceding definition. However, only when the ring A is a field do we obtain a theory analogous to the theory of vector spaces over the reals. This is primarily due to the fact that if A is a field then the relation $\lambda x = \theta$ implies $\lambda = 0$ or $x = \theta$. This implication does not hold if A is merely a ring. The fields mostly used are $\mathbb{R}$ and the field $\mathbb{C}$ of complex numbers. The fact that $\mathbb{R}$ is an ordered field did not intervene in the above definition. However, the existence of an order in $\mathbb{R}$ becomes very useful if we want to define the notion of straight line or of interval in a vector space over the reals. In this case we can also talk about convex sets in X, and it is in this respect that vector spaces over $\mathbb{R}$ play a privileged role among the other vector spaces. Unless otherwise stated, we shall only consider vector spaces over the reals.

Let X be a vector space, and $x_1, x_2, \ldots, x_n$ vectors in X. If $c_1, c_2, \ldots, c_n$ are scalars, then the vector $c_1x_1 + c_2x_2 + \ldots + c_nx_n$ is called a *linear combination* of $x_1, x_2, \ldots, x_n$. Note that linear combinations are finite series.

A set $\{x_1, x_2, \ldots, x_n\}$ of vectors in X is called *independent* iff the relation $c_1x_1 + c_2x_2 + \ldots + c_nx_n = 0$ implies that each $c_i = 0$ $(i = 1, 2, \ldots, n)$. A subset S (finite or infinite) of X is said to be *independent* iff every finite subset of S is independent.

By a *basis* (or Hamel basis) of a vector space X is meant a linearly independent subset B of X with the property that any vector in X can be expressed as a linear combination of some subset of B. It can be shown that every vector space $X \neq \{\theta\}$ admits at least one basis, and that all bases for X have the same cardinal number, called the *dimension* of X. If $X = \{\theta\}$ then it is said to have dimension 0. A vector space is called *finite-dimensional* iff its dimension is 0 or a positive integer, and *infinite-dimensional* otherwise.

Examples of Vector Spaces

1) Take $X = \mathbb{R}^n$, and for $x, y \in \mathbb{R}^n$, $\lambda \in \mathbb{R}$ put:

$$x + y = \langle x_1 + y_1, x_2 + y_2, \ldots, x_n + y_n \rangle$$

$$\lambda x = \langle \lambda x_1, \lambda x_2, \ldots, \lambda x_n \rangle .$$

The case $n = 1$ shows that the real number system can be considered as a vector space over itself.

2) The set of all real-valued functions defined on $[0,1]$ is a vector space over $\mathbb{R}$, if we define $h = f + g$ and $k = \lambda f$ as follows:

$$h(x) = f(x) + g(x),$$

$$k(x) = \lambda f(x), \qquad \text{for every } x \in [0,1].$$

Also the following subsets of the above set, with the functions $+$ and $\cdot$ defined in the same way, are vector spaces over $\mathbb{R}$.

The set of all continuous functions on $[0,1]$.

The set of all infinitely many times differentiable functions on $[0,1]$.

The set of all polynomials with real coefficients and degree less than n. We include here all nonzero constant polynomials (which have degree 0) and the polynomial which is identically zero (this has no degree at all).

The set of all trigonometric polynomials.

3) The set of all holomorphic functions in the complex plane with functions $+$, $\cdot$ defined as in Example 2, is a vector space over the field $\mathbb{C}$ of complex numbers.

4) The L^p spaces considered in Chapter 6 are vector spaces.

It is left to the reader to verify that the spaces given by the above examples are vector spaces.

However, it is not always easy to recognize if a set has the structure of a vector space. As an example, consider the set of all continuous functions on $[0,1]$ whose graphs have a finite length. It is not obvious that the sum of such functions has the same property.

Definition *Two vector spaces* X *and* X' *over the reals (or any other field) are said to be* isomorphic *if and only if there is a one-to-one correspondence* ϕ *between* X *and* X' *such that, for any* $x \in X$, $y \in X$, $\lambda \in \mathbb{R}$, *we have*

$$(i) \quad \phi(x + y) = \phi(x) + \phi(y)$$

and

$$(ii) \quad \phi(\lambda x) = \lambda\phi(x).$$

The function ϕ is called an *isomorphism* between X and X'. It can be shown that condition *(ii)* follows from condition *(i)* if λ is a rational number, and that this is not so if λ is irrational. In other words *(i)* and *(ii)* are independent of each other.

Examples of Isomorphic Vector Spaces

1) Let X be the set of all polynomials $p(x) = a_1x^{n-1} + a_2x^{n-2} + \ldots + a_n$ with real coefficients and of degree less than n. We easily see that the function $\phi: X \to \mathbb{R}^n$ defined by $\phi(p) = \langle a_1, a_2, \ldots, a_n \rangle$ is an isomorphism between X and $\mathbb{R}^n$.

2) Let X_1 be the vector space of all continuous functions on $[0,1]$ and X_2 the vector space of all continuous functions on $[0,1]$ which vanish at $x = 0$, and whose first derivative exists and is continuous on $[0,1]$. Then X_1 and X_2 are isomorphic. The proof is left to the reader.

(Hint: Consider the mapping $f(x) \to \int_0^x f(t)dt$.)

Definition *A vector space* X *is said to be* normed *if and only if there is given a nonnegative real-valued function* $\|\cdot\|$ *(called a* norm *on* X *(*$x \to \|x\|$*)) having the following properties.*

1) $\|x\| = 0$ *if and only if* $x = \theta$.
2) $\|x + y\| \leq \|x\| + \|y\|$ *(triangle inequality).*
3) $\|\lambda x\| = |\lambda|\,\|x\|$, $x \in X$, $\lambda \in \mathbb{R}$.

Definition *Two normed vector spaces* X *and* Y *are said to be* isometrically isomorphic *if and only if there is an isomorphism* ϕ *between* X *and* Y *such that for each* $x \in X$, $\|x\| = \|\phi(x)\|$.

If X is a normed vector space, then the function $d: X \times X \to \mathbb{R}$ defined by $d(x,y) = \|x - y\|$ is evidently a metric on X. Having noticed this fact we can introduce into our discussion the vocabulary of metric spaces: ball, sphere, neighborhood, open, closed, separable, Cauchy sequence, complete, and so forth.

Definition *A normed vector space* X *is called a* Banach space *iff* X *is complete in the metric defined by the norm. The simplest real Banach space is the field of real numbers normed by* $\|x\| = |x|$. Other examples of Banach spaces are listed below. Verifications are left to the reader.

1) The Euclidean n-space, $\mathbb{R}^n$, i.e., the space consisting of all n-tuples of real numbers: $x = \langle x_1, x_2, \ldots, x_n \rangle$ normed by $\|x\| = \left(\sum_{i=1}^{n} x_i^{\,2}\right)^{\frac{1}{2}}$.

2) The L^p spaces considered in Chapter 6 (provided we identify functions which are equal a.e.) normed by $\|f\|_p$.

3) The space $C[a,b]$ of all continuous func-

tions on a closed interval $[a,b]$ with the operations of addition and multiplication by a real number that are usual for functions, normed by $\|f\| = \max_{t\varepsilon[a,b]} |f(t)|$.

4) The space ℓ^2 (Chap. 5, sec. 1) is a Banach space if we define the sum of two elements $x = \langle x_n \rangle$, $y = \langle y_n \rangle$ to be $x + y = \langle x_n + y_n \rangle$ and let $\lambda x = \langle \lambda x_n \rangle$ and $\|x\| = \left(\sum_{n=1}^{\infty} |x_n|^2\right)^{\frac{1}{2}}$.

5) The space of all sequences $x = \langle x_n \rangle$ of real numbers such that $\lim_{n\to\infty} x_n = 0$, with the addition and multiplication defined as in Example 4 above, and the norm $\|x\| = \sup\{|x_n| : 1 \leq n < \infty\}$.

As in the case of real numbers, given a sequence $\langle x_n \rangle$ in a vector space we may form the sequence $\left\langle \sum_{i=1}^{n} x_i \right\rangle_{n=1}^{\infty}$, and call it the series $\sum_{n=1}^{\infty} x_n$. The elements $\sum_{i=1}^{n} x_i$ are called the partial sums of $\sum_{n=1}^{\infty} x_n$. A series $\sum_{n=1}^{\infty} x_n$ in a normed vector space is said to be *summable* (or *convergent*) to an element s of the space, if and only if $\lim_{n\to\infty} \left\| s - \sum_{i=1}^{n} x_n \right\| = 0$. In this case we write $\sum_{n=1}^{\infty} x_n = s$. A series $\sum_{n=1}^{\infty} x_n$ is said to be *absolutely summable* if and only if $\sum_{n=1}^{\infty} \|x_n\| < +\infty$.

1. Proposition *A normed vector space* X *is complete if and only if every absolutely summable series is summable.*

Proof First suppose X is complete and let $\sum_{n=1}^{\infty} x_n$ be a series such that $\sum_{n=1}^{\infty} \|x_n\|$ is convergent. Then for given $\varepsilon > 0$ we can find an $N > 0$ such that $\sum_{n=N}^{\infty} \|x_n\| < \varepsilon$. Also, for $n \geq m \geq N$ we have $\left\| \sum_{i=1}^{n} x_i - \sum_{i=1}^{m} x_i \right\| = \left\| \sum_{m+1}^{n} x_i \right\| \leq \sum_{i=m}^{n} \|x_i\|$ $\leq \sum_{i=N}^{\infty} \|x_i\| < \varepsilon$. This means that $\left\langle \sum_{i=1}^{n} x_i \right\rangle$ is a Cauchy sequence and therefore must converge to an element $s \varepsilon X$, since by assumption X is complete.

Next, suppose that every absolutely summable series in X is summable to some element of X, and let $\langle x_n \rangle$ be a Cauchy sequence in X. Then for each integer k there is an integer n_k such that $n \geq n_k$, $m \geq n_k$ imply $\|x_n - x_m\| < 2^{-k}$. Since the last inequality remains true if we replace n_k by any other integer larger than n_k, we may choose the n_k's such that $n_{k+1} > n_k$. In this case $\langle x_{n_k} \rangle_{k=1}^{\infty}$ is a subsequence of $\langle x_n \rangle$. Put $y_1 = x_{n_1}$, and for $k > 1$ define $y_k = x_{n_{k+1}} - x_{n_k}$. Then since $\|y_k\|$ $< 2^{-k}$ the series $\sum_{k=1}^{\infty} y_k$ is absolutely summable, so that $\sum_{k=1}^{\infty} y_k$ is summable to an element $x \in X$. Hence the sequence $\langle x_{n_k} \rangle$ of the partial sums of this series converges to x. We now prove that the whole sequence $\langle x_n \rangle$ converges to x. Since $\langle x_n \rangle$ is a Cauchy sequence, for given $\varepsilon > 0$ there is an N_1 such that $n \geq N_1$, $m \geq N_1$ imply $\|x_n - x_m\| < \varepsilon/2$. Also, since $\lim_{k\to\infty} x_{n_k} = x$, there is an N_2 such that $k \geq N_2$ imply $\|x_{n_k} - x\| < \varepsilon/2$. Choose a k_o such that $k_o > N_2$ and $n_{k_o} \geq N_1$. Then for $n \geq N_1$ we have:

$$\|x_n - x\| \leq \|x_n - x_{n_{k_o}}\| + \|x_{n_{k_o}} - x\| < (\varepsilon/2) + (\varepsilon/2) = \varepsilon.$$

In other words, for any given $\varepsilon > 0$ there is an N, namely $N = N_1$, such that $n \geq N$ implies $\|x_n - x\| < \varepsilon$. This proves that $\lim_{n\to\infty} x_n = x$.

Definition *A nonempty subset S of a vector space X is called a* subspace *of X or a* linear manifold, *iff S is itself a vector space, relative to the addition and multiplication by scalars which are defined in X. This is equivalent to saying that S is a subspace if and only if $x + y \in S$ and $\lambda x \in S$ whenever x and y are in S and λ is a scalar.*

Clearly the intersection of any nonempty collection of linear manifolds is a linear manifold. It follows that, given a subset E of X there is a smallest linear manifold containing E, namely the intersection of all manifolds which contain E. The set [E] of all linear combinations of all finite subsets of E is clearly a vector space. It is easily verified that [E] is the smallest linear manifold containing E.

8.2 LINEAR TRANSFORMATIONS

Let X and Y be two vector spaces. A *linear transformation* of X into Y is a function A on X into Y such that $A(\alpha x + \beta x) = \alpha Ax + \beta Ay$ for all x and y in X and for all scalars α and β. Note that we often write Ax instead of A(x). In the special case in which Y is the field of scalars of the space X, A is called a *linear functional*. A *linear operator* is a linear transformation of a vector space into itself.

Let A be a linear transformation of a normed vector space X into a normed vector space Y. If there is a number M such that $\|Ax\| \leq M\|x\|$ for all $x \in X$, then A is said to be bounded. The infimum of all such values of M is defined to be the *norm* of A and is denoted by $\|A\|$.

2. Proposition *If A is a bounded linear transformation of a normed vector space X into a normed vector space Y, then*

$$\|A\| = sup\{\|Ax\|: x \in X, \|x\| = 1\}$$

$$= sup\{ \frac{\|Ax\|}{\|x\|} : x \in X, \|x\| \neq 0\}$$

and

$$\|Ax\| \leq \|A\| \cdot \|x\| \quad \textit{for all} \quad x \in X$$

Proof We first notice that since the sets $S_1 = \{\|Ax\|: x \in X, \|x\| = 1\}$, $S_2 = \{\|Ax\|/\|x\|: x \in X, \|x\| \neq 0\}$ of real numbers are equal, we have $\sup S_1 = \sup S_2 = a$. Next we want to prove that $\|A\| = a$. Let $\varepsilon > 0$ be given. Then since $a = \sup S_2$ we can find an element $x_1 \in X$ with $\|x_1\| \neq 0$ such that $\|Ax_1\|/\|x_1\| > a-\varepsilon$ or $\|Ax_1\| > (a-\varepsilon)\|x_1\|$. This implies that $a-\varepsilon < \|A\|$, and since ε is arbitrary we have $a \leq \|A\|$. We now prove that we cannot have $a < \|A\|$. For suppose on the contrary that this inequality holds and let $\|A\| - a = \varepsilon > 0$. Then $a < \|A\| - (\varepsilon/2)$ which implies that the inequality $\|Ax\|/\|x\| \leq a < \|A\| - (\varepsilon/2)$ holds for every $x \neq \theta$ and hence

$$\|Ax\| \leq (\|A\| - (\varepsilon/2))\|x\|$$

holds for every x. This means that $\|A\|$ is not the infimum of all those M for which $\|Ax\| \leq M\|x\|$. It follows from this contradiction that $\|A\| = a$.

One may also prove without difficulty that $\|A\| = \sup\{\|Ax\|: x \in X, \|x\| \leq 1\}$, and that $\|Ax\| \leq \|A\|\|x\|$ for all $x \in X$. The proofs are left to the reader.

3. Proposition *Let A be a linear transformation of a normed vector space X into a normed vector space Y. Then: (a) "A bounded" implies "A uniformly continuous on X." (b) "A continuous at one point of X" implies "A bounded."*

Proof Suppose A bounded, and let $\varepsilon > 0$ be given. If $\|A\| = 0$ then *(a)* is obviously true. So we may assume that $\|A\| > 0$, and choose $\delta = \varepsilon/\|A\|$. For any x and y in X with $\|x - y\| < \delta$ we have $\|Ax - Ay\| \leq \|A\|\|x - y\| < \varepsilon$. This proves *(a)*. To prove *(b)*, assume that A is continuous at $x_0 \in X$. Then there is a $\delta > 0$ such that $\|x - x_0\| < \delta$ implies $\|Ax - Ax_0\| < 1$. We want to prove that there is a positive M such that for every y in X we have $\|Ay\| \leq M\|y\|$. Choose any real number M greater than $1/\delta$, and let y be an arbitrary element of X. If $\|y\| = 0$ then $y = \theta$ so that the inequality $\|Ay\| \leq M\|y\|$ is obviously satisfied. Assume $\|y\| \neq 0$, and put $z = (1/M\|y\|)y + x_0$. Since $\|z - x_0\| = \|(1/M\|y\|)y\| = (1/M) < \delta$, we have $\|Az - Ax_0\| = \|A(z - x_0)\| = \|A(1/M\|y\|)y)\| = (1/M\|y\|)\|Ay\| < 1$ or $\|Ay\| < M\|y\|$. This proves *(b)*.

Let $\mathcal{B}$ be the space of all bounded linear transformations of a normed vector space X into a Banach space Y. We propose now to introduce operations in $\mathcal{B}$ in such a way that with the norm defined above, $\mathcal{B}$ becomes a Banach space. To do this we must define addition and scalar multiplication of the elements of $\mathcal{B}$.

Definition *Let $A \in \mathcal{B}$, $B \in \mathcal{B}$, $x \in X$ and $\lambda \in \mathbb{R}$. Then $A + B$ and λA are defined to be the transformations whose values at x are respectively:*

$$(A + B)x = Ax + Bx \quad \text{and} \quad (\lambda A)x = \lambda Ax.$$

It is easily seen that $A + B$ and λA are both

linear transformations. Furthermore we have

$$\|A+B\| = \sup_{\|x\|=1} \|(A+B)x\| = \sup_{\|x\|=1} \|Ax+Bx\|$$

$$\leq \sup_{\|x\|=1} \{\|Ax\|+\|Bx\|\} \leq \|A\|+\|B\|$$

and

$$\|\lambda A\| = \sup_{\|x\|=1} \|(\lambda A)x\| = \sup_{\|x\|=1} \|\lambda Ax\|$$

$$= |\lambda| \sup_{\|x\|=1} \|Ax\| = |\lambda|\,\|A\|.$$

This proves that $A + B$ and λA are bounded and that $\|\cdot\|$ satisfies the triangle inequality. If $\|A\| = 0$ then the inequality $\|Ax\| \leq \|A\|\|x\| = 0$ implies that A is the bounded linear transformation which maps every element of X onto the θ element of Y. Conversely, if A maps the space X onto the element θ of Y then $\|A\| = 0$. Thus $\|\cdot\|$ satisfies all the requirements of a norm on $\mathcal{B}$.

Having defined $+$, $\cdot$, $\|\cdot\|$ in the space $\mathcal{B}$, we state the following proposition.

4. Proposition $\mathcal{B}$ *is a Banach space.*

Proof That $\mathcal{B}$ is a normed vector space has just been established. All that remains to be proved is that $\mathcal{B}$ is complete. Let $\langle A_n \rangle$ be a Cauchy sequence in $\mathcal{B}$. Then for a given $\varepsilon > 0$ there is a natural number N such that $n,m \geq N$ imply $\|A_n - A_m\| < \varepsilon$. Let x be a fixed element of X. Then, for $n,m \geq N$

$$\|A_n x - A_m x\| = \|(A_n - A_m)x\| \le \|A_n - A_m\|\|x\| \le \varepsilon\|x\|$$

so that $\langle A_n x\rangle$ is a Cauchy sequence in Y. Since Y is complete, there is an element in Y, call it Ax, such that $\lim_{n\to\infty} A_n x = Ax$. This defines a transformation A from X to Y. It remains to prove that A is linear and bounded and that $\|A-A_n\| \to 0$. We have

$$A(x_1+x_2) = \lim_{n\to\infty} A_n(x_1+x_2) = \lim_{n\to\infty} (A_n x_1 + A_n x_2)$$

$$= \lim_{n\to\infty} A_n x_1 + \lim_{n\to\infty} A_n x_2 = Ax_1 + Ax_2$$

and

$$A(\lambda x) = \lim_{n\to\infty} A_n(\lambda x) = \lambda \lim_{n\to\infty} Ax = \lambda Ax$$

which proves that A is linear. To prove that A is bounded, let $\varepsilon > 0$ be given. Then there is an N such that for all $n,m \ge N$ we have $\|A_n - A_m\| < \varepsilon$. In particular for $n \ge N$ and $m = N$ we have $\|A_n\| \le \|A_N\| + \varepsilon$. From the inequalities

$$\|Ax\| \le \|Ax - A_n x\| + \|A_n x\| \text{ and } \|A_n x\| \le \|A_n x - Ax\| + \|Ax\|$$

it follows that

$$\big|\|Ax\| - \|A_n x\|\big| \le \|Ax - A_n x\|$$

which together with the inequality $\|A_n\| \leq \|A_N\| + \varepsilon$ implies that $\|Ax\| = \lim_{n\to\infty} \|A_n x\| \leq (\|A_N\|+\varepsilon)\|x\|$. Thus A is bounded. Finally for $n,m \geq N$ and $x \in X$ we have $\|A_n x - A_m x\| \leq \varepsilon\|x\|$. By letting $m \to \infty$ we get

$$\|(A_n - A)x\| = \|A_n x - Ax\| = \lim_{m\to\infty} \|A_n x - A_m x\| \leq \varepsilon\|x\|$$

or $\|A_n - A\| \leq \varepsilon$. Since ε is arbitrary it follows that $\|A_n - A\| \to 0$ so that $\mathcal{B}$ is complete.

Definition *The space of all bounded linear functionals on a normed vector space* X *is called the* dual *(or conjugate) space of* X, *and is denoted by* X^* *(see end of Chap. 6).*

The following corollary is an immediate consequence of Proposition 4, and the fact that $\mathbb{R}$ is a Banach space.

5. Corollary *The dual space* X^* *of any normed vector space* X *is a Banach space.*

8.3 LINEAR FUNCTIONALS--THE HAHN-BANACH THEOREM

Having the result of Corollary 5 we may consider the following important problem of analysis: *Given a normed vector space* X, *to construct explicitly the space* X^*. An example of a result along this line was given in Chapter 6 where we proved, using the Riesz representation theorem, that for $1 \leq p < +\infty$, the dual $(L^p)^*$ of the space L^p was isometrically isomorphic to the space L^q with $(1/p) + (1/q) = 1$. Another very important example, that we just state without proof, is the following: If X is the space $C[0,1]$ of all continuous functions on $[0,1]$, then X^* is "essentially" the space of all functions α of bounded variation where the linear functional F_α generated by α is defined by the Riemann-Stieltjes integral $F_\alpha f = \int_0^1 f(t)d\alpha(t) \quad (f \in C[0,1])$, the norm of α being equal to the total variation of α in $[0,1]$ (see Taylor [26] p. 397).

Let us now turn our attention to a particular "extension problem" solved by the Hahn-Banach theorem given below.

Generally speaking, in an extension problem one considers a mathematical object (for example, a function) defined in a subspace of a given space, and tries to extend this object to the entire space without losing certain properties. An example of an extension problem was given in Chapter 5, section 8.

The following lemma will be used in the proof of the Hahn-Banach theorem.

6. Lemma *Let T be a subspace of the vector space X. Let p be a real-valued function on X such that*

(i) $p(\lambda x) = \lambda p(x)$ if $\lambda \geq 0$ and $x \in X$.

(ii) $p(x + y) \leq p(x) + p(y)$ for all $x,y \in X$.

Let f be a linear functional on T such that $f(t) \leq p(t)$ for $t \in T$. Let y be an element of X not in T and let U be the subspace of X generated by y and T. Then there exists a linear functional h defined on U such that $h(t) \leq p(t)$ for all t in U, and $h(t) = f(t)$ for all t in T.

Proof Any element in U is of the form $\lambda y + t$ where λ is a scalar and $t \in T$. Therefore the linear functional h will be defined by $h(\lambda y + t) = \lambda h(y) + h(t) = \lambda h(y) + f(t)$, as soon as $h(y)$ is specified in such a way that

(*) $h(\lambda y+t) \leq p(\lambda y+t)$ for all elements $\lambda y+t$ of U.

For t_1 and t_2 arbitrary elements in T we have

$$f(t_1) + f(t_2) = f(t_1 + t_2)$$

$$\leq p(t_1 + t_2) \leq p(t_1 - y) + p(t_2 + y)$$

or

$$-p(t_1 - y) + f(t_1) \leq p(t_2 + y) - f(t_2) .$$

In the last inequality, if t_2 is kept fixed, the set of real numbers $-p(t - y) + f(t)$ that we get, for t running over T, has an upper bound, namely, $p(t_2 + y) - f(t_2)$. It follows that

$$\sup\{-p(t - y) + f(t): t \in T\} \leq p(t_2 + y) - f(t_2).$$

Since the last inequality is true for every $t_2 \in T$, we have:

$$\sup\{-p(t - y) + f(t): t \in T\} \leq$$

$$\leq \inf\{p(t + y) - f(t) : t \in T\}.$$

Define $h(y) = \alpha$, where α is a real number such that

$$\sup\{-p(t - y) + f(t) : t \in T\}$$

$$\leq \alpha \leq \inf\{p(t + y) - f(t) : t \in T\} .$$

The lemma will be proved if we show that for this choice of $h(y)$, (*) holds. For $\lambda = 0$, (*) is obvious. Let $\lambda > 0$. Then

$$h(\lambda y + t) = \lambda h(y) + f(t) = \lambda\alpha + f(t)$$

$$= \lambda\left[\alpha + \frac{1}{\lambda} f(t)\right] = \lambda\left[\alpha + f\left(\frac{t}{\lambda}\right)\right] .$$

From the way that α is defined, we have

$$a \leq p\left(\frac{t}{\lambda} + y\right) - f\left(\frac{t}{\lambda}\right)$$

so that

$$h(\lambda y + t) \leq \lambda\left[\left\{p\left(\frac{t}{\lambda} + y\right) - f\left(\frac{t}{\lambda}\right)\right\} + f\left(\frac{t}{\lambda}\right)\right]$$

$$= \lambda p\left(\frac{t}{\lambda} + y\right) = p(\lambda y + t).$$

Let $\lambda = -\mu < 0$. Then since

$$-\alpha \leq p\left(\frac{t}{\mu} - y\right) - f\left(\frac{t}{\mu}\right),$$

we have

$$h(\lambda y + t) = -\mu\alpha + f(t) = \mu\left(-\alpha + f\left(\frac{t}{\mu}\right)\right)$$

$$\leq \mu\left[\{p\left(\frac{t}{\mu} - y\right) - f\left(\frac{t}{\mu}\right)\} + f\left(\frac{t}{\mu}\right)\right]$$

$$= \mu p\left(\frac{t}{\mu} - y\right) = p(-\mu y + t) = p(\lambda y + t).$$

This proves (*).

7. The Hahn-Banach Theorem *Let* X, T, p, *and* f *be as in Lemma 6. Then there exists a linear functional* F *on* X *such that* $F(x) \leq p(x)$ *for all* x, *and* $f(t) = F(t)$ *for all* t *in* T.

Proof Consider the set $\mathcal{E}$ of all ordered pairs $\langle L,G\rangle$ where L is a linear manifold in X and G is a linear functional defined over L, such that $G(x) \leq p(x)$ for all $x \in L$, and $T \subset L$. On the set $\mathcal{E}$ we define the binary relation $\prec$ as follows. For two elements $\langle L,G\rangle$ and $\langle L',G'\rangle$ in $\mathcal{E}$ we write $\langle L,G\rangle \prec \langle L',G'\rangle$ if and only if $L \subset L'$ and $G(x) = G'(x)$ for all x in L. It is easy to check that $(\mathcal{E},\prec)$ is a partially ordered set, and that $\langle T,f\rangle \in \mathcal{E}$. Then by the Hausdorff maximal principle there is a maximal linearly ordered subset $\{\langle L_\alpha,G_\alpha\rangle\}$ of $\mathcal{E}$, where α runs over a set of indices. Let $U = \bigcup_\alpha L_\alpha$. Then U is also a linear manifold. For if x and y are in U then $x \in L_\alpha$ and

$y \in L_\beta$ for some α and β. The set $\{\langle L_\alpha, G_\alpha\rangle\}$ being linearly ordered we must have either $L_\alpha \subset L_\beta$ or $L_\beta \subset L_\alpha$. Suppose $L_\beta \subset L_\alpha$. Then $x \in L_\alpha$, $y \in L_\alpha$ so that $\lambda x + \mu y$ (λ and μ scalars) is in L_α or $\lambda x + \mu y \in U$. We define a functional F on U by setting $F(x) = G_\alpha(x)$ if $x \in L_\alpha$. Since $\{\langle L_\alpha, G_\alpha\rangle\}$ is linearly ordered, this definition does not depend on the choice of G_α (verify this). We have $F(\lambda x + \mu y) = G_\alpha(\lambda x + \mu y) = \lambda G_\alpha(x) + \mu G_\alpha(y) = \lambda F(x) + \mu F(y)$, which proves that F is linear. Since for every element $\langle L_\alpha, G_\alpha\rangle$ in $\{\langle L_\alpha, G_\alpha\rangle\}$ we have $\langle L_\alpha, G_\alpha\rangle \prec \langle U, F\rangle$, it follows that $\langle U, F\rangle \in \mathcal{E}$. Moreover $\langle U, F\rangle$ is a maximal element of $\mathcal{E}$. For if $\langle L, G\rangle$ is an element such that $\langle U, F\rangle \prec \langle L, G\rangle$, then $\langle L_\alpha, G_\alpha\rangle \prec \langle U, F\rangle \prec \langle L, G\rangle$ implies that $\langle L, G\rangle \in \{\langle L_\alpha, G_\alpha\rangle\}$ by the maximality of $\{\langle L_\alpha, G_\alpha\rangle\}$. Hence $\langle L, G\rangle \prec \langle U, F\rangle$, and so $G = F$. Since $\langle T, f\rangle \prec \langle U, F\rangle$, it follows that $f(t) = F(t)$ for all t in T. It remains to prove that the domain of F is X, i.e., $U = X$. Assume $U \neq X$, and let y be an element of X not in U. Let U' be the linear manifold generated by U and y. Then by Lemma 6 there is a linear functional F' defined on U' such that $F'(x) = F(x)$ for all $x \in U$, and $F'(x) \leq p(x)$ for all $x \in U'$. But this implies that $\langle U, F\rangle \prec \langle U', F'\rangle$ which contradicts the maximality of $\langle U, F\rangle$. The theorem is proved.

The Hahn-Banach theorem is also true for vector spaces over the field of complex numbers. Such vector spaces are called *complex-vector spaces*. A complex-vector space X may be considered to be a vector space over the reals (i.e., a real-vector space). This statement simply means that the set X has all the properties of a real-vector space if we restrict scalar multiplications to be by real scalars only. The following definitions of complex and real linear functionals will be used only in connection with the extension of the Hahn-Banach theorem to complex-vector spaces.

A complex function f on a complex vector space X is called a *complex-linear functional* iff $f(x + y) = f(x) + f(y)$ and $f(\lambda x) = \lambda f(x)$ for all x and y in X, and all complex numbers λ. A real-valued function f defined on a complex (or real) vector space X is called a *real-linear functional* iff the above two equalities hold for all real numbers λ. Theorem 7 is extended as follows.

8. Theorem *Let X be a complex vector space, T a linear subspace, p a real-valued function on X such that $p(x + y) \leq p(x) + p(y)$, and $p(\lambda x) = |\lambda| p(x)$, for all complex numbers λ. Let f be a complex-linear functional on T such that $|f(t)| \leq p(t)$ for all t in T. Then there is a complex-linear functional F defined on X such that $F(t) = f(t)$ for t in T and $|F(x)| \leq p(x)$ for all x in X.*

Proof If f_1 and f_2 are the real and imaginary parts of f so that $f(t) = f_1(t) + if_2(t)$ for every $t \in T$, then f_1 and f_2 are easily seen to be real-linear functionals on the subspace T considered as a real-vector space. Since $f_1(t) \leq$

$\leq |f(t)| \leq p(t)$ we have $f_1(t) \leq p(t)$ for all t in T. The equation $f(it) = if(t)$ together with $f(it) = f_1(it) + if_2(it)$, and $if(t) = i(f_1(t) + if_2(t)) = if_1(t) - f_2(t)$ shows that $f_2(t) = -f_1(it)$, so we can write $f(t) = f_1(t) - if_1(it)$. Now by Theorem 7, considering X and T as real-vector spaces, we can extend f_1 to a real-linear functional G on X, such that $G(x) \leq p(x)$ for all x in X. Define $F(x) = G(x) - iG(ix)$ for $x \in X$. Then F is the complex-linear functional that we are looking for. In fact for x and y in X and the complex number $\lambda = a + ib$ (a,b reals) we have

$$F(x+y) = G(x+y) - iG(i(x+y))$$

$$= G(x) + G(y) - iG(ix) - iG(iy) = F(x) + F(y).$$

$$F(\lambda x) = F(ax+ibx) = G(ax+ibx) - iG(i(ax+ibx))$$

$$= G(ax+ibx) - iG(iax-bx)$$

$$= aG(x) + bG(ix) - i(aG(ix) - bG(x))$$

$$= a(G(x) - iG(ix)) + ib(G(x) - iG(ix))$$

$$= (a+ib)F(x) = \lambda F(x).$$

This proves that F is a complex-linear functional on X. Also for $t \in T$ we have

$$F(t) = G(t) - iG(it) = f_1(t) - if_1(it) = f(t)$$

and

$$|F(t)| = |f(t)| \leq p(t) .$$

It remains to prove that $|F(x)| \leq p(x)$ for all $x \in X$. Let a be the argument of the complex number $F(x)$. We have $F(x) = |F(x)|e^{ia}$ or $|F(x)| = F(x)e^{-ia} = F(xe^{-ia}) = G(xe^{-ia}) - iG(ixe^{-ia})$. Hence, the number $G(xe^{-ia}) - iG(ixe^{-ia})$ being real we must have $G(ixe^{-ia}) = 0$ so that $|F(x)| = G(xe^{-ia}) \leq p(xe^{-ia}) = |e^{-ia}|p(x) = p(x)$. The theorem is proved.

We now give some applications of the Hahn-Banach theorem.

9. Proposition *Let X be a normed vector space and $a \in X$ with $a \neq \theta$. Then there is a bounded linear functional F on X such that $F(a) = \|a\|$ and $\|F\| = 1$.*

Proof Let $S = \{x \in X: x = \lambda a; \lambda \in \mathbb{R}\}$. Then S is a subspace of X containing the element a. Define the linear functional f on S by $f(x) = f(\lambda a) = \lambda\|a\|$, and set $p(y) = \|y\|$ for all y in X. Then X, S, p, and f satisfy the hypothesis of the Hahn-Banach theorem. Therefore there is a linear functional F on X such that $F(x) = f(x)$ for all $x \in S$ and $F(y) \leq \|y\|$ for all y in X. Since $-F(y) = F(-y) \leq \|-y\| = \|y\|$, it follows that $|F(y)| \leq \|y\|$ for all $y \in X$, which means that F is a bounded linear functional on X with $\|F\| \leq 1$. For $y = a$ we have, by definition of f, $F(a) = f(a) = \|a\|$. Thus $\|F\| = 1$.

10. Proposition *Let M be a subspace of the normed vector space X, and y an element of X*

not in M. Suppose that $\|y - t\| \geq \delta$ *for all* $t \in M$. *Then there is a bounded linear functional F on X such that* $\|F\| \leq 1$, $F(y) = \delta$ *and* $F(t) = 0$ *for all* $t \in M$.

Proof Consider the subspace S spanned by M and y. Then any element $s \in S$ is of the form $s = \lambda y + t$ where $t \in M$ and $\lambda \in \mathbb{R}$. This representation of s is unique. Indeed, if $s = \lambda' y + t'$ were another representation of s, then $\lambda y + t = \lambda' y + t'$, or $(\lambda - \lambda')y = t' - t$. Since $y \notin M$ we must have $\lambda = \lambda'$, which implies $t = t'$. Because of the uniqueness of this representation, the functional f defined by $f(s) = \lambda\delta$ is a well-defined mapping from S into the reals. Clearly f is a linear functional and has the properties that $f(t) = 0$ for all $t \in M$, and $f(y) = \delta$. Also for $\lambda \neq 0$, since $(-t/\lambda) \in M$ and $\|\lambda y + t\| = |\lambda| \|y - (-\frac{t}{\lambda})\| \geq |\lambda|\delta$ we have $|\lambda|\delta = |f(s)| \leq \|s\|$. This inequality being also true for $\lambda = 0$, it follows that f is bounded on S with $\|f\| \leq 1$. By the Hahn-Banach theorem we may extend f to a bounded linear functional F on X. Clearly F has the required properties.

8.4 THE NATURAL ISOMORPHISM--REFLEXIVE SPACES

We shall now show that there is a "natural" manner in which a normed linear space X may be considered to be a subspace of X**, where X** = (X*)*. Let x be a fixed element of X. Consider the functional x**, defined on X* by

$$x^{**}(f) = f(x) \ , \quad f \in X^*.$$

Clearly x^{**} is linear. Since $|x^{**}(f)| = |f(x)| \leq \|f\|\|x\|$ x^{**} is also bounded, with $\|x^{**}\| \leq \|x\|$, so that $x^{**} \in X^{**}$. By Proposition 9 there is $f \in X^*$ of norm 1, such that $f(x) = \|x\|$. Hence $\|x^{**}\| = \|x\|$. Consider the mapping $\phi: X \to X^{**}$ defined by $\phi(x) = x^{**}$. Since ϕ is clearly linear and isometric (hence one-to-one), ϕ is an isometric isomorphism of X onto some linear subspace $\phi[X]$ of X^{**}. By Corollary 5, the space X^{**} is a Banach space so that $\overline{\phi[X]}$ being a closed subset of a complete space is itself a Banach space. In other words X is isometrically isomorphic to a dense subset of a Banach space. The mapping ϕ is called the *natural isomorphism* of X into X^{**}.

Definition *If the image of* X *under the above natural isomorphism is all of* X^{**}, *then the space* X *is called* reflexive.

Reflexive spaces are important in analysis.

In Chapter 6 we proved that for $1 < p < \infty$ the dual space of L^p is isometrically isomorphic to the space L^q, with $(1/p) + (1/q) = 1$. Hence for $1 < p < +\infty$, L^p is isometrically isomorphic to $(L^p)^{**}$, which means that L^p is reflexive. Another example of a reflexive space is the $\mathbb{R}^n$ Euclidean space. We leave the proof to the reader. The spaces L^1 and L^∞ are not reflexive. To see that L^1 is not reflexive we essentially need to prove that there are bounded linear functionals on L^∞ which are not generated by functions in L^1. In other words, the Riesz representation theorem is not true for $p = \infty$.

In fact, let $C[0,1]$ be the normed vector space of all continuous functions on $[0,1]$ with $\|x\| = \sup_{t \in [0,1]} |x(t)|$. Then $C[0,1]$ is a subspace of L^∞. Let f be the functional on $C[0,1]$ defined by $f(x) = x(0)$. Clearly f is linear. Since $|f(x)| = |x(0)| \leq 1 \cdot \|x\|$, f is also bounded with $\|f\| \leq 1$. By the Hahn-Banach theorem, f can be extended to a bounded linear functional F on L^∞. Now there is no $g \in L^1$ such that $F(x) = \int_0^1 x(t)g(t)dt$ for all $x \in C(0,1]$. For suppose that such a function g exists. Consider the sequence $\langle x_n \rangle$ in $C[0,1]$ defined by $x_n(t) = 1 - nt$ if $0 \leq t \leq (1/n)$, and $x_n(t) = 0$ if $(1/n) < t \leq 1$. We have $\lim_{n\to\infty} x_n(t) = 0$ for all $t \neq 0$, and by the Lebesgue convergence theorem we get $\lim_{n\to\infty} \int_0^1 x_n(t)g(t)dt = 0$ which is impossible, since for all n, $F(x_n) = x_n(0) = 1$ so that $\lim_{n\to\infty} F(x_n) = 1$. The proof that L^∞ is not reflexive is left to the reader (see Prob. 27).

8.5 THE BOUNDEDNESS OF THE INVERSE TRANSFORMATION

There are some three critical theorems in the theory of Banach spaces which are needed in order to reach a successful maturity. The first such theorem is the Hahn-Banach theorem. The second one, Theorem 13, to be established in this section, concerns the boundedness of the inverse transformation. The third one is Theorem 15, and expresses the so-called "uniform boundedness principle." Here, it is essential that the spaces in question be Banach spaces. The fact that a complete metric space is of second category is used in the proof.

Let A be a one-to-one bounded linear transformation from a Banach space X onto a Banach space Y. Then it is clear that the inverse mapping $A^{-1} : Y \to X$ exists, and is a linear transformation. We shall prove that A^{-1} is also bounded. We first prove the following lemma.

11. Lemma *Let A be a bounded linear transformation of a Banach space X onto a Banach space Y. Then the image of the unit open ball centered on the origin in X contains an open ball centered on the origin in Y.*

Proof Let $S_r = \{x \varepsilon X : \|x\| < r\}$ and $S_r' = \{y \varepsilon Y : \|y\| < r\}$, be open balls with radius r centered on the origin in X and Y respectively. We begin by proving that $\overline{A(S_1)}$ contains some S_r'. Since the mapping A is onto, we have $Y = \bigcup_{n=1}^{\infty} A(S_n)$. Y is a complete metric space, and so it is of second category. Consequently some of the $A(S_n)$, say $A(S_{n_0})$, cannot be nowhere dense, hence $\overline{A(S_{n_0})}$ contains some open ball, say $\{y : \|y-p\| < \eta\}$. But the mapping $y \to y-p$ being a homeomorphism of Y onto itself, we have that $\overline{A(S_{n_0})} - p$ contains the open ball S_η'. Since $\overline{A(S_{n_0})} - p \subset \overline{A(S_{n_0})} -$

$- \overline{A(S_{n_0})} = \overline{A(S_{2n_0})}$ it follows that $S_\eta{}' \subset \overline{A(S_{2n_0})}$.

Also $\overline{A(S_{2n_0})} = \overline{2n_0 A(S_1)} = 2n_0 \overline{A(S_1)}$, which implies that $S_\varepsilon{}' \subset \overline{A(S_1)}$, where $\varepsilon = (\eta/2n_0)$. Next we prove that $S_\varepsilon{}' \subset A(S_3)$. Let $y \in S_\varepsilon{}'$. Since $y \in \overline{A(S_1)}$, there is a $y_1 \in A(S_1)$ such that $\|y-y_1\| < (\varepsilon/2)$, with $y_1 = Ax_1$ for some $x_1 \in S_1$. We next observe that since $S_\varepsilon{}' \subset \overline{A(S_1)}$ we have $S'_{(\varepsilon/2)} \subset \overline{A(S_{(1/2)})}$, so there is a $y_2 \in A(S_{(1/2)})$ such that $\|(y-y_1) - y_2\| < (\varepsilon/4)$ with $y_2 = Ax_2$ for some $x_2 \in S_{(1/2)}$. Continuing in this way, we obtain by induction a sequence $\langle x_n \rangle$ in X such that $x_n \in S_{1/2^{n-1}}$, $n = 1,2,\ldots$, and $\|y - (y_1+y_2+\ldots+y_n)\| < (\varepsilon/2^n)$, with $y_n = Ax_n$. We have $\|x_n\| < (1/2^{n-1})$, and $\sum_{n=1}^{\infty} \|x_n\| < +\infty$. Since X is complete, it follows from Proposition 1, that $\sum_{n=1}^{\infty} x_n$ converges to some $x \in X$. But $\left\|\sum_{k=1}^{n} x_k\right\|$ $\leq \sum_{k=1}^{n} \|x_k\| \leq \sum_{k=1}^{n} (1/2^{k-1}) < 2$, and $\|x\| =$ $\leq \lim_{n\to\infty} \left\| \sum_{k=1}^{n} x_k \right\| \leq 2 < 3$, which means that $x \in S_3$. Since A is continuous, we have $A(x) =$ $A\left(\lim_{n\to\infty} \sum_{k=1}^{n} x_k\right) = \lim_{k\to\infty} A\left(\sum_{k=1}^{n} x_k\right) =$

$\lim_{n\to\infty} (y_1+y_2+\ldots+y_n) = y$, so that $y \in A(S_3)$. This proves that $S_\varepsilon' \subset A(S_3)$, which is equivalent to $S_{(\varepsilon/3)} \subset A(S_1)$. The lemma is proved.

12. Theorem (Open Mapping Theorem) *If X and Y are Banach spaces, and if A is a bounded linear transformation of X onto Y, then the image of any open set in X by A is an open set in Y.*

Proof Let O be any open set in X, and y any point in A[O]. Then there is $p \in O$ such that $y = A(p)$. Since O is an open set there is an open ball S centered on p and contained in O. Since the mapping $x \to x-p$ is a homeomorphism of X onto X, we have that $S - p$ is an open ball centered on the origin in X. It follows by Lemma 11 that $A(S - p)$ contains an open ball, say S', centered on the origin in Y. Clearly $y + S'$ is an open ball centered on y. Furthermore $y + S'$ is an open subset of A[O] since $y + S' \subset y + A(S - p) = A(p) + A(S - p) = A(S) \subset A[O]$. Therefore A[O] is an open set in Y.

13. Theorem *Let A be a one-to-one bounded linear transformation of a Banach space X onto a Banach space Y. Then the inverse transformation A^{-1} from Y onto X is linear and bounded.*

Proof To show that the mapping $A^{-1} : Y \to X$ is continuous we only need to prove that the inverse image of any open set O in X is an open set in Y. But this follows immediately from Theorem 12, since the inverse image of O is A[O] which is an open set in Y. Hence A^{-1} is continuous therefore bounded.

8.6 THE CLOSED GRAPH THEOREM--THE UNIFORM BOUNDEDNESS PRINCIPLE

Let X and Y be vector spaces. Then it is easy to see that the Cartesian product $X \times Y$ is also a vector space if the addition of elements of $X \times Y$ and scalar multiplication are defined by

$$\langle x_1,y_1\rangle + \langle x_2,y_2\rangle = \langle x_1+x_2,\ y_1+y_2\rangle\ ,$$

$$\lambda\langle x,y\rangle = \langle \lambda x,\ \lambda y\rangle.$$

The Cartesian product $X \times Y$ viewed in this way as a vector space is sometimes denoted by $X \oplus Y$. In Chapter 5, section 1, it was pointed out that if $\langle X,d\rangle$ and $\langle Y,\sigma\rangle$ are two metric spaces, then $X \times Y$ is a metric space with the metric ρ defined by

$$\rho(\langle x_1,y_1\rangle,\ \langle x_2,y_2\rangle) = (d^2(x_1,x_2) + \sigma^2(y_1,y_2))^{\frac{1}{2}}\ .$$

It follows that if X and Y are normed vector spaces then $X \times Y$ is also a normed vector space if we take

$$\|\langle x,y\rangle\| = (\|x\|^2 + \|y\|^2)^{\frac{1}{2}}\ .$$

Two other norms we might equally well have taken are

$$\|\langle x,y\rangle\| = \|x\| + \|y\|$$

and

$$\|\langle x,y\rangle\| = \max(\|x\|,\|y\|) \ .$$

All three norms are equivalent on $X \times Y$. The proof is left to the reader (see Prob. 14).

Now let A be a linear transformation of a Banach space X into a Banach space Y. Let $G =$ $\{\langle x,Ax\rangle : x \varepsilon X\}$ be the graph of A. We prove that G_A is a closed subset of $X \oplus Y$ if and only if A has the property that, whenever $\langle x_n\rangle$ is a sequence in X that converges to some point x, and such that $\langle Ax_n\rangle$ converges in Y to a point y, then $y = Ax$.

In fact, suppose that A has the above property and let $\langle x,y\rangle$ be a point of closure of G_A. Then there must exist a sequence of points of G_A, $\langle x_n,Ax_n\rangle$, converging to $\langle x,y\rangle$. This is equivalent to saying that

$$\|x_n-x,\ Ax_n-y\| \to 0 \quad \text{or} \quad \|x_n-x\| + \|Ax_n-y\| \to 0 \ ,$$

which implies that

$$x_n \to x \quad \text{and} \quad Ax_n \to y \ ,$$

which implies that $y = Ax$, or $\langle x,y\rangle = \langle x,Ax\rangle \ \varepsilon \ G_A$. Therefore G_A is closed.

Conversely, suppose G_A is closed and that $x_n \to x$ and $Ax_n \to y$. We must show that $y = Ax$. We have

$$\|\langle x_n, Ax_n\rangle - \langle x,y\rangle\| = \|x_n - x\| + \|Ax_n - y\| \to 0$$

hence

$$\langle x_n, Ax_n\rangle \to \langle x,y\rangle \in \overline{G}_A .$$

Since G_A is closed, $G_A = \overline{G}_A$ so that $\langle x,y\rangle \in G_A$. By the definition of G_A this means that $y = Ax$.

14. Theorem (Closed Graph Theorem) *Let A be a linear transformation on a Banach space X into a Banach space Y. Suppose that the graph G_A of A, is closed in $X \oplus Y$. Then A is bounded.*

Proof For $x \in X$ put $|||x||| = \|x\| + \|Ax\|$. It is clear that $|||\cdot|||$ defines a new norm on X. We first prove that X is a Banach space with respect to the norm $|||\cdot|||$.

Let $\langle x_n\rangle$ be a Cauchy sequence in $\langle X, |||\cdot|||\rangle$. we have

$$|||x_p - x_q||| = \|x_p - x_q\| + \|A(x_p - x_q)\|$$

$$= \|x_p - x_q\| + \|Ax_p - Ax_q\| \to 0 .$$

It follows that $\|x_p - x_q\| \to 0$ and $\|Ax_p - Ax_q\| \to 0$. Hence $\langle x_n\rangle$ and $\langle Ax_n\rangle$ are Cauchy sequences in $\langle X, \|\cdot\|\rangle$ and Y, respectively. Therefore, there are points $x \in X$ and $y \in Y$ such that $\|x_n - x\| \to 0$ and $\|Ax_n - y\| \to 0$. Since G_A is closed, it follows that $y = Ax$ so that

$$|||x_n - x||| = ||x_n - x|| + ||Ax_n - Ax|| \to 0$$

and $\langle X, |||\cdot||| \rangle$ is complete.

The identity map of $\langle X, |||\cdot||| \rangle$ onto $\langle X, ||\cdot|| \rangle$ is a one-to-one bounded linear transformation and so, by Theorem 13, the inverse map from $\langle X, ||\cdot|| \rangle$ onto $\langle X, |||\cdot||| \rangle$ must be bounded. This means that there is a constant M such that, $|||x||| \leq M||x||$ for all $x \in X$. Thus $||x|| + ||Ax|| \leq M||x||$ or $||Ax|| \leq M||x||$ and A is bounded.

We shall now prove the third of the three critical theorems for Banach spaces mentioned in Section 5.8.

15. Theorem (Uniform Boundedness Principle) *Let X be a Banach space and Y be a normed linear space. Let $\mathcal{F}$ be a nonempty set of bounded linear transformations from X into Y. Suppose that for each $x \in X$ there is a constant M_x such that $||Ax|| \leq M_x$ for all $A \in \mathcal{F}$. Then there is a constant M such that $||A|| \leq M$ for all $A \in \mathcal{F}$.*

The theorem is sometimes referred to as the *Banach-Steinhaus Theorem.*

Proof For each $A \in \mathcal{F}$, define the function $f_A: X \to \mathbb{R}$ by $f_A(x) = ||Ax||$. The collection $\{f_A\}$ is a collection of real-valued functions, bounded at each point x of the complete metric space X. It follows from Theorem 28, Chapter 5, that there is an open set O in X, and a constant M' such that for $x \in O$ and all $A \in \mathcal{F}$ we have $f_A(x) = ||Ax|| \leq M'$. Let $y \in O$. Since O is open there is an open ball $S = \{x: ||x-y|| < \delta\}$ such that $S \subset O$. Let

$S_1 = \{x: \|x-y\| \leq \delta_1\}$ with $0 < \delta_1 < \delta$. Then clearly $S_1 \subset S \subset O$. Put $M'/\delta_1 = M$. If z is an arbitrary nonzero point in X, we have $\|(\delta_1/\|z\|)z\| = \delta_1$, so that $(\delta_1/\|z\|)z \in O$. Hence $\|A((\delta_1/\|z\|)z)\| \leq M'$ or $\|A(z)\| \leq M\|z\|$. Thus $\|A\| \leq M$ for all A in $\mathcal{F}$.

8.7 OTHER TOPOLOGIES

We have already pointed out that if in a normed vector space X the distance between two elements x and y is defined to be $d(x,y) = \|x-y\|$, then X becomes a metric space. The topology induced by this metric is called the *norm topology* (or the *strong topology*) of X.

In addition to the *norm topology* other topologies can be defined on a vector space X. Let $\mathcal{F}$ be a collection of linear functionals on X. Then the weakest topology on X such that each f in $\mathcal{F}$ is continuous, is called the *weak topology associated with the collection of linear functionals* $\mathcal{F}$. We can view this topology as follows: If each f in $\mathcal{F}$ has to be continuous, we must have $f^{-1}[O]$ open for every open set O of real numbers. Thus the weak topology associated with the collection $\mathcal{F}$ is the weakest of all topologies in X containing the collection S of all sets $f^{-1}[O]$ where $f \in \mathcal{F}$ and O is an open set of real numbers. We remind the reader (see Chap. 7, sec. 6) that another way to characterize this unique weakest topology containing S is by taking the collection of all arbitrary unions of finite intersections of sets in S. In other words the weak topology associated with $\mathcal{F}$ is the topology that we obtain if we take S as a subbasis. We observe that instead of all open sets of real numbers we can take the topology generated by all $f^{-1}[B]$ where B runs over a basis of the topology of the real numbers. For instance, B may be any open interval of length 2ε, with $\varepsilon > 0$. It follows that a subbasis for the weak topology associated with $\mathcal{F}$ consists of all sets of the form $\{x \in X: |f(x) - f(x_o)| < \varepsilon, f \in \mathcal{F}\}$. A finite intersection of sets of this sort, with a fixed $x_o \in X$ and ε, is denoted by $V(x_o; f_1, f_2, \ldots, f_n, \varepsilon)$. We see that a basis for the weak topology on X, associated with $\mathcal{F}$, consists of all sets of the form $V(x_o; f_1, f_2, \ldots, f_n, \varepsilon)$ where $x_o \in X$, $\varepsilon > 0$, and the $f_i \in \mathcal{F}$.

Suppose now that X is a normed vector space and X* the set of all bounded linear functionals on X. Denote by $\mathcal{O}_w$ the collection of all open sets in the weak topology associated with X* (called the *weak topology* of X), and let $\mathcal{O}$ be the collection of all open sets in the norm topology of X. We prove that the weak topology is coarser than the norm topology. We also say that the norm topology is finer than the weak topology. This means that any set that is open in the weak topology is open in the norm topology. In fact since each $f \in X^*$ is continuous in the norm topology $\mathcal{O}$, each $V(x_o; f_1, f_2, \ldots, f_n, \varepsilon) \in \mathcal{O}$ hence $\mathcal{O}_w \subset \mathcal{O}$.

We shall now show that the topological space $(X, \mathcal{O}_w)$ is a Hausdorff space. Let $x_1, x_2 \in X$,

$x_1 \neq x_2$. By Proposition 9, since $x_1 - x_2 \neq \theta$, there is an $f \in X^*$ such that $f(x_1 - x_2) = \|x_1 - x_2\| \neq 0$. Let us write $|f(x_1) - f(x_2)| = 2\varepsilon$. Then $V(x_1; f,\varepsilon)$ and $V(x_2; f,\varepsilon)$ are neighborhoods of x_1 and x_2 and $V(x_1; f,\varepsilon) \cap V(x_2; f,\varepsilon) = \emptyset$. This proves that $(X, \mathcal{O}_w)$ is a Hausdorff space.

All topological notions pertaining to the space $(X, \mathcal{O}_w)$ will be preceded by the word "weak" or "weakly" to distinguish them from the corresponding notions in the space $(X, \mathcal{O})$. For example, compactness, closure, convergence, open set, and the like are referred to as weak compactness, weak closure, weak convergence, weakly open set, etc., respectively. Also for the sake of emphasis we shall sometimes preface topological properties with the word "strong," for example, strong closure, strongly convergent sequence, etc. According to this terminology a sequence $\langle x_n \rangle$ of elements in $\langle X, \mathcal{O} \rangle$ is said to *converge weakly* to an element x of $\langle X, \mathcal{O} \rangle$ iff $\langle x_n \rangle$ converges to x in the space $\langle X, \mathcal{O}_w \rangle$, which means that every O in $\mathcal{O}_w$ that contains x contains all but a finite number of the x_n. We notice that if a set E is weakly closed, then E is also strongly closed. Indeed, if $\bar{E}^w$ is the weak closure of E, then $E = \bar{E}^w$ and $\complement E \in \mathcal{O}_w$. Since $\mathcal{O}_w \subset \mathcal{O}$, it follows that $\complement E \in \mathcal{O}$ which implies that E is a closed set in $\langle X, \mathcal{O} \rangle$, that is, E is strongly closed. The converse of this statement is not true, however, it is true in the case of a linear manifold as is shown in the following theorem.

16. Theorem *A linear manifold M is weakly closed if and only if it is strongly closed.*

Proof Suppose M is strongly closed and $x \notin M$. This implies that $\inf_{s \in M} \|x-s\| \geq \delta > 0$. By Proposition 10 there is a bounded linear functional F which vanishes on M and does not vanish at x. Now the set $\{y: F(y) \neq 0\}$ is an open set in the weak topology, that is $\{y: F(y) \neq 0\} \in \mathcal{O}_w$. We have $x \in \{y: F(y) \neq 0\}$ and $\{y: F(y) \neq 0\} \cap M = \emptyset$. Hence x is not a point of closure of M in the weak topology. The converse implication is a consequence of the preceding paragraph. The theorem is proved.

We would like to direct the reader's attention to the following statement. Some authors define weak convergence without any reference to the topological space $(X, \mathcal{O}_w)$, as follows: The sequence $\langle x_n \rangle$ from the normed linear space X is said to *converge weakly* to $x_o \in X$, written $x_n \overset{w}{\to} x_o$, if and only if for every $f \in X^*$, $f(x_n) \to f(x_o)$.

We shall now show that this notion of weak convergence coincides with the notion of weak convergence with respect to the topology of $\langle X, \mathcal{O}_w \rangle$. Suppose $x_n \to x$ in the topology $\mathcal{O}_w$. This means that all of the x_n, except for a finite number, belong to $V(x_o; f, \varepsilon)$, or $|f(x_n) - f(x_o)| < \varepsilon$ $(f \in X^*, \varepsilon > 0)$ for all n sufficiently large. But this means simply that $\langle x_n \rangle$ converges weakly in the sense $x_n \overset{w}{\to} x_o$

defined above. On the other hand, assume that $x_n \overset{w}{\to} x_o$ and let O be any weakly open set in the family $\mathcal{O}_w$ containing x_o. Since the sets of the form

$$V(x_o\ ;\ f_1,f_2,\ldots,f_n,\varepsilon)$$

$$= \{x : |f_i(x) - f_i(x_o)| < \varepsilon,\ i = 1,2,\ldots,n\}$$

form a basis for $\mathcal{O}_w$, there must exist $f_1,f_2,\ldots,f_m \in X^*$ such that

$$V(x_o;\ f_1,f_2,\ldots,f_p,\varepsilon) \subset O.$$

But $x_n \overset{w}{\to} x_o$ implies that

$$|f_i(x_n) - f_i(x_o)| < \varepsilon \qquad (i = 1,2,\ldots,p)$$

for sufficiently large n. This means that $x_n \in V(x_o;\ f_1,f_2,\ldots,f_m,\varepsilon)$ for all sufficiently large n. It follows that $\langle x_n\rangle$ converges to x_o with respect to the topology $\mathcal{O}_w$. This proves that the two definitions of weak convergence are equivalent.

Now let X be a normed vector space, and let X^* be the dual space of X. We know by Corollary 5 that X^* is a Banach space. Let ϕ be the natural isomorphism of X into X^{**}, and consider the subset $\phi[X]$ of X^{**}. Then the weak topology associated with the collection $\phi[X]$ of linear functionals on X^*, is called the *weak* topology* on X^*. We see that the weak* topology is not in general the same as the weak topology in X^*, because instead of all X^{**} we considered only the subset $\phi[X]$ of X^{**}. This implies (the proof is left to the reader) that the weak* topology on X^* is weaker than the weak topology on X^*. If X is reflexive then $\phi[X] = X^{**}$, and the weak* topology and the weak topology on X^* will of course coincide.

We recall that by the definition of the mapping ϕ we have for $x \varepsilon X$, and $f \varepsilon X^*$, $\phi_x(f) = f(x)$. Then a typical basis set for the weak* topology is of the form

$$V(f_o; \phi_{x_1}, \phi_{x_2}, \ldots, \phi_{x_n}, \varepsilon)$$

$$= \{f \varepsilon X^*: |\phi_{x_i}(f) - \phi_{x_i}(f_o)| < \varepsilon, i = 1,2,\ldots,n\}$$

$$= \{f \varepsilon X^*: |f(x_i) - f_o(x_i)| < \varepsilon, i = 1,2,\ldots,n\}.$$

We now prove an important theorem which displays an essential feature of the weak topology.

17. Theorem (Alaoglu) *The unit sphere* $S = \{f: \|f\| \leq 1\}$ *of* X^* *is compact in the weak* topology.*

Proof The proof is based on Tychonoff's theorem. If $x \varepsilon X$ consider the closed interval $I_x = [-\|x\|, \|x\|]$ with the induced topology of the real line. Since the space I_x is compact, by Tychonoff's theorem (see Chap. 7, Theorem 24), the Cartesian product $E = \prod_{x \varepsilon X} I_x$ is also compact with respect to the Tychonoff topology. The reader is reminded that E consists of all functions $g : X \to \bigcup_{x \varepsilon X} I_x$ such that $g(x) \varepsilon I_x$, and that the Tychonoff topology for E is the weakest topology for which all the projection mappings are continuous. A typical basis set in this topology is of the form

$$V(g_0;\ x_1, x_2, \ldots, x_n, \varepsilon)$$

$$= \{g \ \varepsilon\ E\colon |g(x_i) - g(x_0)| < \varepsilon,\ i = 1,2,\ldots,n\}$$

where $g_0 \ \varepsilon\ E$, $\varepsilon > 0$, for here $pr_x(g) = g(x)$ (see Chap. 7, sec. 6).

Now for any $f \ \varepsilon\ S$ we have $|f(x)| \leq \|f\| \|x\| \leq \|x\|$, that is $|f(x)| \ \varepsilon\ I_x$. This implies that $S \subset E$. S as a subset of E has the induced topology by the Tychonoff topology of E. A typical basis set in this induced topology is

$$V(g_0;\ x_1, x_2, \ldots, x_n, \varepsilon) \cap S$$

$$= \{g \ \varepsilon\ S\colon |g(x_i) - g_0(x_i)| < \varepsilon,\ i = 1,\ldots,n\}.$$

On the other hand, S as a subset of X^* inherits the topology induced by the weak* topology on X^*, where, as we have seen above, a typical basis set is

$\{f \in X^*: |f(x_i) - (x_o)| < \varepsilon,\ i = 1,2,\ldots,n\}$, which is the same as a typical basis set in the Tychonoff topology. This implies that the two induced topologies on S coincide. To finish the proof it suffices to prove that S is a closed subset of E in the Tychonoff topology. Since S, as a closed subset of the compact space E, is also compact, and since the two induced topologies on S coincide, S is also compact in the weak* topology. To prove that S is closed in E let $g_o \in \overline{S}$, and consider the open set $V(g_o;\ x, y,\ \alpha x + \beta y,\ \varepsilon)$ where $\varepsilon > 0$, $x, y \in X$ and α, β are real numbers. We have

$$V(g_o;\ x, y,\ \alpha x + \beta y,\ \varepsilon) \cap S \neq \emptyset .$$

Let $f \in V(g_o;\ x, y,\ \alpha x + \beta y,\ \varepsilon) \cap S$. We have

$$|g_o(x) - f(x)| < \varepsilon, \quad |g_o(y) - f(y)| < \varepsilon,$$

and

$$|g_o(\alpha x + \beta y) - f(\alpha x + \beta y)| < \varepsilon.$$

Thus

$$|g_o(\alpha x + \beta y) - \alpha g_o(x) - \beta g_o(y)|$$

$$\leq |g_o(\alpha x + \beta y) - f(\alpha x + \beta y)| +$$

$$+ |f(\alpha x + \beta y) - \alpha g_o(x) - \beta g_o(y)|$$

$$= |g_o(\alpha x + \beta y) - f(\alpha x + \beta y)|$$

$$+ |\alpha f(x) + \beta f(y) - \alpha g_o(x) - \alpha g_o(y)|$$

$$\leq |g_o(\alpha x + \beta y) - f(\alpha x + \beta y)| + |\alpha||f(x) - g_o(x)|$$

$$+ |\beta||f(y) - g_o(y)| \leq 3\varepsilon .$$

Since ε is an arbitrary positive number it follows that

$$g_o(\alpha x + \beta y) = \alpha g_o(x) + \beta g_o(y).$$

Similarly for any $x \in X$, there exists an $f_x \in V(g_o;x,\varepsilon) \cap S$, so that $|g_o(x) - f_x(x)| < \varepsilon$ and $|g_o(x)| \leq |f_x(x)| + \varepsilon \leq \|f_x\|\|x\| + \varepsilon \leq \|x\| + \varepsilon$.

Since ε is arbitrary we have $|g_o(x)| \leq \|x\|$. Thus g_o is a bounded linear functional with norm less than or equal to 1. This proves that $g_o \in S$ and S is therefore closed. The theorem is proved.

PROBLEMS

1. Show that the spaces given in Examples 1,2, and 3 of section 8.1, are vector spaces (see p. 388).
2. Show that the vector space of all polynomials with real coefficients and of degree less than n is isomorphic to $\mathbb{R}^n$.

3. Let X_1 be the vector space of all continuous functions on $[0,1]$ and X_2 the vector space of all continuous functions f on $[0,1]$ such that $f(0) = 0$, and whose first derivative exists and is continuous on $[0,1]$. Show that X_1 and X_2 are isometric.
4. If X is a normed vector space, show that the function $d: X \times X \to \mathbb{R}$ defined by $d(x,y) = \|x-y\|$ is a metric on X.
5. Show that the spaces given immediately after the definition of Banach space are Banach spaces.
6. Let X be a normed vector space. Show that:

   (a) The mapping $\langle x,y \rangle \to x + y$ is uniformly continuous on $X \times X$.

   (b) The mapping $\langle \lambda,x \rangle \to \lambda x$ is continuous on $\mathbb{R} \times X$ (λ = a scalar).

   (c) The mapping $x \to \lambda x$ is uniformly continuous on X, (λ = a scalar).
7. Let X be a normed vector space, a be any fixed element of X, and λ be any nonzero fixed scalar. Show that the mappings

$$x \to a + x \quad ; \quad x \to \lambda x$$

   are homeomorphisms of X onto itself. (Hint: Use Problem 6.)
8. Let X be a normed vector space. Let $S(x;r) = \{y: \|y-x\| < r\}$, $S[x;r] = \{y: \|y-x\| \leq r\}$ be respectively the open and closed ball in X. Show that $\overline{S}(x;r) = S[x;r]$. (Hint: Use Theorem 7 of Chapter 5.)
9. Let X be a normed vector space.

(a) Show that if the series $\sum_{n=1}^{\infty} x_n$ ($x_n \in X$ for all n) is convergent, then for any given $\varepsilon > 0$ there is an integer N such that $n \geq N$ and $p \geq 0$ imply

$$\|x_{n+1} + \dots + x_{n+p}\| \leq \varepsilon.$$

(b) Show that if X is a Banach space and if a series $\sum_{n=1}^{\infty} x_n$ satisfies the condition in (a), then $\sum_{n=1}^{\infty} x_n$ is convergent.

10. Let E be a vector subspace of a normed vector space X. Show that $\bar{E}$ is also a vector subspace of X.

11. On the vector space $\mathbb{R}^2$ define the mappings $\|\cdot\|_1$, $\|\cdot\|_2$ and $\|\cdot\|_\infty$ as follows.

For $x = \langle x_1, x_2 \rangle$

$$\|x\|_1 = |x_1| + |x_2|$$
$$\|x\|_2 = \sqrt{|x_1|^2 + |x_2|^2}$$
$$\|x\|_\infty = \max(|x_1|, |x_2|).$$

(a) Show that the mappings $\|\cdot\|_1$, $\|\cdot\|_2$, $\|\cdot\|_\infty$ are norms.

(b) Draw figures to illustrate the closed balls $S[0;1]$ which correspond to each of these norms.

12. Let X be a normed vector space. Show that X is a Banach space if and only if the subset $\{x \in X: \|x\| = 1\}$ considered as a metric subspace of X is complete.

13. Show that in a normed vector space, if $\lim_{n\to\infty} x_n = x$, then $\lim_{n\to\infty} \|x_n\| = \|x\|$.

14. Let X be a normed vector space, and let $\|\cdot\|_1$ and $\|\cdot\|_2$ be two norms on X. We say that $\|\cdot\|_1$ and $\|\cdot\|_2$ are *equivalent norms* if and only if the metric topologies associated with these norms are identical. Show that two norms $\|\cdot\|_1$ and $\|\cdot\|_2$ are equivalent if and only if there exist two constants $a > 0$, $b > 0$ such that

$$a\|x\|_1 \leq \|x\|_2 \leq b\|x\|_1$$

for all x in εX.

15. Let M be a vector subspace of a normed vector space X, and associate with each $x \in X$ the coset

$$x + M = \{x + y : y \in M\}.$$

Denote the set of all cosets of M by X/M.

(a) Show that X/M is a vector space (called the *quotient space*), if we define

$$(x+M)+(y+M) = (x+y)+M; \quad \lambda(x+M) = \lambda x+M$$

for x and $y \in X$ and scalars λ.

(b) Show that if M is a closed subspace of X then the quotient space X/M is a normed vector space, if we define the norm (called the *quotient norm*) of an element by

$$\|x+M\| = \inf\{\|x+y\| : y \in M\}.$$

(c) Show that if X is a Banach space and M is a closed subspace of X then X/M is also a Banach space.

16. Let A be a bounded linear transformation of a normed vector space X into a normed vector space Y. Show that

$$\|A\| = \sup\{\|Ax\| : x \in X, \|x\| \leq 1\}.$$

17. Let A be a bounded linear transformation of a Banach space X into a Banach space Y. Show that if a series $\sum_{n=1}^{\infty} x_n$ is convergent (respectively absolutely convergent) in X, then the series $\sum_{n=1}^{\infty} Ax_n$ is convergent (respectively, absolutely convergent) in Y and that

$$\sum_{n=1}^{\infty} Ax_n = A\left(\sum_{n=1}^{\infty} x_n\right).$$

18. Show that if X is a finite dimensional normed vector space, then every linear transformation of X into a normed vector space Y is bounded.

19. A subset E of a vector space X (not necessarily normed) is said to be *convex* if it has the following geometric property: Whenever $x \in E$, $y \in E$, $\alpha \geq 0$, $\beta \geq 0$, and $\alpha + \beta = 1$, the point $z = \alpha x + \beta y$ lies in E.

As α (or β) runs from 0 to 1 one may visualize z as describing a straight line segment in X, from x to y. E is convex means that E contains the segments between any two of its points.

(a) If X is a normed vector space show that the unit ball (open or closed) is a convex set.

(b) Let S be the set of all points $x = \langle \alpha_1, \alpha_2, \ldots, \alpha_n \rangle$ in ℓ^2 (see Prob. 6, Chap. 6) which satisfy the condition

$$\sum_{n=1}^{\infty} |\alpha_n|^2 n^2 \leq 1.$$

Show that S is convex.

(c) Show that if E is a convex subset of a normed vector space, then its closure $\overline{E}$ is also convex.

(d) Show that the intersection of a collection of convex sets is a convex set.

20. Let X and Y be normed vector spaces. Let $\langle A_n \rangle$ be a sequence of bounded linear transformations from X into Y. Let T be a bounded linear transformation from X into Y. Show that if $A_n \to T$ and $x_n \to x$, then $A_n x_n \to Tx$.

21. Let A be a bounded linear transformation of a normed vector space X into a normed vector space Y. Show that the set $\mathcal{K} = \{x \varepsilon X : Ax = \theta\}$ (where θ is the zero element of Y) is closed. The set $\mathcal{K}$ is called the *kernel* of A.

22. Let X be a normed vector space and M a closed subspace of X. Consider the quotient normed

vector space X/M as defined in Problem 15. Show that the mapping $F: X \to X/M$ defined by $F(x) = x + M$ is a bounded linear transformation of X onto X/M such that $\|F\| = 1$.

23. If X is a normed vector space consisting of more than one point, show that X^* also consists of more than one point.
(Hint: Show that X^* separates points on X: if $x_1 \neq x_2$ in X, then there is $f \in X^*$ such that $f(x_1) \neq f(x_2)$.)

24. Let X be a normed vector space and $x_0 \in X$. Show that if $f(x_0) = 0$ for all f in X^*, then x_0 is the zero vector of X.

25. Let X be a normed vector space. Show that for every $x \in X$ we have

$$\|x\| = \sup\{|f(x)| : f \in X^*, \|f\| = 1\}.$$

26. Let M be a closed vector subspace of a normed vector space X, and let a be a vector not in M. Put $d = \inf\{\|a - m\| : m \in M\}$. Show that there exists an $f \in X^*$ such that $f[M] = 0$, $f(a) = 1$ and $\|f\| = 1/d$.

27. Show that the space L^∞ is not reflexive.

28. (a) Let X be a real normed vector space. Show that if X^* is separable, then X also is separable.
(b) Give an example to show that the converse of the statement in (a) is in general false.
(Hint: Let $\langle f_n \rangle$ be a countable dense set in

X^*. Use the definition of the norm of a bounded linear functional to obtain a sequence $\langle x_n \rangle$ in X such that $\|x_n\| \leq 1$, and $\|f_n(x_n)\| \geq \|f_n\|/2$ for all n. Let M be the set of all linear combinations of the x_n's whose coefficients are rational numbers. Show that $\bar{M} = X$. For the counterexample in (b) consider the spaces L^1 and L^∞. Use Problem 27.)

29. If X is a finite-dimensional normed vector space of dimension n, show that X^* also has dimension n. Show that X is reflexive.

30. Show that a Banach space X is reflexive iff X^* is reflexive.
(Hint: Suppose X is not reflexive. Then the image $\phi[X]$ of X under the canonical mapping $\phi: X \to X^{**}$ is not onto. Consider $\phi[X]$ as a subspace of X. There is a nonzero bounded linear functional F such that $F(x) = 0$ for all $x \in \phi[X]$.)

31. Show that a linear functional f on a normed vector space X is bounded iff the set $\{x \in X : f(x) = 0\}$ is closed.

32. Show that a continuous one-to-one linear transformation of a Banach space X onto a Banach space Y is a homeomorphism.

33. Let X be a vector space which is complete in each of the norms $\|\cdot\|_1$, $\|\cdot\|_2$. Suppose there is a constant C such that

$$\|x\|_1 \leq C\|x\|_2 \quad \text{for all } x \in X.$$

Show that there is another constant B such that $\|x\|_2 \leq B\|x\|_1$ for all $x \in X$. (In other words, show that the norms $\|\cdot\|_1$ and $\|\cdot\|_2$ are equivalent. See Prob. 14.)

34. Let $\langle A_n \rangle$ be a sequence of bounded linear transformations of a Banach space X into a normed vector space Y. Suppose that for each $x \in X$, the sequence $\langle A_n x \rangle$ converges in Y to some element, say Ax. Show that A is a bounded linear transformation.
(Hint: Use Theorem 15.)

35. Show that a nonempty subset E of a normed vector space X is bounded (that is, there is a constant M such that $\|x\| \leq M$ for all $x \in E$) iff $f[E]$ is a bounded set of numbers for each $f \in X^*$.

36. Let X be a normed vector space. Show, by giving an example, that in general the weak* topology on X^* does not coincide with the weak topology on X^*.

37. Show that in a normed vector space if a sequence $\langle x_n \rangle$ converges weakly to x then the sequence $\langle \|x_n\| \rangle$ is bounded.
(Hint: Use Theorem 15.)

38. Let $C[-1,1]$ be the Banach space of all continuous functions on $[-1,1]$ with the supremum norm (i.e., $\|x\| = \sup\{|x(t)| : -1 \leq t \leq 1\}$). Let $\langle \phi_n \rangle_{n=1}^{\infty}$ be a sequence in $C[-1,1]$ such that

(i) $\phi_n(t) \geq 0$ for all n and all $t \in [-1,1]$

(ii) $\phi_n(t) = 0$ for $|t| > 1/n$

(iii) $\int_{-1}^{1} \phi_n(t)dt = 1.$

Define the functionals δ and $\langle f_n \rangle_{n=1}^{\infty}$ on $C[-1,1]$ as follows:

$$\delta(x) = x(0) \quad \text{for} \quad x \in C[-1,1]$$

$$f_n(x) = \int_{-1}^{1} \phi_n(t)x(t)dt \quad \text{for} \quad x \in C[-1,1].$$

(a) Show that the functionals δ and $\langle f_n \rangle$ are bounded linear functionals on $C[-1,1]$.

(b) Show that f_n converges to δ weakly.

CHAPTER 9

Hilbert Space

9.1 INTRODUCTION

The reason why we dedicate a separate chapter to Hilbert space is that this space occupies a special position among the other Banach spaces. One should notice the remarkable harmony and beauty of concepts and proofs throughout the theory of Hilbert space. Perhaps the main reason why this space plays an exceptional role is that there is a natural correspondence between itself and its dual. The theory of Hilbert spaces is well covered by many expository books and it is not in the spirit of this book to give the subject a full coverage. However, because of the importance of the matter, the author believes that a basic analysis book intended mainly for graduate students should contain an introductory treatment and some fundamental results in this field.

9.2 THE DEFINITION AND SOME PROPERTIES

Definition *A complex-vector space* X *is called an* inner product space *(also* pre-Hilbert space) *if and only if there is defined a complex-valued function* (x,y) *on* $X \times X$ *with the following properties:*

(i) $(\lambda_1 x_1 + \lambda_2 x_2, y) = \lambda_1 (x_1, y) + \lambda_2 (x_2, y)$, *where* λ_1, λ_2 *are complex numbers and* $x_1, x_2 \in X$.

(ii) $(x,y) = \overline{(y,x)}$ *(the complex conjugate of* (y,x)*).*

(iii) $(x,x) \geq 0$; $(x,x) = 0$ *if and only if* $x = \theta$.

The reason for the term pre-Hilbert space is that one can pass by the procedure of completion, used in metric spaces, from a pre-Hilbert space to a Hilbert space.

We have defined a complex inner product space. We can also define a real inner product space if we consider spaces over the reals. Then the function (x,y) is real-valued, the scalars λ are real numbers, and $\overline{(y,x)}$ is replaced by (y,x) in (ii). However, it is more convenient in analysis to deal with a complex Hilbert space, for it is only in the complex case that the theory of linear operators assumes its complete form. We shall henceforth restrict all considerations to the complex case.

1. Proposition *If* X *is an inner product space then*

(a) $(x, y+z) = (x,y) + (x,z)$.

(b) $(x, \lambda y) = \overline{\lambda}(x,y)$.

(c) $(\theta, y) = (x, \theta) = 0$.

(d) $(x-y,z) = (x,z) - (y,z).$

$(x,y-z) = (x,y) - (x,z).$

(e) *If* $(x,z) = (y,z)$ *for all* z *in* X *then* $x = y$.

Proof(a): $(x,y+z) = \overline{(y+z,x)} = \overline{[(y,x) + (z,x)]} = \overline{(y,x)} + \overline{(z,x)} = (x,y) + (x,z)$.

(b): $(x,\lambda y) = \overline{(\lambda y,x)} = \overline{[\lambda(y,x)]} = \overline{\lambda(y,x)} = \overline{\lambda}(x,y)$.

(c): $(\theta,y) = (\theta+\theta,y) = (\theta,y) + (\theta,y)$, hence $(\theta,y) = 0$. Similarly, $(x,\theta) = 0$.

(d): $(x-y,z) = (x+(-y),z) = (x,z) + (-y,z) = (x,z) + ((-1)y,z) = (x,z) + (-1)(y,z) = (x,z) - (y,z)$. Similarly $(x,y-z) = (x,y) - (x,z)$.

(e): Suppose $(x,z) = (y,z)$ for all $z \in X$. Then $(x-y,z) = (x,z) - (y,z) = 0$.
Let $z = x - y$. Then $(x-y,x-y) = 0$, and $x - y = \theta$ by property (iii).

Definition *In an inner product space the* norm *of a vector* x *is the nonnegative number* $\|x\|$, *defined by the formula* $\|x\| = \sqrt{(x,x)}$. *With this definition of norm an inner product space becomes a normed vector space in the sense of Chapter 8.*

To establish this assertion we need the following theorem.

2. Theorem *For any vectors* x *and* y *in an inner product space we have*

$$(*) \qquad |(x,y)| \leq \|x\|\cdot\|y\|.$$

The equality holds if and only if x *and* y *are linearly dependent, that is, if and only if there exist scalars* λ *and* μ, *not both zero, such that* $\lambda x + \mu y = 0$.

The above inequality is known as the Cauchy, Cauchy-Schwarz, or Cauchy-Buniakovsky-Schwarz inequality.

Proof Set $A = \|x\|^2$, $B = |(x,y)|$ and $C = \|y\|^2$. Let t be the argument of the complex number (y,x). Then $(y,x) = Be^{it}$ or $B = e^{-it}(y,x)$. If a is an arbitrary real number we have $(x-ae^{-it}y, x-ae^{-it}y)$ $= \|x\|^2 - ae^{it}(x,y) - ae^{-it}(y,x) + a^2\|y\|^2 \geq 0$ and therefore $Ca^2 - 2aB + A \geq 0$. If $C = 0$, then necessarily $B = 0$ for otherwise the last inequality does not hold for $a > A/2B$. But if $B = 0$ then (*) is satisfied. Finally, if $C > 0$ then we take $a = B/C$ and the inequality (*) follows. Next suppose that equality holds in (*). If $C = 0$ then $y = \theta$, and it is clear that x and y are linearly dependent. If $C > 0$ then for $a = B/C$ we have

$$(x-ae^{-it}y, x-ae^{-it}y) = Ca^2 - 2aB + A = 0,$$

which implies $x - ae^{-it}y = \theta$. Hence x and y are linearly dependent.

3. Theorem *Let* X *be an inner product space. Then for any vectors* x *and* y *in* X *we have*

$$\|x + y\| \leq \|x\| + \|y\|.$$

Proof Using the Cauchy inequality, we have

$$\|x+y\|^2 = (x+y,x+y) = (x,x)+(x,y)+(y,x)+(y,y)$$

$$\leq \|x\|^2 + \|x\|\|y\| + \|x\|\|y\| + \|y\|^2 = (\|x\| + \|y\|)^2.$$

Theorem 3 shows that the equality $\|x\| = \sqrt{(x,x)}$ indeed defines a norm in the sense of Chapter 8. For the remaining properties of a norm:

$$\|x\| \geq 0 \quad \text{with} \quad \|x\| = 0 \quad \text{if and only if} \quad x = \theta;$$

$$\|\lambda x\| = |\lambda|\,\|x\|$$

are clearly satisfied. Therefore an inner product space is a normed vector space.

4. Proposition *Let* X *be an inner product space.*

(a) If $x_n \to x$ *and* $y_n \to y$ *then* $(x_n, y_n) \to (x,y)$.

(b) If $\langle x_n \rangle$ *and* $\langle y_n \rangle$ *are Cauchy sequences in* X *then* $\langle (x_n, y_n) \rangle$ *is also a Cauchy sequence of scalars, hence convergent.*

Proof (a) Using the Cauchy-Schwarz inequality we have

$$|(x_n,y_n)-(x,y)| = |(x_n-x,y_n-y)+(x,y_n-y)+(x_n-x,y)|$$

$$\leq \|x_n-x\|\|y_n-y\| + \|x\|\|y_n-y\| + \|x_n-x\|\|y\|.$$

The right-hand side approaches zero as $n \to \infty$.

Proof (b) First observe that since $x_n \to x$ the inequality $\|x_n\| \leq \|x_n-x\|+\|x\|$ shows that the $\|x_n\|$

are bounded. Similarly the $\|y_n\|$ are bounded. We have as before

$$|(x_n,y_n) - (x_m,y_m)|$$

$$\leq \|x_n-x_m\|\|y_n-y_m\| + \|x_m\|\|y_n-y_m\| + \|x_n-x_m\|\|y_m\|,$$

for all m and n, so that the right side approaches 0 as n and $m \to \infty$.

As a consequence of this proposition we observe that if $x_n \to x$ then $\|x_n\| \to \|x\|$. Also if $\langle x_n \rangle$ is a Cauchy sequence, then $\langle \|x_n\| \rangle$ is also a Cauchy sequence.

Many of the classical phenomena that we find in Euclidean geometry are reproduced in an inner product space. We prove here some of these classical theorems.

(*Note:* In this chapter the single word "subspace" means "vector subspace" and not "topological subspace.")

5. Theorem (The Parallelogram Law) *If x and y are any two vectors of an inner product space then*

$$\|x+y\|^2 + \|x-y\|^2 = 2\|x\|^2 + 2\|y\|^2.$$

Proof We have

$$\|x+y\|^2 = (x+y,x+y) = \|x\|^2 + 2R_e(x,y) + \|y\|^2$$

and

$$\|x-y\|^2 = \|x\|^2 - 2R_e(x,y) + \|y\|^2.$$

Theorem 5, in the terminology of the Euclidean plane, asserts that the sum of the squares of the two diagonals of a parallelogram is equal to the sum of the squares of its four sides. If x and y are two vectors of an inner product space X, then x is said to be *orthogonal* to y, in symbol $x \perp y$, iff $(x,y) = 0$. If E is a subset of X, then the notation $x \perp E$ (read: x orthogonal to E) signifies that $x \perp y$ for every y in E.

We denote by $x^{\perp}$ the set of all $y \in X$ that are orthogonal to x. Observe that $x^{\perp}$ is a subspace of X. To see this, let y and z be in $x^{\perp}$. Then $(x,y+z) = (x,y)+(x,z) = 0$ so that $y + z \in x^{\perp}$. Also if α is a scalar we have $(x,\alpha y) = \overline{\alpha}(x,y) = 0$, which means that $\alpha y \in x^{\perp}$.

Now let F, as usual, be the field of scalars. Then it follows from Proposition 4 that the mapping $\phi: X \to F$ defined, for fixed x, by $\phi(y) = (x,y)$ is continuous. Note that the inverse image of $\{\theta\}$ by ϕ is precisely $x^{\perp}$; that is $\phi^{-1}[\{\theta\}] = x^{\perp}$. Since ϕ is continuous, it follows that $x^{\perp}$ is a closed subspace of X.

If M is a subset of X, define $M^{\perp}$ to be the set of all elements y in X that are orthogonal to M. Clearly $M^{\perp} = \bigcap_{x \in M} x^{\perp}$. For $y \in M^{\perp}$ implies $y \perp x$ for all $x \in M$ so that $y \in x^{\perp}$ for all $x \in M^\perp$ and $y \in \bigcap_{x \in M} x^\perp$. Also if $y \in \bigcap_{x \in M} x^\perp$ then $y \perp x$ for all x in M so that $y \in M^\perp$. Since $x^\perp$ is a closed subspace of X it follows that $M^\perp$ is also a closed subspace of X.

Definitions *Let X be a vector space and x and y any two vectors in X. The segment connecting x and y is defined to be the set of all vectors $z = \alpha x + \beta y$ for all α and β such that $\alpha \geq 0$, $\beta \geq 0$, $\alpha + \beta = 1$. A subset E of X is said to be* **convex** *if and only if whenever $x \in E$, $y \in E$, the segment connecting x and y belongs to E.*

A collection $\{x_i\}_{i \in \Lambda}$ of vectors is said to be an **orthogonal system** *if and only if for every two distinct elements i and j of Λ we have $x_i \perp x_j$.*

6. **Theorem (Pythagorean Theorem)** *If $x \perp y$ then*

$$\|x+y\|^2 = \|x\|^2 + \|y\|^2.$$

More generally, if $x_1, x_2, \ldots, x_n$ are pairwise orthogonal then

$$\left\|\sum_{i=1}^n x_i\right\|^2 = \sum_{i=1}^n \|x_i\|^2.$$

Proof Compute.

Warning We know that the converse of the Pythagorean

theorem is true in the Euclidean plane. However, it is not true in an arbitrary inner product space. For example, consider the set of complex numbers $\mathbb{C}$ and define the inner product between two elements $x = x_1 + ix_2$, $y = y_1 + iy_2$ by

$$(x,y) = x\overline{y}.$$

Then clearly $\mathbb{C}$ is an inner product space in which there exist no pairs of orthogonal elements. But still we may find vectors x and y such that

$$\|x+y\|^2 = \|x\|^2 + \|y\|^2.$$

As an example of such vectors we take $x = 1 + i$, $y = 1 - i$.

Remark We have seen that an inner product space is a normed vector space whose norm derives from an inner product. However, *it is erroneous to believe that for every normed vector space X the norm in X is derived from an inner product.* It is possible to prove (Wilansky [28], p. 125) that *any norm which does not obey the parallelogram law is not derived from an inner product.* For example, one verifies that in the Euclidean plane the norm $\|x\| = |x_1| + |x_2|$, for $x = \langle x_1, x_2 \rangle$, does not obey the parallelogram law. Therefore this norm does not derive from an inner product.

Definition *A* Hilbert space *is an inner product space which as a metric space (with respect to the metric induced by its norm) is complete.*

Throughout the rest of this chapter the letter

H will denote a Hilbert space.

Example 1

Let n be a positive integer. Then the space $\mathbb{C}^n(\mathbb{R}^n)$ of all n-tuples of complex numbers (real numbers) with addition and scalar multiplication defined componentwise as usual, and with inner product between $x = \langle x_1, x_2, \ldots, x_n \rangle$ and $y = \langle y_1, y_2, \ldots, y_n \rangle$ given by

$$(x,y) = \sum_{i=1}^{n} x_i \overline{y}_i$$

is a Hilbert space. For we already know that it is a Banach space. It remains to observe that the norm $\|x\| = \left(\sum_{i=1}^{n} |x_i|^2\right)^{\frac{1}{2}}$ is derived from the inner product.

Example 2

The space ℓ^2 of all sequences of complex numbers $\langle x_1, x_2, \ldots \rangle$ with the property $\sum_{i=1}^{\infty} |x_i|^2 < +\infty$ is a Hilbert space with the inner product of the vectors $x = \langle x_1, x_2, \ldots \rangle$, $y = \langle y_1, y_2, \ldots \rangle$ given by $(x,y) = \sum_{i=1}^{\infty} x_i \overline{y}_i$ (see Chap. 5, p. 218).

Example 3

The Banach space $L^2(a,b)$ of all Lebesgue measurable complex-valued functions (see end of Chap. 3) f on the closed and bounded interval $[a,b]$ such that $\int_a^b |f|^2 dx < +\infty$ is a Hilbert space if the inner product for f and g in $L^2(a,b)$ is defined by

$$(x,y) = \int_a^b f(t)\overline{g(t)}dt.$$

We shall recall again that $L^2(a,b)$ should be regarded as a space of equivalence classes of functions under the equivalence relation in $L^2(a,b)$ defined by "$f \sim g$ iff $f = g$ a.e."

Example 4

The vector space X of all continuous complex-valued functions on $[-1,1]$ is an inner product space if

$$(f,g) = \int_{-1}^{1} f(t)\overline{g(t)}dt,$$

but is not a Hilbert space.

Proof Clearly X is an inner product space. To show that X is not a Hilbert space we prove X is not complete. Let $\langle x_n \rangle$ be a sequence of elements in X defined as follows:

$$x_n(t) = \begin{cases} 0 & \text{if } -1 \le t \le 0 \\ nt & \text{if } 0 < t < \frac{1}{n} \\ 1 & \text{if } \frac{1}{n} \le t < 1 \end{cases}$$

By elementary calculus we easily see that $\langle x_n \rangle$ is a Cauchy sequence in X. Assume now that there is an element x in X such that $\|x_n - x\| \to 0$, or equivalently that

$$\int_{-1}^{1} |x_n(t) - x(t)|^2 dt \to 0, \quad \text{as} \quad n \to \infty.$$

Since the integrand is nonnegative, this implies that

$$\int_{-1}^{0} |x_n(t)-x(t)|^2 dt = \int_{-1}^{0} |x(t)|^2 dt \to 0.$$

In other words $\int_{-1}^{0} |x(t)|^2 dt = 0$. Since x is continuous on $[-1,0]$ it follows that $x(t) = 0$ for $-1 \le t \le 0$. On the other hand for $0 < \varepsilon < 1$, and $n > 1/\varepsilon$ we have

$$\int_{\varepsilon}^{1} |x_n(t)-x(t)|^2 dt = \int_{\varepsilon}^{1} |1-x(t)|^2 dt.$$

But the first member of this equality approaches zero as $n \to \infty$. Thus

$$\int_{\varepsilon}^{1} |1-x(t)|^2 dx = 0$$

which implies that $x(t) = 1$ on $[\varepsilon,1]$. Since ε was arbitrary, it follows that $x(t) = 1$ for $0 < t \le 1$. Thus $x(t) = 0$ if $-1 \le t \le 0$ and

$x(t) = 1$ if $0 < t \leq 1$, which contradicts the assumption that x is continuous.

7. Proposition *Let E be a nonempty closed convex set in a Hilbert space H. Then there exists a unique element* a *in E such that*

$$\|a\| \leq \|x\| \quad \textit{for every} \quad x \in E.$$

Proof Set $\alpha = \inf\{\|x\|: x \in E\}$. Then there is a sequence $\langle x_n \rangle$ in E such that $\lim_{n\to\infty} \|x_n\| = \alpha$. By the parallelogram law (Theorem 5) we have

$$\|x_n - x_m\|^2 = 2\|x_n\|^2 + 2\|x_m\|^2 - \|x_n + x_m\|^2$$

$$= 2\|x_n\|^2 + 2\|x_n\|^2 - 4\left\|\frac{x_n+x_m}{2}\right\|^2.$$

Since E is convex, $(x_n+x_m)/2$ belongs to E, so that $\|(x_n+x_m)/2\| \geq \alpha$. Hence

$$\|x_n - x_m\|^2 \leq 2\|x_n\|^2 + 2\|x_m\|^2 - 4\alpha^2.$$

Since the right side of this inequality approaches zero as $n,m \to \infty$ it follows that $\langle x_n \rangle$ is a Cauchy sequence. Hence there is an element a such that $x_n \to a$. Since E is closed a belongs to E. Clearly $\lim_{n\to\infty} \|x_n\| = \alpha = \|a\|$.

To prove the uniqueness of the vector a assume that there is an element b in E such that $\|a\| = \|b\| = \alpha$. Then by the parallelogram law we have as before

$$\|a-b\|^2 \leq 2\|a\|^2 + 2\|b\|^2 - 4\alpha^2 = 0$$

which implies $a = b$.

8. Proposition *Let H be a Hilbert space and M a closed vector subspace of H. Then there exist a unique pair of mappings P_1 and P_2 called the projections of H onto M and $M^\perp$ respectively*

$$P_1 : H \to M;$$

$$P_2 : H \to M^\perp$$

such that

(1) $\quad x = P_1 x + P_2 x \quad$ *for all* $x \in H$.

Furthermore, P_1 and P_2 have the following properties

(a) If $x \in M$ then $P_1 x = x$, $P_2 x = \theta$; if $x \in M^\perp$ then $P_1 x = \theta$, $P_2 x = x$.

(b) $\|x-P_1 x\| = \inf\{\|x-y\|: y \in M\}$, $x \in H$

(c) $\|x\|^2 = \|P_1 x\|^2 + \|P_2 x\|^2$

(d) The mappings P_1 and P_2 are linear.

Proof First observe that for any $x \in H$ the set $\{x+y: y \in M\} = x + M$ is convex and closed. To see this let $x + y_1$ and $x + y_2$ be two elements in $x + M$. Then for $\alpha \geq 0$, $\beta \geq 0$, $\alpha + \beta = 1$ we have

$\alpha(x + y_1) + \beta(x + y_2) = \alpha x + \beta x + \alpha y_1 + \beta y_2 = x + \alpha y_1 + \beta y_2$. Since M is a subspace and $y_1, y_2 \in M$ we have $\alpha y_1 + \beta y_2 \in M$, so that $x + \alpha y_1 + \beta y_2 \in x + M$. This proves that $x + M$ is convex. Next let z be a point in the closure of $x + M$. Then there exists a sequence $\langle y_n \rangle$ in M such that

$$\|(z-x) - y_n\| = \|z - (x+y_n)\| \to 0.$$

This means that $z - x$ is a point in the closure of M. Since M is closed we have $z - x \in M$ or $z \in x + M$. This proves that $x + M$ is closed.

We now define the mappings P_1 and P_2. For each $x \in H$ consider the closed and convex set $x + M$. Define P_2x to be the unique element of smallest norm in $x + M$ given by Proposition 7. Next define P_1x to be the element $x - P_2x$. Then since $P_2x \in x + M$ it follows that $P_1x \in M$. Thus P_1 is a mapping from H into M. To prove that P_2 is a mapping of H into $M^{\perp}$ we must prove that $(P_2x, y) = 0$ for all elements in M. We may assume that $y \neq \theta$, for otherwise there is nothing to prove. Set $t = y/\|y\|$. Since P_2x is the smallest element in norm of $x + M$, and since for every scalar α, $P_2x - \alpha t$ also belongs to $x + M$, we have

$$(P_2x,P_2x) = \|P_2x\|^2 \leq \|P_2x-\alpha t\|^2 = (P_2x-\alpha t,P_2x-\alpha t).$$

It follows that

$$|\alpha|^2 - \bar{\alpha}(P_2x,t) - \alpha(t,P_2x) \geq 0.$$

If we choose $\alpha = (P_2x,t)$ we get $|(P_2x,t)| \leq 0$ so that $(P_2x,t) = 0$ or $(P_2x,y) = 0$. Thus P_2 maps H into $M^\perp$, and clearly $x = P_1x + P_2x$.

To prove the uniqueness of the mappings P_1 and P_2 first observe that $M \cap M^\perp = \{\theta\}$. For if $t \in M \cap M^\perp$ then $(t,t) = 0$ which implies $t = \theta$. Next suppose that for $x \in H$ we have $x = x_1 + x_2$ with $x_1 \in M$ and $x_2 \in M^\perp$. Then $x_1 + x_2 =$ $P_1x + P_2x$ or $x_1 - P_1x = P_2x - x_2$. Since $x_1 - P_1x \in M$ and $P_2x - x_2 \in M^\perp$ both $x_1 - P_1x$ and $P_2x - x$ belong to $M \cap M^\perp$ so that,

$$x_1 = P_1x \quad \text{and} \quad x_2 = P_2x.$$

This proves the uniqueness of the mappings P_1, P_2.

We now prove the properties (a) through (d).

(a) If $x \in M$ we have $x = x + \theta = P_1x+P_2x$ which implies $P_1x = x$, $P_2x = \theta$, since P_1, P_2 are unique. Similarly if $x \in M^\perp$, then $P_1x = \theta$ and $P_2x = x$.

(b) It follows from (1) and the definition of P_2x.

(c) It follows from the Pythagorean theorem.

(d) We have $x = P_1x + P_2x$, $y = P_1y + P_2y$. Thus

$$\begin{aligned}\alpha x + \beta y &= \alpha(P_1x + P_2x) + \beta(P_1y + P_2y)\\ &= (\alpha P_1x + \beta P_1y) + (\alpha P_2x + \beta P_2y).\end{aligned}$$

Since $\alpha P_1x + \beta P_1y \in M$ and $\alpha P_2x + \beta P_2y \in M^{\perp}$ we have

$$P_1(\alpha x + \beta y) = \alpha P_1x + \beta P_1y$$

and

$$P_2(\alpha x + \beta y) = \alpha P_2x + \beta P_2y.$$

Hence P_1 and P_2 are linear.

9. Corollary *If M is a proper closed subspace of H then there exists an $x_o \in H$ such that $x_o \neq \theta$ and $x_o \perp M$.*

Proof Let $y \in H - M$ and take $x_o = P_2y$. Then $x_o \in M^{\perp}$. Also $x_o \neq \theta$. For suppose on the contrary $x_o = \theta$. Then $y = P_1y + P_2y = P_1y + \theta = P_1y$, so that $y \in M$ which contradicts $y \in H - M$.

9.3 ORTHONORMAL SYSTEMS

A set of vectors $\mathcal{S}$ in an inner product space X is said to be an orthonormal system if and only if $\mathcal{S}$ is an orthogonal system and for each e in $\mathcal{S}$ we have $\|e\| = 1$.

If X contains only the zero vector then it has no orthonormal systems.

Example 5

Consider the Hilbert space $L^2(0,2\pi)$ (Ex. 3). One easily verifies that the functions

$$\{e_n : n = 0, \pm 1, \pm 2, \ldots\}$$

defined by $e_n(x) = e^{inx}/\sqrt{2\pi}$, form an orthonormal system in $L^2(0,2\pi)$. For any function f in $L^2(0,2\pi)$ the *Fourier coefficients* of f are the numbers

$$c_n = (f, e_n) = \frac{1}{2\pi} \int_0^{2\pi} f(x) e^{-inx}\, dx$$

with $n = 0, \pm 1, \pm 2, \ldots$.

10. Proposition *Let* $\{e_1, e_2, \ldots, e_n\}$ *be an orthonormal system in an inner product space. Then for every vector* x *in the space*

$$(i) \quad \left\| x - \sum_{k=1}^{n} (x, e_k) x_k \right\|^2 = \|x\|^2 - \sum_{k=1}^{n} |(x, e_k)|^2$$

hence

(ii) $\sum_{k=1}^{n} |(x,e_k)|^2 \leq \|x\|^2$ *(Bessel's inequality).*

Proof Let $\alpha_1, \alpha_2, \ldots, \alpha_n$ be arbitrary complex numbers. By Theorem 6 we have

$$\left\|\sum_{k=1}^{n} \alpha_k e_k\right\|^2 = \sum_{k=1}^{n} \|\alpha_k e_k\|^2 = \sum_{k=1}^{n} |\alpha_k|^2.$$

Also

$$\left\|x - \sum_{k=1}^{n} \alpha_k e_k\right\|^2 = \left(x - \sum_{k=1}^{n} \alpha_k e_k,\ \ x - \sum_{k=1}^{n} \alpha_k e_k\right)$$

$$= \|x\|^2 - \sum_{k=1}^{n} \alpha_k \overline{(x,e_k)} - \sum_{k=1}^{n} (x,e_k)\bar{\alpha}_k + \sum_{k=1}^{n} \alpha_k \bar{\alpha}_k$$

$$= \|x\|^2 - \sum_{k=1}^{n} |(x,e_k)|^2 + \sum_{k=1}^{n} |(x,e_k) - \alpha_k|^2.$$

If we set $\alpha_k = (x,e_k)$ then (i) follows.

Remark (a) Observe that in the proof of Proposition 10 the choice $\alpha_k = (x,e_k)$ minimizes

$\left\|x - \sum_{k=1}^{n} \alpha_k e_k\right\|$ and provides a best approximation of x by a linear combination of the vectors $e_1, e_2, \ldots, e_n$.

(b) Let x be any vector and let

$y = \sum_{k=1}^{n} (x,x_k) e_k$. Set $z = x - y$. Then clearly

$(z,e_k) = 0$ for $k = 1,2,\ldots,n$. Thus we have a decomposition of x, namely $x = y + z$ where y is a linear combination of $e_1,e_2,\ldots,e_n$ and $z \perp y$. It is easy to see that such a decomposition is unique.

Let $\mathcal{S}$ be an orthonormal system in an inner product space X, and let x be any element in X. Then the scalars $\{(x,e)\}_{e\in S}$ are called the *Fourier coefficients* of x relative to the orthonormal system $\mathcal{S}$.

11. Proposition *If $\mathcal{S}$ is an orthonormal system in an inner product space X, and if x is any vector in X, then the set $E = \{e \in \mathcal{S} : (x,e) \neq 0\}$ is countable.*

Proof For each positive integer n set

$$A_n = \{e \in \mathcal{S}: |(x,e)|^2 > \|x\|^2/n\}.$$

Then clearly $E = \bigcup_{n=1}^{\infty} A_n$. But it follows from Bessel's inequality that each set A_n cannot contain more than $n - 1$ vectors of $\mathcal{S}$. Therefore E is countable.

The inequality (ii) in Proposition 10 can be given the following loose geometric interpretation: The sum of the squares of the components of a vector in various perpendicular directions does not exceed the square of the length of the vector itself. The reader could verify this in the three dimensional Euclidean space. Bessel's inequality is a special case of a more general inequality with the same name. We first introduce a new notation. Let $\{x_\alpha : \alpha \in \Lambda\}$ be a collection of nonnegative real numbers, not ordered in any way, and where the indexing set Λ can be even uncountable. Under these conditions the symbol

$$\sum_{\alpha\in\Lambda} x_\alpha$$

denotes the supremum of the set of all finite sums $x_{\alpha_1} + x_{\alpha_2} + \ldots + x_{\alpha_n}$ where $\alpha_1, \alpha_2, \ldots, \alpha_n$ are members of Λ.

12. Theorem (Bessel's Inequality) *If* $\{e_\alpha : \alpha \in \Lambda\}$ *is an orthonormal system in a Hilbert space* H, *then*

$$\sum_{\alpha\in\Lambda} |(x, e_\alpha)|^2 \leq \|x\|^2 \qquad (1)$$

for every vector x *in* H.

Proof As in Proposition 11 we write $E = \{e_\alpha : (x, e_\alpha) \neq 0\}$. Then E is countable. If E is empty then (1) obviously holds. If E is finite then E can be written in the form $E = \{e_1, e_2, \ldots, e_n\}$ for some positive integer n. Then

$$\sum_{\alpha\in\Lambda} |(x, e_\alpha)|^2 = \sum_{i=1}^{n} |(x, e_i)|^2$$

and (1) follows from (ii) of Proposition 10.

Finally assume that E is denumerable (that is countably infinite) and let the elements in E be arranged in a definite order

$$E = \langle e_1, e_2, \ldots, e_n, \ldots \rangle.$$

Then it follows from (ii) of Proposition 10 that $\sum_{n=1}^{\infty} |(x,e_n)|^2$ converges. Thus every series obtained from this by rearranging its terms also converges and all such series have the same sum. This implies that

$$\sum_{\alpha\varepsilon\Lambda} |(x,e_\alpha)|^2 = \sum_{n=1}^{\infty} |(x,e_n)|^2 \leq \|x\|^2.$$

The theorem is proved.

13. Corollary *If* $\langle e_n \rangle$ *is a countably infinite orthonormal system in* X, *then for every* x *in* X $(x, e_n) \to 0$ *as* $n \to \infty$.

Observe that if $x_n \to \theta$, then for each $x \varepsilon X$ we have $(x,x_n) \to 0$. The above corollary shows that the converse of this last statement is not true, since the x_n's can be an orthonormal system, so that $\|x_n\| = 1$.

14. Theorem *Let* H *be a Hilbert space and let* $\{e_n \colon n = 1, 2, \ldots\}$ *be a countably infinite orthonormal system in* H. *Let* $\langle \alpha_n \rangle$ *be a sequence of real or complex numbers. Then*

(a) $\sum_{n=1}^{\infty} \alpha_n e_n$ *is summable if and only if* $\sum_{n=1}^{\infty} |\alpha_n|^2 < +\infty$.

(b) If $\sum_{n=1}^{\infty} \alpha_n e_n$ *is summable to the element* x *of* H, *then* $\alpha_n = (x, e_n)$.

Proof (a) Consider the partial sums $S_n = \sum_{k=1}^{n} \alpha_k e_k$ and suppose $n > m$. We have

$$\|S_n - S_m\|^2 = \left\|\sum_{k=m+1}^{n} \alpha_k e_k\right\|^2 = \sum_{k=m+1}^{n} |\alpha_k|^2.$$

If $\sum_{n=1}^{\infty} \alpha_n e_n$ is summable then $\sum_{k=m+1}^{n} |\alpha_k|^2$ converges to zero so that $\sum_{n=1}^{\infty} |\alpha_n|^2$ is convergent. Conversely, if $\sum_{n=1}^{\infty} |\alpha_n|^2$ converges it follows that $\langle S_n \rangle$ is a Cauchy sequence, and since H is complete $\sum_{n=1}^{\infty} \alpha_n e_n$ is summable to some element in H.

Proof (b) Consider $S_n = \sum_{k=1}^{n} \alpha_k e_k$. For $i < n$ we have $\alpha_i = (S_n, e_i)$. By Proposition 4 we get $\alpha_i = \lim_{n\to\infty} (S_n, e_i) = (x, e_i)$.

The following proposition is a direct corollary of Theorem 14 applied to the space $L^2(0,2\pi)$ with the orthonormal system $\{e_n: n = 0,1,2,...\}$ as in Example 5. However, we state this proposition in a form of theorem, because of its great importance.

15. Theorem (Riesz-Fisher) *If* $c_n (n = 0, \pm 1, \pm 2, ...)$ *are given complex numbers for which* $\sum_{n=-\infty}^{\infty} |c_n|^2$ *converges, then there exists a function* f *in* $L^2(0,2\pi)$ *whose Fourier coefficients are the* c_n*'s.*

Definition *An orthonormal system* $\mathcal{S}$ *in an inner product space is said to be* complete *iff there exists no orthonormal system of which* $\mathcal{S}$ *is a proper subset.*

Complete orthonormal systems are frequently called *total orthonormal systems* or *orthonormal bases*. We now prove that *an orthonormal system in an inner product space* X *is complete if and only if the condition* $x \perp \mathcal{S}$ *implies* $x = \theta$.

To see this suppose $\mathcal{S}$ is complete and let x be an element in X such that $x \perp \mathcal{S}$. Then x must be the zero element θ. For if $x \neq \theta$ then $\mathcal{S} \cup \{(x/\|x\|\}$ is an orthonormal system containing $\mathcal{S}$ properly. But this contradicts the assumption that $\mathcal{S}$ is complete.

Conversely, suppose that $y \perp \mathcal{S}$ implies $y = \theta$. Then $\mathcal{S}$ must be complete. For if not, there must exist an orthonormal system A properly containing $\mathcal{S}$. In this case let $y \in A - \mathcal{S}$. We have $y \perp \mathcal{S}$ and $\|y\| = 1$. Hence $y \neq \theta$, which contradicts our assumption.

16. Proposition *Let X be an inner product space and E an orthonormal system in X. Then there exists a complete orthonormal system containing E.*

Proof Let $\mathcal{F}$ be the collection of all orthonormal systems in X containing E. Then $\mathcal{F}$ is partially ordered by set inclusion. This means that if A and B are members of $\mathcal{F}$, then $A \prec B$ if and only if $A \subset B$. The collection $\mathcal{F}$ is nonempty since it contains E. Hence by Hausdorff maximal principle (Chap. 1, sec. 10) $\mathcal{F}$ contains a maximal linearly ordered subset $T = \{A_\alpha : \alpha \in \Lambda\}$. It is clear that for any α, $A_\alpha \subset \bigcup_{\alpha\in\Lambda} A_\alpha$. Furthermore, $E \subset \bigcup_{\alpha\in\Lambda} A_\alpha$. We claim that $\bigcup_{\alpha\in\Lambda} A_\alpha$ is a complete orthonormal system. If x_1 and x_2 are in $\bigcup_{\alpha\in\Lambda} A_\alpha$, then $x_1 \in A_1$ and $x_2 \in A_2$ for some A_1 and $A_2 \in T$. Since T is linearly ordered we have either $A_1 \subset A_2$ or $A_2 \subset A_1$. Suppose $A_1 \subset A_2$, so that $x_1 \in A_2$ and $x_2 \in A_2$. Since A_2 is orthonormal $x_1 \perp x_2$ if $x_1 \neq x_2$ or $(x_1, x_2) = 1$ if $x_1 = x_2$. Thus $\bigcup_{\alpha\in\Lambda} A_\alpha$ is an orthonormal system. To prove that $\bigcup_{\alpha\in\Lambda} A_\alpha$ is complete, suppose on the contrary that

$\bigcup_{\alpha\varepsilon\Lambda} A_\alpha$ is a proper subset of an orthonormal system $\mathcal{S}$. It is clear that $\mathcal{S} \notin T$, and that $\mathcal{S}$ contains every member of T. But if this is the case, then the set $T \cup \{\mathcal{S}\}$ is a linearly ordered set in $\mathcal{F}$ containing T. This contradicts the maximality of T.

Remark (a) Proposition 16 asserts that a nontrivial inner product space X (i.e., $X \neq \{\theta\}$) contains a complete orthonormal system. For if $x \varepsilon X$ with $x \neq \theta$, then $E = \{(x/\|x\|)\}$ is an orthonormal system, so that by the above proposition there is a complete orthonormal system containing E.

Remark (b) As we mentioned earlier a complete orthonormal system is often called an orthonormal basis. However, it is worthwhile to notice that *in a Hilbert space an infinite complete orthonormal system is never a basis (a Hamel basis).* To see this let $\mathcal{S}$ be a complete orthonormal system in H. Then since $\mathcal{S}$ is infinite it contains an infinite sequence of distinct elements, say $\langle e_n \rangle_{n=1}^{\infty}$. The series

$$\sum_{n=1}^{\infty} \frac{1}{n^2} e_n$$

being absolutely summable, it is summable to some element x of H. By (b) of Theorem 14 we have $1/n^2 = (x, e_n)$. On the other hand if $\mathcal{S}$ were a basis it would be possible to express x as a linear combination of elements in $\mathcal{S}$, that is

$$x = \lambda_\alpha e_\alpha + \ldots + \lambda_\mu e_\mu$$

where $e_\alpha, \ldots, e_\mu \in \mathcal{S}$ and $\lambda_\alpha, \ldots, \lambda_\mu$ are scalars. Thus

$$\frac{1}{n^2} = (x, e_n) = (\lambda_\alpha e_\alpha + \ldots + \lambda_\mu e_\mu, e_n) = 0$$

which clearly is not true. Hence $\mathcal{S}$ is not a basis.

17. Theorem *Let $\{e_\alpha : \alpha \in \Lambda\}$ be an orthonormal system in a Hilbert space H. Then the following conditions are equivalent to one another.*

(a) $\{e_\alpha\}$ is a complete orthonormal system in H.

(b) The set A of all linear combinations of elements of $\{e_\alpha\}$ is dense in H.

(c) For every $x \in H$ we have

$$\|x\|^2 = \sum_{\alpha\in\Lambda} |(x, e_\alpha)|^2$$

(d) Let $x \in H$, $y \in H$ and $E = \{e_\alpha : (x, e_\alpha)(y, e_\alpha) \neq 0\}$.

Let the vectors in E be arranged in an arbitrary but definite order:

$$E = \langle e_1, e_2, \ldots, e_n, \ldots \rangle.$$

Then

$$(x,y) = \sum_{n=1}^{\infty} (x,e_n)\overline{(y,e_n)}.$$

(*Note:* The inequality $2|(x,e_\alpha)(y,e_\alpha)| \leq |(x,e_\alpha)|^2 + |(y,e_\alpha)|^2$ together with Bessel's inequality ensure that the series in (d) converges absolutely so that its sum is independent of the order in which the vectors in E have been arranged. The formula given in (c) is called *Parseval's identity.*)

Proof (a) ⇒ (b) We know that A is the smallest subspace of H which contains all vectors e_α. Let $\bar{A}$ be the closure of A. We prove that $\bar{A}$ is also a subspace. For let $x,y \in \bar{A}$. Then there exist sequences $\langle x_n \rangle$ and $\langle y_n \rangle$ in A such that $x_n \to x$ and $y_n \to y$. Hence $x_n + y_n \to x + y$ so that $x + y \in \bar{A}$. Also since $\alpha x_n \in A$ and $\alpha x_n \to \alpha x$ we have that $\alpha x \in \bar{A}$. This proves that $\bar{A}$ is a subspace. Assume now that A is not dense in H. This means that $\bar{A}$ is a proper subset of H. Hence, by Corollary 9, there exists a nonzero vector x_0 such that $x_0 \perp \bar{A}$. But this implies that x_0 is a nonzero vector orthogonal to all vectors in $\{e_\alpha\}$, which contradicts the assumption that $\{e_\alpha\}$ is a complete orthonormal system. Hence (a) ⇒ (b).

Proof (b) ⇒ (c) If $x = \theta$ then (c) trivially holds. So let $x \neq \theta$ and ε such that $0 < \varepsilon < \|x\|$. Since A is assumed to be dense in H, it is possible to find vectors $e_{\alpha_1}, e_{\alpha_2}, \ldots, e_{\alpha_n}$ in A, and scalars $\lambda_1, \lambda_2, \ldots, \lambda_n$ such that

$$\|x - (\lambda_1 e_{\alpha_1} + \lambda_2 e_{\alpha_2} + \dots + \lambda_n e_{\alpha_n})\| < \varepsilon.$$

By Remark (a) following Proposition 10 the last inequality will hold (or even be improved) if the λ_k's are replaced by the (x, e_{α_k})'s $(k = 1, 2, \dots, n)$. It follows then

$$\|x\| \leq \left\| \sum_{k=1}^{n} (x, e_{\alpha_k}) e_{\alpha_k} \right\| + \varepsilon$$

or

$$(\|x\| - \varepsilon)^2 \leq \left\| \sum_{k=1}^{n} (x, e_{\alpha_k}) e_{\alpha_k} \right\|^2$$

$$= \sum_{k=1}^{n} |(x, e_{\alpha_k})|^2 \leq \sum_{\alpha \varepsilon \Lambda} |(x, e_\alpha)|^2.$$

Since ε was arbitrary we get the inequality

$$\|x\|^2 \leq \sum_{\alpha \varepsilon \Lambda} |(x, e_\alpha)|^2$$

which combined with Bessel's inequality gives (c).

Proof (c) ⇒ (d) Let $\langle e_n \rangle$ be an arbitrary arrangement of the vectors in E, and let us denote by $\hat{x}$ and $\hat{y}$ the sequence $\langle (x, e_n) \rangle$ and $\langle (y, e_n) \rangle$ respectively. Then since

$$\sum_{n=1}^{\infty} |(x,e_n)|^2 < +\infty;$$

$$\sum_{n=1}^{\infty} |(y,e_n)|^2 < +\infty$$

both $\hat{x}$ and $\hat{y}$ belong to the Hilbert space ℓ^2 considered in Example 2. Hence (c) and (d) can be written $(x,x) = (\hat{x},\hat{x})$ and $(x,y) = (\hat{x},\hat{y})$ respectively. Then if (c) holds we get

$$(1) \qquad (x + \lambda y, x + \lambda y) = (\hat{x} + \lambda\hat{y}, \hat{x} + \lambda\hat{y})$$

for every value of the scalar λ. If we take $\lambda = 1$ and $\lambda = i$, we get respectively from (1)

$$R_e(x,y) = R_e(\hat{x},\hat{y}) \quad \text{and} \quad \mathcal{I}(x,y) = \mathcal{I}(\hat{x},\hat{y})$$

which shows that (x,y) and $(\hat{x},\hat{y})$ have the same real and imaginary part, so that they are equal. Hence (d) holds.

Proof (d) ⇒ (a) Suppose on the contrary that (d) does not imply (a). This means there exists a nonzero element $x_o \in H$ such that $x_o \perp e_\alpha$ for all $\alpha \in \Lambda$. If we take $x = y = x_o$ in (d) we get

$$0 \neq (x,y) = \|x_o\|^2 = \sum_{n=1}^{\infty} (x_o,e_n)\overline{(x_o,e_n)} = 0,$$

since $(x_o,e_\alpha) = 0$ for all $\alpha \in \Lambda$. A contradiction. Hence (d) ⇒ (a).

Let us now consider again the space $L^2(0,2\pi)$ (see Ex. 3 and Ex. 5). We already have seen that the system $\{e_n: n = 0,\pm 1,\pm 2,\ldots\}$ where $e_n(x) = e^{inx}/\sqrt{2\pi}$ is an orthonormal system in $L^2(0,2\pi)$. It is an event of great importance in the theory of Fourier series that this system is complete. For a proof see [29]. By Theorem 17 the completeness of $\{e_n\}$ is equivalent to saying that for each $f \in L^2(0,2\pi)$ we can write Parseval's identity

$$\sum_{n=-\infty}^{\infty} |c_n|^2 = \int_0^{2\pi} |f(x)|^2 dx$$

where

$$c_n = \frac{1}{2\pi}\int_0^{2\pi} f(x)e^{-inx}dx.$$

Also, again by Theorem 17, the function f can be expanded in a series as follows

$$f(x) = \frac{1}{\sqrt{2\pi}}\sum_{n=-\infty}^{\infty} c_n e^{inx}.$$

We emphasize that the above expression should not be interpreted as stating that the series converges to $f(x)$ pointwise. It says that the partial sums of the series converge to f in norm. In other words if we set

$$f_n(x) = \frac{1}{\sqrt{2\pi}} \sum_{k=-n}^{n} c_k e^{ikx}$$

then

$$\|f_n - f\| \to 0 \quad \text{as} \quad n \to +\infty.$$

9.4 THE DUAL SPACE OF A HILBERT SPACE

As we pointed out in the introduction of this chapter, the main reason why a Hilbert space H plays an exceptional role among the other Banach spaces is that there is a natural correspondence between H and its dual H^*. Our purpose in this section is to clarify this point by proving that there is a one-to-one mapping of H onto H^* which is an isometry.

For a fixed $y \varepsilon H$ define the functional F_y on H by $F_y(x) = (x,y)$ for all $x \varepsilon H$. Then F_y is linear:

$$F_y(x_1+x_2) = (x_1+x_2,y) = (x_1,y) + (x_2,y)$$

$$= F_y(x_1) + F_y(x_2)$$

and

$$F_y(\lambda x) = (\lambda x,y) = \lambda(x,y) = \lambda F_y(x).$$

Also F_y is bounded. By the Cauchy-Schwarz

inequality we have

$$|F_y(x)| = |(x,y)| \le \|x\|\cdot\|y\|,$$

which shows that F_y is bounded and $\|F_y\| \le \|y\|$.

Next we prove that $\|F_y\| = \|y\|$. Clearly if $y = \theta$ then this holds. Suppose $y \neq \theta$. Then by Proposition 2 of Chapter 8 we have

$$\|F_y\| = \sup\{|F_y(x)|: \|x\| = 1\}$$

$$\ge \left|F_y\left(\frac{y}{\|y\|}\right)\right| = \left|\left(\frac{y}{\|y\|},y\right)\right| = \|y\|.$$

This inequality combined with the inequality $\|F_y\| \le \|y\|$, obtained earlier, provides $\|F_y\| = \|y\|$.

We have so far proved that every vector $y \in H$ gives rise to an element $F_y \in H^*$ such that $\|y\| = \|F_y\|$. In other words the mapping $y \to F_y$ is a norm preserving mapping of H into H^*. The following theorem shows that every functional in H^* arises in just this way.

18. Theorem (F. Riesz) *Let H be a Hilbert space and let F be a continuous linear functional on H ($F \in H^*$). Then there exists a unique vector y in H such that*

$$\text{(1)} \qquad F(x) = (x,y)$$

for every x in H.

Proof of existence If F is the zero functional then we choose $y = \theta$. Hence assume that $F \neq 0$. Define

$$M = \{x \in H: F(x) = 0\}.$$

Then if $x_1, y_2 \in M$ and α, β scalars we have

$$F(\alpha x_1 + \beta y_2) = \alpha F(x_1) + \beta F(y_2) = 0.$$

Thus M is a subspace of H. Also notice that M is the inverse image of the closed set $\{0\}$ by the continuous mapping F, so that M is a closed subspace of H. Furthermore, since $F \neq 0$, we have for some $x \in H$, $F(x) \neq 0$ which shows that M is a proper closed subspace of H. By Corollary 9, there exists a nonzero element $x_o \in H$ such that $x_o \perp M$. Put $y = \alpha x_o$ where $\alpha = \overline{F(x_o)}/\|x_o\|^2$. Then $y \in M^\perp$ and $F(y) = \|y\|^2$. Let x be any element in H. Set

$$x_1 = x - \frac{F(x)}{\|y\|^2} y \quad \text{and} \quad x_2 = \frac{F(x)}{\|y\|^2} y.$$

We have $x_1 \in M$ since

$$F(x_1) = F(x) - F(x) \frac{F(y)}{\|y\|^2} = F(x) - F(x) = 0.$$

Hence

$$(x,y) = (x_1 + x_2,y) = (x_1,y) + (x_2,y) = (x_2,y)$$

$$= \left(\frac{F(x)}{\|y\|^2}\, y, y\right) = \frac{F(x)}{\|y\|^2}\,\|y\|^2 = F(x)$$

which proves the existence of a vector y such that (1) holds.

Proof of uniqueness of y Suppose there is a $y' \in H$ such that $F(x) = (x,y')$ for all $x \in H$. This implies that $(x,y') = (x,y)$ or $(x,y'-y) = 0$ for all $x \in H$. For $x = y'-y$ we get $\|y'-y\|^2 = 0$ or $y' = y$.

The result of Theorem 18 shows that the norm-preserving mapping of H into H^* defined by

$$(2) \qquad y \to F_y \quad \text{with} \quad F_y(x) = (x,y)$$

is a mapping of H onto H^*.

Observe that if H is complex, the mapping (2) is not linear since

$$F_{\alpha y_1+\beta y_2}(x) = (x,\alpha y_1 + \beta y_2) = \bar{\alpha}F_{y_1}(x) + \bar{\beta}F_{y_2}(x)$$

so that

$$F_{\alpha y_1+\beta y_2} = \bar{\alpha}F_{y_1} + \bar{\beta}F_{y_2}.$$

But in any case we have

$$\|F_{y_1} - F_{y_2}\| = \|F_{y_1-y_2}\| = \|y_1-y_2\|.$$

which proves that (2) is an isometry.

Before we close this chapter we wish to add a few words concerning isomorphic Hilbert spaces. In Chapter 2, section 3, the general concept of isomorphic mathematical systems was discussed. In the case of Hilbert spaces an isomorphism ϕ of a Hilbert space H_1 onto a Hilbert space H_2 is defined to be a one-to-one linear mapping of H_1 onto H_2, which also preserves inner products: $(\phi(x), \phi(y)) = (x,y)$ for all x and y in H_1.

The following theorem (we omit the proof) tells us that there are essentially only two types of separable Hilbert spaces. We recall that H is said to be separable iff it contains a countable dense subset.

19. Theorem *Let H be a separable Hilbert space.*

(a) If H has finite dimension it is isomorphic with $\mathbb{C}^n$ (Ex. 1).

(b) If H is infinite dimensional, it is isomorphic with ℓ^2 (Ex. 2).

PROBLEMS

1. Show that the spaces given in the Examples 1, 2, and 3 of section 9.2 are Hilbert spaces.
2. In the Euclidean space $\mathbb{R}^2$ define the norm of an element $x = \langle x_1, x_2 \rangle$ by $\|x\| = |x_1| + |x_2|$. Show that this norm does not obey the parallelogram law.

3. Let H be a Hilbert space and x_0 a fixed element of H. Show that the mappings

$$x \to (x,x_0); \quad x \to (x_0,x); \quad x \to \|x\|$$

are all continuous on H.

4. Let M be a vector subspace of a Hilbert space H. Show that $M = (M^{\perp})^{\perp}$ iff M is closed. Show that if S is a subset of H then $S^{\perp}$ is a closed subspace of H.

5. Show that a Hilbert space is separable if and only if it contains a complete orthonormal system which is at most countable.

6. Let $\langle e_n \rangle$ be a countable orthonormal system in a Hilbert space H. Show that the set $\{e_n: n \in N\}$ is closed and bounded but not compact. Let Q be the set of all elements x in H of the form

$$x = \sum_{n=1}^{\infty} c_n e_n, \quad \text{with} \quad |c_n| \le \frac{1}{n}.$$

Prove that Q (called the *Hilbert cube*) is compact.

7. Let S be a measurable subset of the interval $[0,2\pi]$. Show that

$$\lim_{n\to\infty} \int_S e^{inx}dx = 0.$$

8. Let S be a nonempty subset of a Hilbert space H. Show that if $[S]$ is the set of all linear combinations of vectors in S then $\overline{[S]} = H$ iff $S^{\perp}$ is the set consisting of the zero vector of H.

9. Show that if $\langle x_n \rangle$ is a sequence of vectors in a Hilbert space H such that every $x \in H$ is a linear combination of finitely many of the x_n, then H is finite dimensional.

10. Show that if H_1 and H_2 are Hilbert spaces, then one of them is isomorphic to a subspace of the other.

11. Show that in the Hilbert space ℓ^2 the system of vectors $\langle 1,0,\ldots \rangle$, $\langle 0,1,0,\ldots \rangle$, $\langle 0,0,1,0,\ldots \rangle,\ldots,$ is a complete orthonormal system.

12. Show that in a Hilbert space, $\|x_n\| \to \|x\|$ and $(x_n,x) \to \|x\|$ imply $x_n \to x$.

13. (Gram-Schmidt orthonormalization procedure.) Let $\langle y_n \rangle_{n=1}^{\infty}$ be a sequence of linearly independent vectors in an inner product space. Show that there exists an orthonormal system $\langle x_n \rangle_{n=1}^{\infty}$ such that for all n, the space $[x_1,x_2,\ldots,x_n]$ generated by the vectors $x_1,\ldots,x_n$, is identical to the space $[y_1,\ldots,y_n]$ generated by the vectors $y_1,\ldots,y_n$. (Hint: Let $x_1 = \|y_1\|^{-1}y_1$. Assume that orthonormal vectors $x_1,x_2,\ldots,x_{n-1}$ are already de-

fined in such a way that $[x_1, x_2, \ldots, x_k] = [y_1, y_2, \ldots, y_k]$ for $k = 1, 2, \ldots, n-1$. Set $z = y_n - \sum_{k=1}^{n-1} (y_n, x_k) x_k$ and define $x_n = \|z\|^{-1} z$. Show that $[x_1, x_2, \ldots, x_n] = [y_1, y_2, \ldots, y_n]$.)

14. In the Hilbert space $L^2(-1,1)$ (Ex. 3) consider the sequence $S = \langle 1, t, t^2, \ldots \rangle$. Show that the vectors in S are linearly independent. Show that the orthonormalization of S by the Gram-Schmidt procedure (Prob. 13) yields the sequence of polynomials

$$\left\langle c_n \cdot \frac{d^n (t^2-1)^n}{dt^n} \right\rangle \qquad (n = 0, 1, 2, \ldots),$$

where c_n are certain positive constants. These are the so-called *Legendre polynomials*.

15. Let x_1, x_2, x_3 and x_4 be four vectors in an inner product space X. Show that

$$\|x_1 - x_3\| \cdot \|x_2 - x_4\| \leq \|x_1 - x_2\| \cdot \|x_3 - x_4\| + \|x_2 - x_3\| \cdot \|x_1 - x_4\|.$$

When does equality hold? (Hint: You may suppose $x_1 = \theta$. Consider the mapping $x \to x/\|x\|$ $(x \neq \theta)$ of X into X.)

16. Let F be a bounded linear functional on a Hilbert space H. Let M be the kernel of F defined by

$$M = \{x \in H: \ F(x) = 0\}.$$

Prove that if $M \neq H$, then M is a vector space of dimension 1.

17. Let F be a linear functional on a Hilbert space H, and let $M = \{x \in H: F(x) = 0\}$ be the kernel of F. Show that if F is not bounded then $\overline{M} = H$.

18. Let f be a bounded linear functional on a subspace M of a Hilbert space H. Prove that f has a unique norm preserving extension to a bounded linear functional F on H, and that F vanishes on $M^{\perp}$.
(Hint: There is an $m_o \in M$ such that $f(m) = (m,m_o)$ for all m in M. Also there is a $y \in H$ such that $F(z) = (z,y)$ for all $z \in H$. We have $F(m_o) = f(m_o) = \|m_o\|^2 = (m_o,y) \leq \|m_o\|\|y\|$. Since $\|y\| = \|m_o\|$, it follows $|(m_o,y)| = \|m_o\|\|y\|$. Hence m_o and y are linearly dependent.)

19. Show that if H is a Hilbert space then H is reflexive.
(Hint: We must prove that the canonical mapping $\phi: H \to H^{**}$ [defined by $x \to F_x$ with $F_x(f) = f(x)$ for all $f \in H^*$] is onto H^{**}. Let $F \in H^{**}$. We want to show that there is a

$y \in H$ such that $F_y = F$. Define $g: H \to \mathbb{C}$ [$\mathbb{C}$ is the field of complex numbers] by $g(x) = \overline{F(f_x)}$ where $f_x \in H^*$ is the image of x under the mapping (2) given in Theorem 18. Show that g is in H^*. Then by Theorem 18 there is a $y \in H$ such that $g(z) = (z,y)$ for all $z \in H$. We prove that $F_y = F$. Let f be any element in H^*. Since the mapping (2) of Theorem 18 is onto H^*, there must exist an element $t \in H$ such that $f = f_t$, so that $f(z) = f_t(z) = (z,t)$ for all $z \in H$. We have $F_y(f) = F_y(f_t) = f_t(y) = (y,t)$. Also $F(f) = F(f_t) = \overline{g(t)} = \overline{(t,y)} = (y,t)$. Hence $F_y = F$.)

CHAPTER 10

Measure and Integration

10.1 INTRODUCTION

The classical theory of measure and integration, developed in the direction laid out by Lebesgue (beginning of the twentieth century) deals mainly with functions defined on the real line, and with sets of real numbers. As we have seen in Chapter 3 the Lebesgue integral generalizes in a fruitful way the familiar Riemann integral and it is relieved of some of the serious defects that the Riemann integral possesses. Later on, by the last half of the thirties, an abstract axiomatic treatment appeared to be the necessary and useful way of developing the theory of measure and integration, so that its essential structure could be perceived and the theory be applied to more general spaces than the Euclidean spaces. Although the generalized theory did not have the full force of the original one, the ideas about abstract measure and integration had found important applications in probability, and more precisely in making explicit the notion of "random variable" so important in this subject. Also integration theory in abstract spaces played an important role in the spectral theory of linear operators, in harmonic analysis on topological groups, which is an extension of classical harmonic analysis, and in many other branches of mathematics.

10.2 MEASURABLE FUNCTIONS AND MEASURE

To see how the notion of abstract measure arises from an intuitive geometrical point of view, consider the problem of formulating this notion on an arbitrary set S, as an abstraction of the concepts of length, volume, mass, and so forth. The most natural way is to define a function μ that assigns to suitable subsets E of S a number $\mu(E)$ (or in short μE), and to call this number the *measure* of E. Then we face questions such as (a) what are the "suitable sets"? (b) what are the characteristic properties of the function μ? Being acquainted with the Lebesgue measure, which is a generalization of the notion of length of an interval, we may answer question (a) by requiring that the "suitable sets" form a σ-algebra of subsets of S, and answer question (b) by demanding that μ have all the properties of the Lebesgue measure. With the above discussion in mind, we proceed now to a more systematic presentation of the subject.

Definition *(a) A* measurable space *is an ordered pair* $\langle X, \mathcal{B} \rangle$ *consisting of a set* X *and a* σ*-algebra of subsets of* X. *The members of* $\mathcal{B}$ *are called the* measurable sets *in* X.

To simplify the terminology, if no confusion is possible we shall often refer to a measurable space X instead of $\langle X, \mathcal{B} \rangle$.

(b) If X *is a measurable space,* Y *is a topological space, and* f *is a function from* X *into* Y*, then* f *is said to be* measurable *if and only if* $f^{-1}[O]$ *is a measurable set in* X *for every open set* O *in* Y*.*

The reader should realize that the definition of a measurable function depends on the σ-algebra $\mathcal{B}$ of measurable sets chosen in X. For example if $\mathcal{B}$ is taken to be the collection of all subsets of X then clearly every function $f: X \to Y$ is measurable. In particular, if X is a topological space and $\mathcal{B}$ is the collection of Borel sets of X (i.e., the smallest σ-algebra containing all open sets of X), and if for every open set O in Y the set $f^{-1}[O]$ is a Borel set, then we say that f is *Borel measurable*. For example, it follows from the definition of a continuous function that if f is continuous, then it is Borel measurable. If X is the real line $\mathbb{R}$ and $\mathcal{M}$ is the σ-algebra of measurable sets as described in Chapter 3, then an extended real-valued function on $\mathbb{R}$ is called *Lebesgue measurable* if and only if $f^{-1}[O]$ belongs to $\mathcal{M}$ for every open set in the range of f. We shall see (Theor. 1(c)) that this definition of measurable function is equivalent to the one given in Chapter 3.

If $\langle X, \mathcal{B} \rangle$ is a measurable space and if f is a measurable function on X, then f is said to be $\mathcal{B}$*-measurable*.

1. Theorem *Let* $\langle X, \mathcal{B} \rangle$ *be a measurable space,* Y *a topological space, and* f *a mapping on* X *into* Y*. Let* $\mathcal{A}$ *denote the collection of all subsets* S *of* Y *such that* $f^{-1}[S]$ *is a measurable set of* X*,*

i.e.,

$$\mathcal{A} = \{S \subset Y : f^{-1}[S] \in \mathcal{B}\}.$$

Then

(a) $\mathcal{A}$ is a σ-algebra of subsets of Y.

(b) If f is measurable, and E is a Borel set in Y, then $f^{-1}[E] \in \mathcal{B}$.

(c) If Y is the extended real line, i.e., $Y = [-\infty, \infty]$, and if for every real number α we have $f^{-1}[(\alpha, \infty)] \in \mathcal{B}$, then f is measurable.

Proof (a) First we prove that if $S \in \mathcal{A}$ then $\complement S \in \mathcal{A}$. We have $f^{-1}[S] \in \mathcal{B}$. Since $f^{-1}[\complement S] = \complement f^{-1}[S]$ and $\complement f^{-1}[S] \in \mathcal{B}$, we also have $\complement S \in \mathcal{A}$.

Next let $\langle S_n \rangle$ be a sequence of sets in $\mathcal{A}$. We prove that $\bigcup S_n \in \mathcal{A}$. We have $f^{-1}\left[\bigcup S_n\right] = \bigcup f^{-1}[S_n]$. Since for each n, $f^{-1}[S_n]$ belongs to $\mathcal{B}$, and since $\mathcal{B}$ is a σ-algebra, we have $\bigcup f^{-1}[S_n] \in \mathcal{B}$. Hence $\bigcup S_n \in \mathcal{A}$.

Proof (b) Let $\mathcal{A}$ be as in the hypothesis of the theorem. If O is any open set in Y, since f is measurable we have $f^{-1}[O] \in \mathcal{B}$. This means that $\mathcal{A}$ contains all open sets of Y, and since $\mathcal{A}$ is a σ-algebra, it follows that $\mathcal{A}$ contains the smallest σ-algebra containing the open sets, that is $\mathcal{A}$ contains the Borel sets.

Proof (c) Set $\mathcal{N} = \{E \subset [-\infty, +\infty]: f^{-1}[E] \in \mathcal{B}\}$. Then it follows from part (a) that $\mathcal{N}$ is a σ-algebra of subsets of $[-\infty,+\infty]$. We now recall that $[-\infty,+\infty]$ being the compactification of the real line by adjoining the objects $-\infty$ and $+\infty$ (see Chap. 7, sec. 9), every open set in $[-\infty,+\infty]$ is a countable union of segments of the type $[-\infty,\alpha)$, $(\alpha,+\infty]$, $[-\infty,\beta) \cap (\alpha,+\infty]$ with α and β real numbers. By assumption $(\alpha,+\infty] \in \mathcal{N}$ for every real α. Since

$$[-\infty,\alpha] = \complement\{(\alpha,+\infty]\}$$

$$[-\infty,\alpha) = \bigcup_{n=1}^{\infty} [-\infty,\ \alpha - \frac{1}{n}\ ] = \bigcup_{n=1}^{\infty} \complement\{(\alpha - \frac{1}{n},\ +\infty]\}$$

it follows, from the fact that $\mathcal{N}$ is a σ-algebra, that the sets $[-\infty,\alpha)$ and $[-\infty,\beta) \cap (\alpha,+\infty]$ also belong to $\mathcal{N}$. Hence for every open set O in $[-\infty,+\infty]$, we have $f^{-1}[O] \in \mathcal{B}$ which proves that f is measurable.

Remark Part (c) of Theorem 1 shows that an extended real-valued function f defined on the real line is measurable if and only if the set $\{x: f(x) > \alpha\}$ is measurable for every real number α. This agrees with the definition of a measurable function given in Chapter 3.

2. Theorem *Let f and g be extended real-valued measurable functions and let c be a real number. Then the functions $f + c$, cf, $f + g$, $f \cdot g$ are also measurable. Furthermore, if $\langle f_n \rangle$ is a sequence of extended real-valued measurable functions, then the functions*

$$\sup_n f_n \ , \quad \inf_n f_n$$

are both measurable, and consequently

$$\overline{lim}_{n\to\infty} f_n \quad and \quad \underline{lim}_{n\to\infty} f_n$$

are also measurable.

The proof of this theorem is similar to the proofs of Theorems 12 and 13 of Chapter 13, and is left to the reader.

We also observe, as we did in Chapter 3, that the conventions $(+\infty) + (-\infty) = 0$ and $0\cdot(\pm\infty) = 0$ are used here.

A function ϕ on a measurable space X is said to be *simple* if it is measurable and its range is a finite set of real numbers. If $a_1, a_2, \ldots, a_n$ are the distinct values of a simple function ϕ and $A_i = \{x\colon \phi(x) = a_i\}$ for $i = 1,2,\ldots,n$, then

$$\phi = \sum_{i=1}^{n} a_i \chi_{A_i}$$

where χ_{A_i} is the characteristic function of the set A_i.

3. Theorem *Let X be a measurable space and let $f\colon X \to [0,\infty]$ be measurable. Then there exists a sequence $\langle f_n \rangle$ of nonnegative simple measurable functions with $f_{n+1} \geq f_n$ such that*

$$lim_{n\to\infty} f_n(x) = f(x)$$

for every $x \in X$ (see Chap. 3, Theor. 15).

Definition *A* **measure** *on a measurable space* $\langle X, \mathcal{B} \rangle$ *is a function* μ *defined for all sets of* $\mathcal{B}$, *whose range is in* $[0,\infty]$, *and which has the following properties.*

(a) $\mu(\emptyset) = 0$.

(b) If $\{E_n\}$ *is a countable collection of pairwise disjoint members of* $\mathcal{B}$ *then*

$$\mu\left(\bigcup_{n=1}^{\infty} E_n\right) = \sum_{n=1}^{\infty} \mu(E_n).$$

Property *(b)* of μ is referred to by saying that μ is *countably additive*.

Definition *A* **measure space** $\langle X, \mathcal{B}, \mu\rangle$ *is a measurable space* $\langle X, \mathcal{B} \rangle$ *together with a measure* μ *defined on* $\mathcal{B}$. *If no confusion is possible we shall refer to a measure space* X *instead of* $\langle X, \mathcal{B}, \mu\rangle$.

4. **Theorem** *Let* μ *be a measure on a measurable space* $\langle X, \mathcal{B} \rangle$. *Then*

(a) $\mu\left(\bigcup_{i=1}^{n} E_i\right) = \sum_{i=1}^{n} \mu(E_i)$ *if* $E_1, E_2, \ldots, E_n$ *are pairwise disjoint members of* $\mathcal{B}$.

(b) If $A \in \mathcal{B}$, $B \in \mathcal{B}$, *and* $A \subset B$, *then* $\mu(A) \leq \mu(B)$.

(c) Let $\langle E_n \rangle$ *be a sequence of members of* $\mathcal{B}$ *with* $E_n \subset E_{n+1}$ *for each* n. *Then if* $E = \bigcup_{n=1}^{\infty} E_n$ *we have* $\mu(E) = \lim_{n\to\infty} \mu(E_n)$.

(d) Let $\langle E_n \rangle$ *be a sequence of members of* $\mathcal{B}$ *with* $E_n \supset E_{n+1}$ *for each* n, *and* $\mu(E_1) < +\infty$. *Then*

if $E = \bigcap_{n=1}^{\infty} E_n$ *we have* $\lim_{n\to\infty} \mu(E_n) = \mu(E)$.

Proof The proofs of *(a)* and *(b)* being simple, are left to the reader.

Proof (c) Define the sequence $\langle A_n \rangle$ by setting $A_1 = E_1$ and $A_n = E_n - E_{n-1}$ for $n \geq 2$. Then the A_n are pairwise disjoint members of $\mathcal{B}$. Also, we easily see that

$$E = \bigcup_{n=1}^{\infty} A_n \quad \text{and} \quad E_n = \bigcup_{i=1}^{n} A_i .$$

Using part *(a)* and the countable additivity property of μ we get

$$\mu(E_n) = \sum_{i=1}^{n} \mu(A_i) \quad \text{and} \quad \mu(E) = \sum_{n=1}^{\infty} \mu(A_n).$$

Thus the $\mu(E_n)$ are the partial sums of the series $\sum_{n=1}^{\infty} \mu(A_n)$ so that

$$\mu(E) = \lim_{n\to\infty} \mu(E_n).$$

Proof (d) Define $B_n = E_1 - E_n$ for $n = 1,2,\ldots$.

Then $B_n \subset B_{n+1}$, and $E_1 - E = \bigcup_{n=1}^{\infty} B_n$. Since $E_1 = E_n \cup B_n$ and $\mu(E_1) < +\infty$, we have $\mu(B_n) = \mu(E_1) - \mu(E_n)$. By applying part *(c)* to the sequence

$\langle B_n \rangle$ we obtain

$$\mu(E_1) - \mu(E) = \mu(E_1 - E) = \lim_{n\to\infty} \mu(B_n)$$

$$= \mu(E_1) - \lim_{n\to\infty} \mu(E_n)$$

or

$$\mu(E) = \lim_{n\to\infty} \mu(E_n).$$

In general, for a given measurable space X the construction of a "useful" measure on X is not an easy task. Besides the Lebesgue measure discussed in Chapter 3, here are two other examples of measures. The proofs, being simple, are left to the reader.

Example (a)

Let X be an arbitrary set and let $\mathcal{B}$ be the collection of all subsets of X. For any $E \in \mathcal{B}$ define $\mu(E) = +\infty$ if E is an infinite set, and let $\mu(E)$ be the number of elements in E if E is finite. The measure μ is called the *counting measure* on X.

Example (b)

Let X and $\mathcal{B}$ be as in Example (a), and let $a \in X$. For $E \in \mathcal{B}$, define $\mu(E) = 1$ if $a \in E$, and $\mu(E) = 0$ if $a \notin E$.

10.3 INTEGRATION

In this section X will be an arbitrary non-empty set, $\mathcal{B}$ a σ-algebra of subsets of X, and μ a measure on $\mathcal{B}$.

Definition *Let* ϕ *be a real-valued measurable nonnegative simple function on* X, *of the form*

$$\phi = \sum_{i=1}^{n} a_i \chi_{A_i}$$

where $a_1, a_2, \ldots, a_n$ *are the distinct values of* ϕ *and* $A_i = \{x : \phi(x) = a_i\}$, $i = 1, 2, \ldots, n$. *Let* $E \in \mathcal{B}$. *Then the* **integral** *of* ϕ *over* E *is defined by*

$$\int_E \phi d\mu = \sum_{i=1}^{n} a_i \mu(A_i \cap E).$$

(*Note*: If for some i we have $a_i = 0$, and $\mu(A_i \cap E) = +\infty$, then using the convention $0 \cdot (+\infty) = 0$ we put $a_i \mu(A_i \cap E) = 0$.)

Definition *Let* $f: X \to [0, \infty]$ *be an extended real-valued nonnegative measurable function, and* $E \in \mathcal{B}$. *Then the* **integral** *of* f *over* E *is defined by*

$$\int_E f d\mu = \sup\{\int_E \phi d\mu\}$$

where the supremum is taken over all simple measurable functions ϕ, *such that* $0 \leq \phi \leq f$.

(*Note*: The reader should observe that if f is a simple function, then the two definitions given above provide the same number as the integral of f.)

Several of the following theorems abstract the most important properties of the Lebesgue integral.

Often the proofs, being identical to those of the corresponding theorems in Chapter 3, will, therefore, be omitted.

5. Theorem *Let f and g be nonnegative extended real-valued measurable functions on X, let A, B and E be measurable sets, and c a nonnegative real number. Then*

(a) If $f \leq g$ then $\int_E f d\mu \leq \int_E g d\mu$

(b) If $A \subset B$ then $\int_A f d\mu \leq \int_B f d\mu$

(c) $\int_E (cf) d\mu = c \int_E f d\mu$

(d) If $f(x) = 0$ for all $x \in E$ then $\int_E f d\mu = 0$

(e) If $\mu(E) = 0$ then $\int_E f d\mu = 0$

(f) $\int_E f d\mu = \int_X f \cdot \chi_E d\mu$.

The proof is a direct consequence of the definition of the integral. We prove only part *(a)* and leave the rest to the reader.

Proof (a) We have

$$\int_E f d\mu = \sup\{\int_E \phi d\mu\} \quad ; \quad \int_E g d\mu = \sup\{\int_E s d\mu\}$$

where ϕ and s are simple nonnegative measurable functions such that $\phi \leq f$ and $s \leq g$. Put

$$S_1 = \{\phi : 0 \leq \phi \leq f\}, \quad S_2 = \{s : 0 \leq s \leq g\}.$$

Since $f \leq g$ we have $S_1 \subset S_2$. This implies

$$\sup_{\phi \varepsilon S_1} \{\int_E \phi d\mu\} \leq \sup_{s \varepsilon S_2} \{\int_E s d\mu\}$$

and so

$$\int_E f d\mu \leq \int_E g d\mu.$$

6. Theorem (Monotone Convergence Theorem) *Let* $\langle f_n \rangle$ *be a sequence of measurable nonnegative extended real-valued functions on* X *such that* $f_n \leq f_{n+1}$ *for* $n = 1,2,\ldots,$ *and* $\lim_{n\to\infty} f_n(x) = f(x)$ *for every* $x \varepsilon X$. *Then* f *is measurable and*

$$\lim_{n\to\infty} \int_X f_n d\mu = \int_X f d\mu.$$

7. Theorem *Let* $f_n \colon X \to [0,\infty]$ *be measurable for* $n = 1,2,\ldots$. *Define*

$$f(x) = \sum_{n=1}^{\infty} f_n(x).$$

Then

$$\int_X f d\mu = \sum_{n=1}^{\infty} \int_X f_n d\mu$$

(see Chap. 3, Prop. 29).

8. Theorem (Fatou's Lemma) *Let* $f_n \colon X \to [0,\infty]$ *be measurable for* $n = 1,2,\ldots$. *Then*

$$\int_X (\underline{lim}_{n\to\infty} f_n) d\mu \leq \underline{lim}_{n\to\infty} \int_X f_n d\mu$$

(see Chap. 3, Theor. 30).

9. Theorem *Let* $f\colon X \to [0,\infty]$ *be measurable and consider the mapping*

$$\phi\colon \mathcal{B} \to [0,\infty]$$

defined by

$$\phi(E) = \int_E f d\mu \quad \textit{for} \quad E \in \mathcal{B}.$$

Then ϕ *is a measure on the measurable space* $\langle X, \mathcal{B} \rangle$. *Also for every measurable nonnegative extended real-valued function* g *we have*

$$(1) \qquad \int_X g d\phi = \int_X g f d\mu.$$

Proof Let $\langle E_n \rangle$ be a sequence of pairwise disjoint measurable sets of X. Set $E = \bigcup_{n=1}^{\infty} E_n$. Then

$$f \cdot \chi_E = \sum_{n=1}^{\infty} f\chi_{E_n}.$$

To see this let $x \in X$. If x is not in E then

$$(f\chi_E)(x) = 0 = \sum_{n=1}^{\infty} (f\chi_{E_n})(x).$$

If x is in E, then since the E_n are pairwise disjoint, there must exist a n_o such that $x \in E_{n_o}$ and $x \notin E_n$ for $n \neq n_o$. Hence, for this x

$$(f\chi_E)(x) = f(x)\chi_E(x) = f(x)\cdot 1$$

$$= f(x)\cdot\chi_{E_{n_o}}(x) = \sum_{n=1}^{\infty}(f\chi_{E_n})(x)$$

so that

$$f\chi_E = \sum_{n=1}^{\infty} f\chi_{E_n}.$$

We have

$$\phi(E) = \int_E f d\mu = \int_X f\chi_E d\mu;$$

$$\phi(E_n) = \int_{E_n} f d\mu = \int_X f\chi_{E_n} d\mu.$$

If we apply Theorem 7 to the sequence $\langle f\chi_{E_n}\rangle$ we get

$$\phi(E) = \sum_{n=1}^{\infty}\phi(E_n).$$

So far we have proved that ϕ is a nonnegative countably additive extended real-valued function on $\mathcal{B}$. To see that ϕ is a measure it remains to prove that $\phi(\emptyset) = 0$. But this follows immediately from the definition of ϕ.

To prove *(1)* observe that *(1)* holds if g is the characteristic function of a measurable set E. Hence *(1)* holds for every simple measurable function g, and the general case follows from the monotone convergence theorem.

(*Note*: Theorem 9 tells us that every measurable function $f: X \to [0,\infty]$ defines a measure on $\langle X, \mathcal{B} \rangle$. It is a fact of great importance that the converse of Theorem 9, called the *Radon-Nikodym theorem*, is also true. It will be proved later [see Theor. 21].)

If f is an arbitrary extended real-valued function on X, the functions

$$f^+ = \max\{f,0\} \quad ; \quad f^- = \max\{-f,0\}$$

are called the *positive* and *negative parts* of f, respectively. We have

$$|f| = f^+ + f^- \quad \text{and} \quad f = f^+ - f^-.$$

Clearly if f is measurable, then f^+ and f^- are measurable too (see Prob. 2).

Definition (a) *A nonnegative measurable function $f: X \to [0,\infty]$ is said to be* integrable *if and only if*

$$\int_X f d\mu < +\infty .$$

(b) *An arbitrary measurable function $f: X \to [-\infty,\infty]$ is said to be integrable if and only if both f^+ and f^- are integrable. In this case*

$\int_E f d\mu$, *the integral of* f *over a measurable set* E, *is defined by*

$$\int_E f d\mu = \int_E f^+ - \int_E f^- .$$

The integral $\int_E f d\mu$ *is written sometimes as* $\int_E f(x) d\mu(x)$, $(x \in E)$.

10. Theorem *Let* f *and* g *be integrable functions and let* a *and* b *be real numbers. Then* $af + bg$ *is integrable, and*

$$\int_X af + bg = a \int_X f + b \int_X g .$$

(see Chap. 3, Theor. 33.)

Remark Theorem 10 shows that the integrable functions form a vector space, and that the integral is a linear functional on this space.

The reader is already familiar with the concept of "almost everywhere" (abbreviated a.e.). For example, in Chapter 3 a function f was said to vanish a.e. if and only if the set $\{x: f(x) \neq 0\}$ was of Lebesgue measure zero. The same concept can be introduced in the case of an abstract measure space. Let $\langle X, \mathcal{B}, \mu \rangle$ be a measure space and let P be a property that a point x in X may or may not have. For example, if f is a given function on X, P may be the property that $f(x) = 0$. Let S be any measurable set in X, and let S_0 be the set of all points in S that do not have the property P, or, equivalently, the set of all points in S where P does not hold. Then we say that P holds on S a.e. if and only if $\mu(S_o) = 0$. Clearly the concept of "almost everywhere" depends on the measure μ. For if μ_1 and μ_2 are two different measures on $\langle X, \mathcal{B} \rangle$, then it might occur that $\mu_1(S_o) = 0$ and $\mu_2(S_o) \neq 0$, which would mean that P holds a.e. on S with respect to the measure μ_1 but not with respect to the measure μ_2. For this reason, whenever clarity requires, we shall specify the measure μ with respect to which P holds a.e. and we shall write "P holds a.e. $[\mu]$ on S."

Sets of measure zero are, so to speak, negligible in integration. This means that if an integrable function is modified on a set of measure zero then it remains integrable and the value of its integral does not change. More precisely, if S is a measurable set of X, if S_o is a measurable subset of S with $\mu S_o = 0$, and if f is an integrable function on S, then

$$\int_S f d\mu = \int_{S-S_o} f d\mu.$$

To see this, observe that $S = (S - S_o) \cup S_o$. Then $S - S_o$ and S_o being disjoint, we have

$$\int_S f d\mu = \int_{S-S_o} f d\mu + \int_{S_o} f d\mu = \int_{S-S_o} f d\mu,$$

for $\int_{S_o} fd\mu = 0$, since $mS_o = 0$.

It follows from the above discussion that if f and g are integrable functions on a measurable set S, and if $f(x) = g(x)$ a.e. $[\mu]$, then

$$\int_S fd\mu = \int_S gd\mu.$$

We recall that in the theory of Lebesgue measure if a set is of measure zero then every subset of S is measurable and of measure zero. In other words the σ-algebra $\mathcal{M}$ of the Lebesgue measurable sets has the property that it contains all the subsets of all sets of measure zero. However, this is not true in every measure space $\langle X, \mathcal{B}, \mu\rangle$. That is, it may happen that a set S in $\mathcal{B}$ with $\mu(S) = 0$ has a subset $S_o \subset S$ which is not even measurable, i.e., $S_o \notin \mathcal{B}$.

If this last phenomenon does not occur the consequences are important enough to justify a name for this property of the measure. We give the following definition.

Definition *Let* $\langle X, \mathcal{B}, \mu\rangle$ *be a measure space. Then the measure* μ *is said to be* complete *iff* $\mathcal{B}$ *contains all subsets of all sets of measure zero; i.e., if* $S \in \mathcal{B}$, $\mu(S) = 0$ *and* $S_o \subset S$, *then* $S_o \in \mathcal{B}$.

If μ *is a complete measure then* $\langle X, \mathcal{B}, \mu\rangle$ *is called a* complete measure space.

If μ is not a complete measure, the following theorem shows that it is possible to enlarge $\mathcal{B}$, and extend the definition of μ so as to obtain a complete measure.

Borel measure is an example of an incomplete measure. For details see Gelbaum and Olmsted [10], Example 17, page 98.

11. Theorem *Let* $\langle X, \mathcal{B}, \mu\rangle$ *be a measure space. Let* $\mathcal{N}$ *be the class of all sets* A *such that to each* A *corresponds some* $E \in \mathcal{B}$ *with* $A \subset E$ *and* $\mu(E) = 0$. *Let* $\bar{\mathcal{B}}$ *be the class of all sets of the form* $E \cup A$ *with* $E \in \mathcal{B}$ *and* $A \in \mathcal{N}$. *Then*

(a) $\bar{\mathcal{B}}$ *is a* σ*-algebra containing* $\mathcal{B}$.

(b) If $E_1 \cup A_1$ *and* $E_2 \cup A_2$ $(E_1, E_2 \in \mathcal{B}$ *and* $A_1, A_2 \in \mathcal{N})$ *are two representations of the same set in* $\bar{\mathcal{B}}$, *then* $\mu(E_1) = \mu(E_2)$.

(c) The function $\bar{\mu}: \bar{\mathcal{B}} \to \bar{\mathbb{R}}$ *(where* $\bar{\mathbb{R}}$ *is the extended real number system) defined by*

$$\bar{\mu}(E \cup A) = \mu(E).$$

is a complete measure on $\bar{\mathcal{B}}$.

Proof (a) Obviously $\mathcal{B} \subset \bar{\mathcal{B}}$ since $\emptyset \in \mathcal{N}$. Next we verify the defining properties of a σ-algebra. Since $X \in \mathcal{B}$ we have $X \in \bar{\mathcal{B}}$. We have for $E_n \in \mathcal{B}$ and $A_n \in \mathcal{N}$ $(n = 1,2,\ldots)$, $\bigcup_n (E_n \cup A_n) = \left(\bigcup_n E_n\right) \cup \left(\bigcup_n A_n\right)$ so that $\bigcup_n (E_n \cup A_n) \in \bar{\mathcal{B}}$, since $\bigcup_n E_n \in \mathcal{B}$ and $\bigcup_n A_n \in \mathcal{N}$. To see that $\bar{\mathcal{B}}$ is closed under the formation of differences, we first observe that if $E \in \mathcal{B}$ and $A \in \mathcal{N}$, then $E-A \in \bar{\mathcal{B}}$. For suppose $A \subset F$ where $F \in \mathcal{B}$ and $\mu(F) = 0$. Let

$M = F - A$. Then $E - A = (E - F) \cup (E \cap M) \in \bar{\mathcal{B}}$. If $E_1, E_2 \in \mathcal{B}$ and $A_1, A_2 \in \mathcal{N}$, it is not difficult to show that $(E_1 \cup A_1) - (E_2 \cup A_2) = \{(E_1 - E_2) - A_2\} \cup \{(A_1 - A_2) - E_2\}$. This result combined with the fact that $E - A \in \bar{\mathcal{B}}$ shows that $\bar{\mathcal{B}}$ is closed under differences. Therefore $\bar{\mathcal{B}}$ is a σ-algebra.

Proof (b) We have seen that

$$(E_1 \cup A_1) - (E_2 \cup A_2)$$

$$= \{(E_1 - E_2) - A_2\} \cup \{(A_1 - A_2 - E_2)\}.$$

Thus if $E_1 \cup A_1 = E_2 \cup A_2$, then $(E_1 - E_2) - A_2 = \emptyset$ which implies that $E_1 - E_2 = (E_1 - E_2) \cap A_2$, and so $\mu(E_1 - E_2) = 0$. Also since $E_1 = (E_1 \cap E_2) \cup (E_1 - E_2)$ we get $\mu(E_1) = \mu(E_1 \cap E_2)$. Exchanging the roles of E_1 and E_2 we also have $\mu(E_2) = \mu(E_1 \cap E_2)$ so that $\mu(E_1) = \mu(E_2)$.

Proof (c) It is now easy to prove that the well-defined function $\bar{\mu}$ is a complete measure on $\bar{\mathcal{B}}$. The details of the proof are left to the reader.

The measure $\bar{\mu}$ derived from μ in the above theorem is called *completion* of μ. The measure $\bar{\mu}$ is a minimal extension of μ. By this we mean that if μ_1 is a complete measure on a σ-algebra $\mathcal{B}_1$ such that $\mathcal{B} \subset \mathcal{B}_1$ and $\mu_1(E) = \mu(E)$ if $E \in \mathcal{B}$, then $\bar{\mathcal{B}} \subset \mathcal{B}_1$ and $\mu_1(S) = \bar{\mu}(S)$ when $S \in \bar{\mathcal{B}}$. To see this suppose $A \in \mathcal{N}$ with $A \subset F$, $F \in \mathcal{B}$, $\mu(F) = 0$. Then $\mu_1(F) = 0$ and so $A \in \mathcal{B}_1$, $\mu_1(A)$

$= 0$, because μ_1 is complete. Now an arbitrary element $E \cup A$ of $\bar{\mathcal{B}}$ belongs to $\mathcal{B}_1$. Furthermore $\mu_1(E) \leq \mu_1(E \cup A) \leq \mu_1(E) + \mu_1(A) = \mu_1(A)$, so that $\mu_1(E \cup A) = \mu_1(E) = \mu(E) = \overline{\mu}(E \cup A)$.

(*Note*: Since every measure can be completed, we shall assume from now on that any given measure is complete.)

12. Theorem (Lebesgue's Dominated Convergence Theorem) *Let* $\langle f_n \rangle$ *be a sequence of measurable functions such that*

$$\lim_{n\to\infty} f_n(x) = f(x) \quad \text{a.e. on } X.$$

Suppose there exists an integrable function g *such that*

$$|f_n(x)| \leq g(x) \quad (n = 1,2,\ldots) \quad \text{for all } x \in X.$$

Then f *is integrable and*

$$\lim_{n\to\infty} \int_X |f_n - f| = 0 \quad ; \quad \lim_{n\to\infty} \int_X f_n = \int_X f$$

(see Chap. 3, Theor. 34).

13. Theorem (Bounded Convergence Theorem) *Let* $\langle f_n \rangle$ *be a sequence of measurable functions defined on a measurable set* E *of finite measure. Suppose that there exists a real number* M *such that*

$|f_n(x)| \leq M$ *for all* n *and all* $x \in E$. *Suppose also that*

$$\lim_{n\to\infty} f_n(x) = f(x)$$

for each $x \in E$. *Then*

$$\lim_{n\to\infty} \int_E f_n = \int_E f$$

(see Chap. 3, Theor. 35).

14. Theorem *(a) Let* $f: X \to [0,\infty]$ *be measurable, and suppose that* $\int_E fd\mu = 0$ $(E \in \mathcal{B})$. *Then* $f = 0$ *a.e.* $[\mu]$.

(b) If f *is integrable on* X *and if* $\int_E fd\mu = 0$ *for every measurable set* E, *then* $f = 0$ *a.e.* $[\mu]$.

Proof (a) Put $E_n = \{x \in E: f(x) > (1/n)\}$ $(n = 1, 2,\ldots)$. Then

$$\frac{1}{n}\,\mu(E_n) \leq \int_{E_n} fd\mu \leq \int_E fd\mu = 0,$$

and so $\mu(E_n) = 0$. We have $\{x \in E: f(x) > 0\} = \bigcup_{n=1}^{\infty} E_n$ so that $\mu\{x \in E: f(x) > 0\} = 0$. Hence $f = 0$ a.e.

Proof (b) We have $f = f^+ - f^-$. Consider the set $S = \{x: f(x) > 0\}$. Then by hypothesis, we have

$$0 = \int_S f d\mu = \int_S f^+ d\mu - \int_S f^- d\mu = \int_S f^+ d\mu$$

and by part *(a)*, we have $f^+ = 0$ a.e. $[\mu]$.

We find in a similar manner that $f^- = 0$ a.e. $[\mu]$. Hence $f = 0$ a.e. $[\mu]$.

10.4 SIGNED MEASURES

Let $\langle X, \mathcal{B}, \mu\rangle$ be a measure space and let f be any integrable function on X. Then, by Theorem 9, the function $\phi: \mathcal{B} \to [0,\infty]$ defined by

$$\phi(E) = \int_E |f| d\mu \qquad (E \in \mathcal{B})$$

is a measure on the measurable space $\langle X, \mathcal{B}\rangle$.

Consider now the function ν defined on $\mathcal{B}$ by

$$\nu(E) = \int_E f d\mu \qquad (E \in \mathcal{B}).$$

Clearly, since f is integrable ν is seen to be a countably additive real-valued ($\nu(E)$ is finite) set function on $\mathcal{B}$. The function ν is an example of what we shall call *signed measure* (see definition below). Observe that $\nu(E)$ may be negative.

Our purpose is to investigate, up to a certain degree, the nature of a signed measure. There are three aspects of this investigation. First we shall prove (see Theor. 16) that every signed measure can be expressed as the difference of two measures. For example, in the case of the signed measure ν considered above we can write (see Prob. 14) $\nu = \nu^+ - \nu^-$ where ν^+ and ν^- are the measures defined by

$$\nu^+(E) = \int_E f^+ d\mu \quad , \quad \nu^-(E) = \int_E f^- d\mu \ .$$

The measure ϕ then can be expressed as $\phi = \nu^+ + \nu^-$. But if a signed measure is not given in terms of another measure μ, and a function f, then the problem is how to find ν^+ and ν^- directly from ν.

The second stage of the investigation is to determine when a signed measure is expressible as an integral, as in the case with ν. Here the key result is the *Radon-Nikodym theorem* (see Theor. 21).

The third aspect of our study is to show that under some rather general conditions any signed measure can be expressed as the sum of a so-called "absolutely continuous" signed measure and a "singular" signed measure (Theor. 22).

Definition *Let* $\langle X, \mathcal{B} \rangle$ *be a measurable space. Then a function* $\nu : \mathcal{B} \to [-\infty, +\infty]$ *is said to be a* signed measure *on* $\langle X, \mathcal{B} \rangle$ *if and only if the following conditions are satisfied:*

(a) At most one of the values $-\infty, +\infty$ *is in the range of* ν.

(b) $\nu(\emptyset) = 0$

(c) ν *is countably additive: For any sequence* $\langle E_n \rangle$ *of disjoint measurable sets* $(E_n \in \mathcal{B})$ *we have*

$$\nu\left(\bigcup_{n=1}^{\infty} E_n\right) = \sum_{n=1}^{\infty} \nu(E_n) \ .$$

Remark Observe that if ν is a signed measure, and if $\nu\left(\bigcup_{n=1}^{\infty} E_n\right)$ is finite, then the series $\sum_{n=1}^{\infty} \nu(E_n)$ converges absolutely. For since the union of the sets E_n is not changed if the subscripts are permuted, every arrangement of the series must converge to the same number $\nu\left(\bigcup_{n=1}^{\infty} E_n\right)$ and so by a theorem due to Riemann, the series must converge absolutely.

If $\nu\left(\bigcup_{n=1}^{\infty} E_n\right)$ is not finite, then equality in *(c)* is taken to mean that the series properly diverges.

Clearly a measure is also a signed measure, but a signed measure is not necessarily a measure.

A signed measure ν is not in general monotone. That is $E \in \mathcal{B}$, $F \in \mathcal{B}$, $E \subset F$ does not always imply $\nu(E) \leq \nu(F)$.

A measure or a signed measure ν is called *σ-finite* if and only if each measurable set E is expressible as a countable union of measurable sets $E_1, E_2, \ldots, E_n, \ldots,$ such that $\nu(E_n)$ is finite for each n.

A measure space $\langle X, \mathcal{B}, \mu \rangle$ is said to be a *σ-finite measure space* if and only if μ is a σ-finite measure.

Definition *Let* ν *be a signed measure on a measurable space* $\langle X, \mathcal{B} \rangle$. *We define the function* $|\nu| : \mathcal{B} \to [0, +\infty]$ *by*

$$|\nu|(E) = \sup \sum_{i=1}^{n} |\nu(E_i)| \qquad (E \in \mathcal{B})$$

where the supremum is taken over all possible finite collections $\{E_1, E_2, \ldots, E_n\}$ *of disjoint sets with* $E_i \in \mathcal{B}$, *and* $E_i \subset E$ *for* $i = 1,2,\ldots,n$. *Then* $|\nu|(E)$ *is called the* total variation *of* ν *on* E. *The function* $|\nu|$ *is called the total variation of* ν.

15. Theorem *If* ν *is a signed measure on a measurable space* $\langle X, \mathcal{B} \rangle$ *then the total variation* $|\nu|$ *of* ν *is a measure on* $\langle X, \mathcal{B} \rangle$.

Proof From the definition of $|\nu|$ follows that $|\nu|(\emptyset) = 0$. Let $\langle E_n \rangle$ be a sequence of disjoint measurable sets in X. Put $E = \bigcup_{n=1}^{\infty} E_n$. Let $A_1, A_2, \ldots, A_k$ be any finite collection of disjoint measurable sets in E and let $A_{in} = A_i \cap E_n$ $(i = 1,2,\ldots,k)$. We have

$$A_i = \bigcup_{n=1}^{\infty} A_{in} \quad , \quad \nu(A_i) = \sum_{n=1}^{\infty} \nu(A_{in}) \; .$$

Also

$$\sum_{i=1}^{k} |\nu(A_i)| \le \sum_{n=1}^{\infty} \left(\sum_{i=1}^{k} |\nu|(A_{in}) \right) \le \sum_{n=1}^{\infty} |\nu|(E_n),$$

and so

$$|\nu|(E) \leq \sum_{n=1}^{\infty} |\nu|(E_n).$$

We now prove the reverse of this last inequality. We assume $|\nu|(E) < +\infty$, for otherwise there is nothing to prove. Since $E_n \subset E$ the definition of $|\nu|$ shows that $|\nu|(E_n)$ is also finite. Let $\varepsilon > 0$ be given. Then there is a finite collection of disjoint measurable sets in E_n, say $\{E_{in}\}_{i=1}^{p}$, such that

$$\sum_{i=1}^{p} |\nu(E_{in})| > |\nu|(E_n) - \frac{\varepsilon}{2^n}.$$

For any arbitrary positive integer M we have

$$\sum_{n=1}^{M} |\nu|(E_n) < \sum_{n=1}^{M} \left[\sum_{i=1}^{p} |\nu(E_{in})| + \frac{\varepsilon}{2^n}\right] \leq |\nu|(E)+\varepsilon.$$

Since M and ε were arbitrary we get

$$\sum_{n=1}^{\infty} |\nu|(E_n) \leq |\nu|(E).$$

Hence $|\nu|(E) = \sum_{n=1}^{\infty} |\nu|(E_n)$, and so $|\nu|$ is a measure.

Definition *If* ν *is a signed measure on a measurable space* $\langle X, \mathcal{B}\rangle$, *then the functions* ν^+ *and* ν^- *are defined on* $\mathcal{B}$ *as follows:*

$$\nu^+(E) = \sup\{\nu(A) : A \in \mathcal{B},\ A \subset E\}$$

$$\nu^-(E) = -\inf\{\nu(A) : A \in \mathcal{B},\ A \subset E\}\ .$$

ν^+ *and* ν^- *are called* the positive and negative variations of ν *respectively.*

Clearly for each $E \in \mathcal{B}$ we have $\nu^+(E) \geq 0$, $\nu^-(E) \geq 0$. Since $-\nu$ is also a signed measure, by applying these definitions to $-\nu$ we obtain $\nu^- = (-\nu)^+$.

16. Theorem *Let* ν *be a signed measure on a measurable space* $\langle X, \mathcal{B}\rangle$. *Then*

(a) ν^+ *and* ν^- *are measures on* $\mathcal{B}$.

(b) If for some $E \in \mathcal{B}$, $\nu^+(E) = +\infty$, *then* $\nu(E) = +\infty$. *If for some* $E \in \mathcal{B}$, $\nu^-(E) = +\infty$, *then* $\nu(E) = -\infty$, *and so one of the functions* ν^+, ν^- *has no infinite values.*

(c) $\nu = \nu^+ - \nu^-$

(d) $|\nu| = \nu^+ + \nu^-$.

Proof (a) Clearly $\nu^+(\emptyset) = 0$. Let $\langle E_n\rangle$ be a sequence of pairwise disjoint measurable sets, with

$E = \bigcup_{n=1}^{\infty} E_n$. If $A \in \mathcal{B}$, $A \subset E$, then

$A = \bigcup_{n=1}^{\infty} (A \cap E_n)$ so that

$$\nu(A) = \sum_{n=1}^{\infty} \nu(A \cap E_n) \leq \sum_{n=1}^{\infty} \nu^+(E_n)$$

which implies that

$$\nu^+(E) \leq \sum_{n=1}^{\infty} \nu^+(E_n) .$$

To prove the reverse inequality assume that $\nu^+(E) < +\infty$, for otherwise there is nothing to prove. Then, as we pointed out in the proof of the preceding theorem, $\nu^+(E_n) < +\infty$. Let $\varepsilon > 0$ be given. Then there is $A_n \varepsilon \mathcal{B}$, $A_n \subset E_n$ such that $\nu(A_n) > \nu^+(E_n) - (\varepsilon/2^n)$. Then $A = \bigcup_{n=1}^{\infty} A_n$ is in $\mathcal{B}$, $A \subset E$, and $\nu(A) \leq \nu^+(E)$. We have

$$\sum_{n=1}^{\infty} \nu^+(E_n) < \sum_{n=1}^{\infty} \{\nu(A_n) + \frac{\varepsilon}{2^n}\}$$

$$= \nu(A) + \varepsilon \leq \nu^+(E) + \varepsilon .$$

Since ε was arbitrary we get $\nu^+(E) \geq \sum_{n=1}^{\infty} \nu^+(E)$. This proves that ν^+ is a measure. Since $\nu^- = (-\nu)^+$ we also conclude that ν^- is a measure.

Proof (b) Assume that for some $E \varepsilon \mathcal{B}$ we have $\nu^+(E) = +\infty$. Then there exists an $E_1 \varepsilon \mathcal{B}$, $E_1 \subset E$ with $\nu^+(E_1) > 1$. Since $\nu^+(E) = \nu^+(E_1) + \nu^+(E - E_1) = +\infty$, for at least one of the sets E_1 and $E - E_1$ the value of ν^+ is $+\infty$. Let F_1 be the one of these sets for which $\nu^+(F_1) = +\infty$. Putting $F_0 = E$, and proceeding by induction we obtain sequences $\langle E_n \rangle$ and $\langle F_n \rangle$ such that $E_n \subset F_{n-1}$, $\nu(E_n) > n$, and F_n is either E_n or $F_{n-1} - E_n$ the choice being such as to insure that $\nu^+(F_n) = +\infty$. We have $F_{n+1} \subset F_n$. Hence if $m < n$, and we have both $F_m = F_{m-1} - E_m$ and $F_n = F_{n-1} - E_n$, then $E_n \cap E_m = \emptyset$. This implies that $\nu\left(\bigcup E_n\right) = +\infty$, since $\nu(E_n) > n$ for each n. Now since $E = \left(\bigcup E_n\right) \cup \left(E - \bigcup E_n\right)$ it follows that $\nu(E) = \nu\left(\bigcup E_n\right) + \nu\left(E - \bigcup E_n\right) = +\infty$, for only one of the values $+\infty$, $-\infty$ is assumed by ν.

We now have two more possibilities to consider.

It may happen that $F_n = F_{n-1} - E_n$ for an infinite number of values of n, say $n_1 < n_2 < \ldots$. Then put $F = \bigcup E_{n_i}$. We have $\nu(F) = \sum \nu(E_{n_i}) \geq \sum n_i = +\infty$. Since $F \subset E$ we have

$$\nu(E) = \nu(F) + \nu(E - F),$$

and as before since only one of the values $+\infty$, $-\infty$ can be assumed by ν we must have $\nu(E) = \infty$.

The other possibility is that we might have $F_n \neq E_n$ for at most a finite number of values of n. In this case if for some n, $\nu(F_n) = +\infty$, then as in the previous case we have $\nu(E) = +\infty$. Otherwise we have $n < \nu(F_n) < +\infty$ for all sufficiently large n, and one easily proves (see part (d) of Theor. 4) that $\nu\left(\bigcap F_n\right) = +\infty$, which again implies as before that

$\nu(E) = +\infty$. Now using the fact $\nu^- = (-\nu)^+$ we conclude that if $\nu^-(E) = +\infty$ for some $E \in \mathcal{B}$, then $\nu(E) = -\infty$.

Proof (c) From the definitions it is clear that

$$-\nu^-(E) \leq \nu(E) \leq \nu^+(E).$$

It follows from these inequalities that if $\nu(E) = +\infty$, then $\nu(E) = \nu^+(E) - \nu^-(E)$, and part *(c)* is proved in this case. Assume now that $\nu(E)$ is finite. Then, for every measurable set A contained in E, we have $\nu(E) = \nu(A) + \nu(E-A)$ which implies that $\nu(A)$ is finite. Also if $A \subset E$, $A \in \mathcal{B}$ we have $E - A \subset E$ so that $-\nu^-(E) \leq \nu(E-A) \leq \nu^+(E)$. Now $\nu(A) = \nu(E) - \nu(E-A)$ and so $\nu(A) \leq \nu(E) + \nu^-(E)$. Hence $\nu^+(E) \leq \nu(E) + \nu^-(E)$. Similarly we obtain $-\nu^-(E) \geq \nu(E) - \nu^+(E)$. Since $\nu(E)$ is finite we have that $\nu^+(E)$ and $\nu^-(E)$ are finite, and the last two inequalities prove that $\nu(E) = \nu^+(E) - \nu^-(E)$.

Proof (d) Let $E \in \mathcal{B}$, and let α be an extended real number such that $\alpha < |\nu|(E)$. Let $E_1, E_2, \ldots, E_n$ be measurable disjoint subsets of E such that $\alpha < \sum_{i=1}^{n} |\nu(E_i)|$. Let F be the union of the E_i's for which $\nu(E_i) \geq 0$ and let G be the union of the E_i's for which $\nu(E_i) < 0$. Then $\nu(F) \leq \nu^+(E)$, $\nu(G) \geq -\nu^-(E)$, and so $\alpha < \nu(F) - \nu(G) \leq \nu^+(E) + \nu^-(E)$ which implies $|\nu|(E) \leq \nu^+(E) + \nu^-(E)$.

To prove the reverse of this inequality we assume that $|\nu|(E)$ is finite, for otherwise there is nothing to prove. For each measurable subset A

of E we have (by the definition of $|\nu|$) $|\nu(A)| \leq |\nu|(E)$. It follows that $\nu^+(E)$ and $\nu^-(E)$ are finite. If $A \subset E$, $A \in \mathcal{B}$ we have

$$2\nu(A) = \nu(A)+\nu(E)-\nu(E-A)$$

$$2\nu(A) \leq \nu(E)+|\nu(A)|+|\nu(E-A)| \leq \nu(E)+|\nu|(E).$$

Hence

$$2\nu^+(E) \leq \nu(E)+|\nu|(E).$$

By part *(c)* it follows that $\nu^+(E)+\nu^-(E) \leq |\nu|(E)$. This proves *(d)*.

The representation $\nu = \nu^+ - \nu^-$ of ν is called the *Jordan decomposition* of ν.

Let ν be a signed measure on a measurable space $\langle X, \mathcal{B} \rangle$, and suppose that there exist disjoint measurable sets A and B such that $X = A \cup B$, and such that for every set E in $\mathcal{B}$ we have $\nu(E \cap A) \geq 0$ and $\nu(E \cap B) \leq 0$. Then such a decomposition of X into the sets A and B is called a *Hahn decomposition* of X relative to ν.

Observe that *if* $X = A \cup B$ *is a Hahn decomposition then the first set is always the one for which* $\nu(E \cap A) \geq 0$.

The following proposition shows that there is a relationship between a Hahn decomposition of a space X relative to a signed measure ν, and the Jordan decomposition of ν.

17. Proposition *Let* ν *be a signed measure on a measurable space* $\langle X, \mathcal{B} \rangle$, *and let* $X = A \cup B$

be a Hahn decomposition of X *relative to* ν. *Then for each measurable set* E *we have*

$$\nu^+(E) = \nu(E \cap A) \quad ; \quad \nu^-(E) = -\nu(E \cap B) \; .$$

Proof Let $F \in \mathcal{B}$, and $F \subset E \cap A$. Then $F = F \cap A$ so that $\nu(F) \geq 0$. By the definition of ν^- we have $-\nu^-(E \cap A) \geq 0$, which implies that $\nu^-(E \cap A) = 0$ because ν^- is a nonnegative function.

In a similar way we obtain $\nu^+(E \cap B) = 0$. We have $E = (E \cap A) \cup (E \cap B)$, and so $\nu^+(E) = \nu^+(E \cap A)$. Since $\nu = \nu^+ - \nu^-$ we have $\nu(E \cap A) = \nu^+(E \cap A) - \nu^-(E \cap A) = \nu^+(E \cap A)$. Hence $\nu^+(E) = \nu(E \cap A)$. We obtain the formula $\nu^-(E) = -\nu(E \cap B)$ in a similar manner.

An immediate consequence of Proposition 17 is that if $X = A \cup B$ is a Hahn decomposition of X, then $\nu^-(A) = \nu^+(B) = 0$.

Example

Let $\langle X, \mathcal{B}, \mu\rangle$ be a measure space and let $f: X \to [-\infty,\infty]$ be an integrable function. Then one can verify (see Prob. 14) that the function ν defined by

$$\nu(E) = \int_E f d\mu \qquad (E \in \mathcal{B})$$

is a signed measure, and that

$$\nu^+(E) = \int_E f^+ d\mu \quad , \quad \nu^-(E) = \int_E f^- d\mu$$

$$|\nu| = \nu^+ + \nu^- \quad , \quad |\nu|(E) = \int_E |f|d\mu.$$

Set $A = \{x: f(x) \geq 0\}$, $B = \{x: f(x) < 0\}$. Then $X = A \cup B$ is a Hahn decomposition of X. For clearly $A \cap B = \emptyset$. Let $E \in \mathcal{B}$. If $x \in A$, then $f^-(x) = 0$, and if $x \in \mathcal{B}$, then $f^+(x) = 0$. This implies that $\nu(E \cap A) = \nu^+(E \cap A) \geq 0$, and $\nu(E \cap B) = -\nu^-(E \cap B) \leq 0$. Hence $X = A \cup B$ is a Hahn decomposition of X relative to ν.

The above example shows that *a Hahn decomposition may not be unique.* Indeed $X = A' \cup B'$, where $A' = \{x: f(x) > 0\}$, $B' = \{x: f(x) \leq 0\}$ is also a Hahn decomposition of X.

18. Theorem *Let* ν *be a signed measure on a measurable space* $\langle X, \mathcal{B} \rangle$. *Then there is a Hahn decomposition of* X *relative to* ν.

Proof We may, without loss of generality, assume that the value $+\infty$ is not in the range of ν. For if $+\infty$ is in the range of ν we may consider the signed measure $-\nu$. Then $+\infty$ will certainly not be in the range of $-\nu$. Now if $X = B \cup A$ were a Hahn decomposition of X relative to $-\nu$, then $X = A \cup B$ would be a Hahn decomposition for ν. Hence, assume that $\nu(E) < +\infty$ for each $E \in \mathcal{B}$. Since $X \in \mathcal{B}$, this implies $\nu(X) < +\infty$. By Theorem 16 we then have $\nu^+(X) < +\infty$. Since $\nu^+(X)$ is finite, it follows from the definition of ν^+, that for each n there is a set $E_n \in \mathcal{B}$ such that

$$+\infty > \nu^+(X) \geq \nu(E_n) > \nu^+(X) - \frac{1}{2^n} .$$

Using the Jordan decomposition of ν and the obvious fact $\nu^+(E_n) \leq \nu^+(X)$, we obtain $\nu^-(E_n) < 1/2^n$.

We now prove that $\nu^+(X - E_n) \leq 1/2^n$, by showing that for every measurable subset E of $X - E_n$ we have $\nu(E) < 1/2^n$. Indeed, since $E \subset X - E_n$ and $E \cap E_n = \emptyset$ we have $\nu(E \cup E_n) = \nu(E) + \nu(E_n) \leq \nu^+(X)$, which implies $\nu(E) \leq \nu^+(X) - \nu(E_n) < 1/2^n$. Hence $\nu^+(X - E_n) \leq 1/2^n$. Put

$$A = \bigcap_{k=1}^{\infty}\left(\bigcup_{n=k}^{\infty} E_n\right),$$

$$B = X - A = \bigcup_{k=1}^{\infty}\left(\bigcap_{n=k}^{\infty} (X - E_n)\right).$$

We claim that the Hahn decomposition sought by the theorem is $X = A \cup B$. First we prove that $\nu^+(B) = \nu^-(A) = 0$. Using the inequality $\nu^-(E_n) < 1/2^n$ obtained earlier, we get

$$\nu^-(A) \leq \nu^-\left(\bigcup_{n=k}^{\infty} E_n\right) \leq \sum_{n=k}^{\infty} \nu^-(E_n) < \frac{1}{2^{k-1}}$$

which implies $\nu^-(A) = 0$. Also, observing that B is the union of the nondecreasing sequence $\left\langle \bigcap_{n=k}^{\infty} (X - E_n) \right\rangle_{k=1}^{\infty}$, we get by part (c) of Theorem

4 that

$$\lim_{k\to\infty} \nu^+\left(\bigcap_{n=k}^{\infty} (X - E_n)\right) = \nu^+(B) .$$

Hence $\nu^+(B) = 0$, since $\nu^+(X - E_n) \leq 1/2^n$.

Next, suppose $E \in \mathcal{B}$. Then $E \cap A \subset A$ so that $0 \leq \nu^-(E \cap A) \leq \nu^-(E) = 0$, which implies that $\nu(E \cap A) = \nu^+(E \cap A) \geq 0$. Similarly we find $\nu(E \cap B) \leq 0$.

10.5 ABSOLUTE CONTINUITY

Let $\langle X, \mathcal{B}, \mu\rangle$ be a measure space and consider the signed measure ν, defined earlier by $\nu(E) = \int_E f d\mu$ $(E \in \mathcal{B})$ where f is an integrable function on X.

We see that whenever $\mu(E) = 0$ we also have $\nu(E) = 0$. This observation leads to the following definition.

Definition *(a) Let* μ *be a measure on a measurable space* $\langle X, \mathcal{B}\rangle$ *and let* ν *be any measure or signed measure on* $\langle X, \mathcal{B}\rangle$. *Then we say that* ν *is* absolutely continuous *with respect to* μ, *and write*

$$\nu << \mu$$

iff $\nu(E) = 0$ *for every* $E \in \mathcal{B}$ *for which* $\mu(E) = 0$.

(b) If ν *is a measure or a signed measure on* $\langle X, \mathcal{B}\rangle$, *and if there is an* $A \in \mathcal{B}$ *such that* $\nu(E) = \nu(A \cap E)$ *for every* $E \in \mathcal{B}$, *then we say that* ν

is concentrated *on* A. *This simply means that* $\nu(E) = 0$ *if* $E \cap A = \emptyset$.

(c) Suppose that ν_1 *and* ν_2 *are measures or signed measures and suppose that there exist disjoint sets* A_1 *and* A_2 *such that* ν_1 *is concentrated on* A_1, *and* ν_2 *is concentrated on* A_2. *Then we say that* ν_1 *and* ν_2 *are mutually singular and we write* $\nu_1 \perp \nu_2$.

19. Proposition *Let* $\langle X, \mathcal{B} \rangle$ *be a measurable space, let* ν, ν_1, ν_2, *be measures or signed measures on* $\langle X, \mathcal{B} \rangle$, *and let* μ *be a measure on* $\langle X, \mathcal{B} \rangle$. *Then*

(a) If ν *is concentrated on a measurable set* E, *then* $|\nu|$ *is also concentrated on* E.

(b) If $\nu_1 \perp \nu_2$ *then* $|\nu_1| \perp |\nu_2|$.

(c) If $\nu_1 \perp \mu$ *and* $\nu_2 \perp \mu$, *then* $\nu_1 + \nu_2 \perp \mu$.

(d) If $\nu_1 << \mu$ *and* $\nu_2 << \mu$, *then* $\nu_1 + \nu_2 << \mu$.

(e) If $\nu << \mu$, *then* $|\nu| << \mu$.

(f) If $\nu_1 << \mu$ *and* $\nu_2 \perp \mu$, *then* $\nu_1 \perp \nu_2$.

(g) If $\nu << \mu$ *and* $\nu \perp \mu$, *then* $\nu = 0$.

Proof (a) Let $S \in \mathcal{B}$ with $S \cap E = \emptyset$. Let $\{S_1, S_2, \ldots, S_n\}$ be a finite disjoint collection of measurable subsets of S. We have $\nu(S_i) = 0$ for $1 \leq i \leq n$, and so $\sum_{i=1}^{n} |\nu(S_i)| = 0$. This implies that $|\nu|(S) = 0$, which proves that $|\nu|$ is concentrated on E.

Proof (b) Let ν_1 and ν_2 be concentrated on E_1 and E_2 respectively, with $E_1 \cap E_2 = \emptyset$. Then by *(a)*, $|\nu_1|$ and $|\nu_2|$ are concentrated on E_1 and E_2 respectively, and so $|\nu_1| \perp |\nu_2|$.

Proof (c) $\nu_1 \perp \mu$ implies the existence of disjoint sets A_1, B_1 in $\mathcal{B}$ such that ν_1 is concentrated on A_1 and μ on B_1. Similarly, $\nu_2 \perp \mu$ implies the existence of disjoint sets A_2, B_2 in $\mathcal{B}$ such that ν_2 is concentrated on A_2 and μ on B_2. Set

$$E = A_1 \cup A_2 \quad \text{and} \quad F = B_1 \cap B_2 .$$

Clearly $E \cap F = 0$. We prove that $\nu_1 + \nu_2$ is concentrated on E, and μ on F. Let $M \varepsilon \mathcal{B}$, $N \varepsilon \mathcal{B}$, and such that $M \cap E = \emptyset$ and $N \cap F = \emptyset$. Then

$$\nu_1(M) = \nu_2(M) = \mu(N) = 0.$$

Hence

$$(\nu_1 + \nu_2)(M) = \mu(N) = 0,$$

which proves that $\nu_1 + \nu_2 \perp \mu$.

Proof (d) Let $E \varepsilon \mathcal{B}$ with $\mu(E) = 0$. Then $\nu_1 << \mu$ and $\nu_2 << \mu$ imply $\nu_1(E) = \nu_2(E) = 0$. Hence $(\nu_1 + \nu_2)(E) = 0$, which proves that $\nu_1 + \nu_2 << \mu$.

Proof (e) Suppose $E \varepsilon \mathcal{B}$ with $\mu(E) = 0$. Let $E_1, E_2, \ldots, E_n$ be disjoint measurable subsets of E. Then since μ is a measure we have $\mu(E_i) = 0$ $(1 \leq i \leq n)$. Since $\nu << \mu$ we also have $\nu(E_i) = 0$

$(1 \leq i \leq n)$. Hence $\sum_{i=1}^{n} |\nu(E_i)| = 0$, which implies $|\nu|(E) = 0$.

Proof (f) Since $\nu_2 \perp \mu$, there exists an $E \varepsilon \mathcal{B}$ on which ν_2 is concentrated and such that $\mu(E) = 0$. Since μ is a measure, for every measurable subset A of E we also have $\mu(A) = 0$. Since $\nu_1 << \mu$, it follows that for every A in E we have $\nu_1(A) = 0$. This means that ν_1 is concentrated on the complement of E, and so $\nu_1 \perp \nu_2$.

Proof (g) By part *(f)* we have $\nu \perp \nu$. But it is clear that the only measure which is singular with respect to itself is the measure 0. Hence $\nu = 0$.

20. Lemma *Let* λ *and* μ *be measures on a measurable space* $\langle X, \mathcal{B}\rangle$. *Suppose that* $\lambda << \mu$, $\mu(X) < +\infty$ *and* $\lambda(X) > 0$. *Then there exists* $\alpha > 0$ *and* $S \varepsilon \mathcal{B}$ *such that* $\mu(S) > 0$, *and such that* $\nu = \lambda - \alpha\mu$ *is a signed measure on* $\langle X, \mathcal{B}\rangle$ *with* $\nu^-(S) = 0$.

Proof The assumption $\lambda(X) > 0$ simply means that λ is not the zero measure. Since $\mu(X)$ is finite we have, for any fixed $\varepsilon > 0$, that $\lambda - \varepsilon\mu$ is a signed measure. Take $\varepsilon_n = 1/n$ $(n = 1,2,\ldots)$ and let $X = A_n \cup B_n$ be a Hahn decomposition of X relative to the signed measure $\nu_n = \lambda - \varepsilon_n\mu$. Put

$$A = \bigcup_{n=1}^{\infty} A_n, \qquad B = \bigcap_{n=1}^{\infty} B_n.$$

Since $B_n = X - A_n$, we have that $A = X - B$. Also, by Proposition 17, $B \subset B_n$ implies $\nu_n(B) = -\nu_n^-(B) \leq 0$, and so $0 \leq \lambda(B) \leq \varepsilon_n \mu(B)$. Since $\mu(B) < +\infty$ and ε_n can be made arbitrarily small we have $\lambda(B) = 0$. It follows that $\lambda(X) = \lambda(A) + \lambda(B) = \lambda(A) > 0$. Since $\lambda << \mu$, it follows that $\mu(A) > 0$ which implies that for some n_0 we have $\mu(A_{n_0}) > 0$. This proves the lemma, for we can take $S = A_{n_0}$ and $\alpha = \varepsilon_{n_0}$.

We are now ready to prove the following important theorem, which is the converse of Theorem 9.

21. Theorem (Radon-Nikodym) *Let $\langle X, \mathcal{B}, \mu \rangle$ be a measure space where μ is a σ-finite measure. Let ν be a signed measure on $\langle X, \mathcal{B} \rangle$ such that ν assumes only finite values, and such that $\nu << \mu$. Then there exists a function $f: X \to \mathbb{R}$ such that*

$$\int_X |f| d\mu < +\infty \quad \textit{and} \quad \nu(E) = \int_E f d\mu$$

for each $E \in \mathcal{B}$. The function f is unique in the sense that if g is another function which satisfies the theorem, then $f = g$ a.e. $[\mu]$.

Proof We proceed by steps. First we prove the theorem under the additional assumptions that $\mu(X) < +\infty$, and ν is a measure. Next we drop the restriction $\mu(X) < +\infty$, but we still assume that ν is a measure. Finally we prove the theorem in its general form. The uniqueness of the function f is an immediate consequence of part *(b)* of Theorem 14.

Step 1 Assume $\mu(X) < +\infty$, and that ν is a measure. Let $\mathcal{F}$ be the collection of all integrable, nonnegative functions f on the measure space $\langle X, \mathcal{B}, \mu\rangle$ such that $\int_E f d\mu \leq \nu(E)$ for each E.

Then $\mathcal{F}$ is not empty since it contains the zero function. Set

$$c = \sup\{ \int_X f d\mu : f \in \mathcal{F} \}.$$

Since $\nu(X)$ is assumed to be finite, the number c is finite too. Now, there is a sequence $\langle f_n \rangle$ of functions in $\mathcal{F}$ such that

$$\lim_{n\to\infty} \int_X f_n d\mu = c .$$

Define $g_n = \max\{f_1, f_2, \ldots, f_n\}$. It is clear that for each n, g_n is a nonnegative integrable function on X. We prove that $g_n \in \mathcal{F}$ for each n.

For $E \in \mathcal{B}$, consider the finite sequence $\langle E_i \rangle_{i=1}^{n}$ defined as follows:

$$E_1 = \{x : g_n(x) = f_1(x)\} \cap E.$$

$$E_i = \{x : g_i(x) = f_i(x)\} \cap (E - E_1 - E_2 - \ldots - E_{i-1}),$$

$$(i = 2, 3, \ldots, n).$$

It is clear that the E_i's are disjoint measurable

sets with $E = E_1 \cup E_2 \cup \ldots \cup E_n$. Also for $x \in E_i$ we have $g_n(x) = f_i(x)$. Hence

$$\int_E g_n d\mu = \int_{E_1} f_1 d\mu + \int_{E_2} f_2 d\mu + \ldots + \int_{E_n} f_n d\mu = \nu(E),$$

which proves that $g_n \in \mathscr{F}$. Let $g = \sup\{f_n : n = 1,2,\ldots\}$. Then g is measurable by Theorem 2. Moreover, the sequence $\langle g_n \rangle$ is nondecreasing, and it is clear that

$$\lim_{n\to\infty} g_n = g.$$

Hence by the Lebesgue dominated convergence theorem we get

$$\lim_{n\to\infty} \int_E g_n d\mu = \int_E g d\mu \leq \nu(E),$$

which proves that $g \in \mathscr{F}$. Since $f_n \leq g_n$, we conclude that

$$c = \int_X g d\mu \quad .$$

Now define a function f by

$$f(x) = \begin{cases} g(x) & \text{if } g(x) < +\infty \\ 0 & \text{if } g(x) = +\infty \end{cases}$$

Then $g(x) = f(x)$ a.e. $[\mu]$, which implies that f is integrable on $\langle X, \mathcal{B}, \mu \rangle$, and that

$$\int_E f d\mu = \int_E g d\mu \qquad (E \in \mathcal{B}).$$

More particularly for $E = X$ we get

$$c = \int_X g d\mu = \int_X f d\mu .$$

Define $\lambda \colon \mathcal{B} \to \mathbb{R}$ by

$$\lambda(E) = \nu(E) - \int_E f d\mu \qquad (E \in \mathcal{B}).$$

It is easily seen that λ is an absolutely continuous measure with respect to ν. If we can show that λ is the zero measure, then the first step of the theorem will be proved. So suppose on the contrary that λ is not the zero measure. This implies that $\lambda(X) > 0$. Now the hypotheses in the Lemma 20 are satisfied. So there is an $\alpha > 0$ and $S \in \mathcal{B}$ such that $\mu(S) > 0$, $\lambda - \alpha\mu$ is a signed measure, and $(\lambda - \alpha\mu)^-(S) = 0$. The last equality implies that for each $E \in \mathcal{B}$, $\alpha\mu(E \cap S) \leq \lambda(E \cap S)$. Consider the function $h = f + \alpha\chi_S$ where χ_S is the characteristic function of S. For any $E \in \mathcal{B}$, we have

$$\int_E h d\mu = \int_E f d\mu + \alpha\mu(E \cap S) \leq \int_E f d\mu + \lambda(E \cap S) =$$

$$= \int_E f d\mu + \nu(E \cap S) - \int_{E \cap S} f d\mu = \int_{E-S} f d\mu + \nu(E \cap S)$$

$$\leq \nu(E - A) + \nu(E \cap A) = \nu(E).$$

Thus the function h satisfies the inequality

$$\int_E h\,d\mu \leq \nu(E) \qquad (E \in \mathcal{B})$$

and in particular $\int_X h\,d\mu \leq \nu(X)$, which means that h belongs to the collection $\mathcal{F}$, since $h \geq 0$. But on the other hand we have

$$\int_X h\,d\mu = \int_X f\,d\mu + \alpha\mu(S) = c + \alpha\mu(S) > c$$

which contradicts the fact, just established, that $h \in \mathcal{F}$. Hence $\lambda = 0$, and so $\nu(E) = \int_E f\,d\mu$ for $E \in \mathcal{B}$.

Step 2 We now prove the theorem only under the assumption that ν is a measure. Since μ is σ-finite we can express $X = \bigcup_{n=1}^{\infty} X_n$ where the X_n are disjoint measurable sets such that $\mu(X_n) < +\infty$, for each n. Let $\mathcal{B}_n$ denote all members of $\mathcal{B}$ which belong to X_n. Then $\mathcal{B}_n$ is clearly a σ-algebra of subsets of X_n. If we still denote by μ the restriction of μ to $\mathcal{B}_n$, then for each n, $\langle X_n, \mathcal{B}_n, \mu\rangle$ is a measure space which satisfies the hypotheses of Step 1. Hence there is a nonnegative integrable function $f_n\colon X_n \to \mathbb{R}$ such that

$$\nu(E) = \int_E f_n d\mu \quad \text{for} \quad E \in \mathcal{B}_n .$$

Define $f: X \to \mathbb{R}$ by $f(x) = f_n(x)$ if $x \in X_n$. Let $E \in \mathcal{B}$, and put $E_n = X_n \cap E$. Then $E = \bigcup_{n=1}^{\infty} E_n$ with $E_n \in \mathcal{B}_n$. We have

$$\int_E f d\mu = \sum_{n=1}^{\infty} \int_{E_n} f d\mu$$

$$= \sum_{n=1}^{\infty} \int_{E_n} f_n d\mu = \sum_{n=1}^{\infty} \nu(E_n) = \nu(E)$$

which proves the theorem in the case where ν is a measure.

Step 3 We now prove the theorem in its general form. Let $\nu = \nu^+ - \nu^-$ be the Jordan decomposition of ν. By Proposition 19, part (e), the assumption $\nu << \mu$ implies that $|\nu| << \mu$, which clearly implies $\nu^+ << \mu$ and $\nu^- << \mu$. By Step 2 there are nonnegative integrable functions f_1 and f_2 such that for each $E \in \mathcal{B}$ we have

$$\nu^+(E) = \int_E f_1 d\mu \quad , \quad \nu^-(E) = \int_E f_2 d\mu$$

which proves that the function $f = f_1 - f_2$ is the sought for function in the theorem, since for $E \in \mathcal{B}$ we have

$$\nu(E) = \nu^+(E) - \nu^-(E)$$

$$= \int_E f_1 d\mu - \int_E f_2 d\mu = \int_E f d\mu \ .$$

The function f given by Theorem 21 is called the *Radon-Nikodym derivative of* ν *with respect to* μ. It is denoted by $[d\nu/d\mu]$.

Remark Steps 1 and 2 in the proof of Theorem 21 show that if in addition ν is assumed to be a measure, then the Radon-Nikodym derivative is a nonnegative function.

22. Theorem *Let* $\langle X, \mathcal{B}, \mu\rangle$ *be a measure space where* μ *is a* σ*-finite measure. Let* ν *be a* σ*-finite signed measure on* $\langle X, \mathcal{B}\rangle$. *Then we can find* σ*-finite measures* ν_1 *and* ν_2 *on* $\langle X, \mathcal{B}\rangle$ *such that* $\nu = \nu_1 + \nu_2$, $\nu_1 << \mu$ *and* $\nu_2 \perp \mu$. *The signed measures* ν_1 *and* ν_2 *are unique.*

The pair ν_1 and ν_2, is called the *Lebesgue decomposition* of ν.

Proof We proceed by steps.

Step 1 Suppose in addition that $\mu(X) < +\infty$, $\nu(X) < +\infty$, and that ν is a measure. Then $\mu + \nu$ is also a measure, and clearly $\nu << \mu + \nu$. By the Remark made after Theorem 21 there exists a nonnegative real-valued integrable function on X with respect to $\mu + \nu$ such that for each $E \in \mathcal{B}$, we have

$$\nu(E) = \int_E f d(\mu + \nu).$$

Clearly f is integrable with respect to both measures μ and ν, and so

$$\nu(E) = \int_E f d\mu + \int_E f d\nu.$$

Put

$$A_1 = \{x: f(x) = 1\}; \quad A_2 = \{x: f(x) > 1\};$$

$$A = A_1 \cup A_2; \quad B = \{x: f(x) < 1\}$$

and

$$E_n = \{x: f(x) \geq 1 + \frac{1}{n}\} \quad (n = 1,2,\ldots).$$

Then we easily see that

$$A \cup B = X, \quad A \cap B = \emptyset \text{ and } A_2 = \bigcup_{n=1}^{\infty} E_n.$$

We have

$$\nu(E_n) = \int_{E_n} f d\mu + \int_{E_n} f d\nu \geq \left(1 + \frac{1}{n}\right)[\mu(E_n) + \nu(E_n)]$$

which implies that $\mu(E_n) = \nu(E_n) = 0$, and so $\mu(A_2) = \nu(A_2) = 0$. We also have

$$\nu(A_1) = \int_{A_1} f d\mu + \int_{A_1} f d\nu = \mu(A_1) + \nu(A_1)$$

so that $\mu(A_1) = 0$. This, combined with $\mu(A_2) = 0$,

implies $\mu(A) = 0$.

For any $E \in \mathcal{B}$, define ν_1 and ν_2 by

$$\nu_1(E) = \nu(E \cap B) \quad , \quad \nu_2(E) = \nu(E \cap A).$$

Clearly ν_1 and ν_2 are measures on $\mathcal{B}$ and $\nu = \nu_1 + \nu_2$. We prove that $\nu_1 << \mu$. Let $E \in \mathcal{B}$ with $\mu(E) = 0$. Then $\mu(E \cap B) = 0$ so that

$$\nu(E \cap B) = \int_{E \cap B} f d\mu + \int_{E \cap B} f d\nu = \int_{E \cap B} f d\nu.$$

This implies that

$$0 = \nu(E \cap B) - \int_{E \cap B} f d\nu$$

$$= \int_{E \cap B} 1 \cdot d\nu - \int_{E \cap B} f d\nu = \int_{E \cap B} (1 - f) d\nu.$$

Since $1 - f(x) > 0$ whenever $x \in B$, we conclude that $\nu(E \cap B) = \nu_1(E) = 0$. Finally since for every $E \in \mathcal{B}$ we have $\nu_2(E \cap B) = 0$, it follows that $\nu_2 \perp \mu$.

Step 2 We now drop the assumptions $\mu(X) < +\infty$ and $\nu(X) < +\infty$, but we retain the additional assumption that ν is a measure. The hypothesis that μ and ν are σ-finite measures implies the existence of a sequence $\langle X_n \rangle$ of pairwise disjoint measurable sets such that $X = \bigcup_{n=1}^{\infty} X_n$ with $\mu(X_n) < +\infty$ and $\nu(X_n) < +\infty$. For since μ is σ-finite there is a sequence $\langle S_i \rangle$ of sets in $\mathcal{B}$ such that $X = \bigcup_{i=1}^{\infty} S_i$ with $\mu(S_i) < +\infty$. We may take the S_i's pairwise disjoint, for if they are not we may replace the $\langle S_i \rangle$ by $\langle T_i \rangle$ where $T_1 = S_1$ and $T_i = S_i - (T_1 \cup T_2 \cup \ldots \cup T_{i-1})$ for $i \geq 2$. Now since ν is a σ-finite measure each S_i can be written $S_i = \bigcup_{k=1}^{\infty} E_k^{(i)}$, where the $E_k^{(i)}$'s are pairwise disjoint measurable sets with $\nu(E_k^{(i)}) < +\infty$ for each i and k. It is clear now that for the sequence $\langle X_n \rangle$ we can take any arrangement of the $E_k^{(i)}$'s. We now proceed as in Step 1. For each X_n there is a decomposition $X_n = A_n \cup B_n$ such that $A_n \cap B_n = \emptyset$, $\mu(A_n) = 0$, and such that for each $E \in \mathcal{B}$,

$$\nu(E \cap B_n) = 0 \quad \text{if} \quad \mu(E \cap B_n) = 0.$$

Put

$$A = \bigcup_{n=1}^{\infty} A_n, \quad B = X - A.$$

Then $\mu(A) = 0$. Suppose now that $\mu(E) = 0$ for some $E \in \mathcal{B}$. Then $\mu(E \cap B_n) = 0$, and so $\nu(E \cap B_n) = 0$. Since the X_n are disjoint we have $B = \bigcup_{n=1}^{\infty} B_n$. Hence $E \cap B = \bigcup_{n=1}^{\infty} (E \cap B_n)$ so that

$\nu(E \cap B) = 0$. Define now ν_1 and ν_2 as in Step 1 by

$$\nu_1(E) = \nu(E \cap B), \quad \nu_2(E) = \nu(E \cap A) \quad \text{for} \quad E \in \mathcal{B}.$$

Then $\nu = \nu_1 + \nu_2$. Clearly since ν is σ-finite, ν_1 and ν_2 are σ-finite too.

Step 3 We now prove the theorem without any additional assumption. Let $\nu = \nu^+ - \nu^-$ be the Jordan decomposition of ν. Since ν^+ and ν^- are measures, by Step 2 there are two sets A_1 and A_2 such that $\mu(A_1) = \mu(A_2) = 0$ and such that if $B_1 = X - A_1$ and $B_2 = X - A_2$ we have $\nu^+(E \cap B_1)=0$ whenever $\mu(E) = 0$ and $\nu^-(E \cap B_2) = 0$ whenever $\mu(E) = 0$. Put $A = A_1 \cup A_2$ and $B = B_1 \cap B_2 = X-A$. If for an $E \in \mathcal{B}$ we have $\mu(E) = 0$ then $\nu(E \cap B) = \nu^+(E \cap B_1 \cap B_2) - \nu^-(E \cap B_1 \cap B_2) = 0$ because $\mu(E \cap B_2) = \mu(E \cap B_1) = 0$. Since $\mu(A) = 0$, we have what is necessary in this case.

Step 4 To prove that ν_1 and ν_2 are unique suppose that $\nu = \nu_3 + \nu_4$ is another decomposition of ν of the same kind, as $\nu = \nu_1 + \nu_2$. Since ν is σ-finite, the uniqueness of the ν_1 and ν_2 will be proved if we prove that

$$\nu_1(E) = \nu_3(E) \quad \text{and} \quad \nu_2(E) = \nu_4(E)$$

for every measurable set such that $\mu(E)$ and $\nu(E)$

are finite. But if we assume that all the values of μ and ν are finite, then we can write $\nu_1 - \nu_3 = \nu_4 - \nu_2$. We also have $\nu_1 - \nu_3 << \mu$ and $\nu_4 - \nu_2 \perp \mu$. By parts (c), (d), and (g) of Proposition 19 we conclude that

$$\nu_1 - \nu_3 = \nu_4 - \nu_2 = 0.$$

The theorem is proved.

10.6 THE L^p SPACES

A theory similar to the one developed in Chapter 6 can be given if the underlying space is an abstract measure space $\langle X, \mathcal{B}, \mu \rangle$. If p is a positive real number, we denote by $L^p(\mu)$ the collection of all measurable functions $f: X \to \overline{\mathbb{R}}$ ($\overline{\mathbb{R}}$ is the extended real line) such that $\int_X |f|^p d\mu < \infty$. Two functions, f and g in $L^p(\mu)$, are considered as identical if and only if $f = g$ a.e. $[\mu]$. A measurable function $f: X \to \overline{R}$ is called *essentially bounded* (with respect to μ) if and only if there is some positive real number M such that $\mu\{x: |f(x)| > M\} = 0$. The infimum of all such numbers M is called the *essential supremum of* f.

As we did in Chapter 6, we shall only be concerned with L^p spaces for values $p \geq 1$ exclusively.

For $1 \leq p < \infty$ and $f \in L^p(\mu)$ we set

$$\|f\|_p = \{ \int_X |f|^p d\}^{\frac{1}{p}}$$

and

$$\|f\|_\infty = \text{essential supremum of } f.$$

Theorems 1, 2, 3 and Proposition 4 of Chapter 6, concerning respectively the Hölder and Minkowski inequalities, the completeness of the L^p spaces, and bounded linear functionals on L^p, remain true in this case and the proofs are identical to those given in Chapter 6.

The simple functions play an important role in $L^p(\mu)$.

23. Proposition *Let S be the class of all simple functions on a measure space $\langle X, \mathcal{B}, \mu\rangle$ such that for each $\phi \in S$*

$$\mu(\{x\colon \phi(x) \neq 0\}) < +\infty.$$

Then if $1 \leq p < \infty$, S is dense in $L^p(\mu)$.

Proof Clearly S is a subclass of $L^p(\mu)$, since every simple function which vanishes outside of a set of finite measure belongs to $L^p(\mu)$. Let $f \in L^p(\mu)$. We first assume $f \geq 0$. By Theorem 3 there is a nondecreasing sequence $\langle f_n\rangle$ of nonnegative simple functions such that $f_n \to f$ for every $x \in X$. Since $0 \leq f_n \leq f$ we have $f_n \in L^p(\mu)$, and so $f_n \in S$. We have $|f - f_n|^p \leq f^p$, and Theorem

12 shows that

$$\lim_{n\to\infty} \|f - f_n\|_p = 0.$$

In the general case, if f is any function in $L^p(\mu)$ we write $f = f^+ - f^-$ and apply the above result to f^+ and f^-. This proves that f is in the L^p-closure of S, which means that S is dense in L^p.

24. Lemma *Let* $\langle X, \mathcal{B}, \mu\rangle$ *be a measure space with* $\mu(X) < +\infty$. *Let* $g \in L^1(\mu)$, *and let* p *and* q *be extended real numbers such that* $1 \leq p < +\infty$ *and* $(1/p) + (1/q) = 1$. *Furthermore, suppose there exists a constant* M *such that*

$$\left|\int_X g\phi d\mu\right| \leq M\|\phi\|_p$$

for all simple functions ϕ. *Then* $g \in L^q(\mu)$.

Proof We first consider the case $p = 1$. We prove that $g \in L^\infty(\mu)$ with $\|g\|_\infty \leq M$. Let $E = \{x: |g(x)| \geq M + \varepsilon\}$ with $\varepsilon > 0$, and set $\phi = (\operatorname{sgn} g)\chi_E$. Then $\|\phi\|_1 = \mu(E)$. Also

$$M\mu(E) = M\|\phi\|_1 \geq \left|\int_X g\phi d\mu\right|$$

$$= \int_E |g| d\mu \geq (M + \varepsilon)\mu(E).$$

But this inequality holds if and only if $\mu(E) = 0$. Hence, since $\varepsilon > 0$ was arbitrary, we conclude that

$g \in L^{\infty}(\mu)$ and $\|g\|_{\infty} \leq M$.

Suppose now $p > 1$ and consider the nonnegative function $|g|^q$. By Theorem 3, there is a nondecreasing sequence $\langle f_n \rangle$ of nonnegative measurable simple functions such that $f_n \to |g|^q$ as $n \to \infty$. Put $\phi_n = (f_n)^{1/p} \cdot (\operatorname{sgn} g)$. Then ϕ_n is a simple function, and considered as an element of $L^p(\mu)$ it has norm

$$\|\phi_n\|_p \leq \left(\int_X f_n d\mu\right)^{\frac{1}{p}} .$$

We have, since $(1/p) + (1/q) = 1$,

$$f_n = f_n^{\frac{1}{q}+\frac{1}{p}} \leq f_n^{\frac{1}{p}}|g| = \phi_n g$$

and so

$$\int_X f_n d\mu \leq \int_X \phi_n g d\mu \leq M\|\phi_n\|_p \leq M\left(\int_X f_n d\mu\right)^{\frac{1}{p}} .$$

This implies

$$\left(\int_X f_n d\mu\right)^{\frac{1}{q}} \leq M , \int_X f_n d\mu \leq M^q .$$

By the monotone convergence theorem we obtain

$$\int_X |g|^q d\mu \leq M^q$$

which proves that $g \in L^q(\mu)$.

We now prove the Riesz representation theorem, whose proof was postponed in Chapter 6, Theorem 5.

25. Theorem (Riesz Representation Theorem) *Let* $\langle X, \mathcal{B}, \mu\rangle$ *be a measure space and let* μ *be* σ*-finite. Let* p *and* q *be extended real numbers such that* $1 \leq p < +\infty$, *and* $(1/p) + (1/q) = 1$. *Let* F *be a bounded linear functional on* $L^p(\mu)$. *Then there exists a unique* $g \in L^q(\mu)$ *such that*

$$F(f) = \int_X fg\,d\mu \quad \text{for all} \quad f \in L^p(\mu) .$$

We have also $\|F\| = \|g\|_q$.

Proof We first prove the theorem under the additional assumption $\mu(X) < \infty$, and then we proceed to the general case.

Step 1 Suppose $\mu(X) < \infty$, and let $E \in \mathcal{B}$. Then clearly the characteristic function χ_E of E belongs to $L^p(\mu)$. Define the mapping

$$\nu : \mathcal{B} \to \mathbb{R}$$

by

$$\nu(E) = F(\chi_E) \qquad (E \in \mathcal{B}).$$

We have

$$\|\chi_E\|_p = \left(\int_X |\chi_E|^p d\mu\right)^{\frac{1}{p}} = [\mu(E)]^{\frac{1}{p}}$$

and so

$$(1) \qquad |\nu(E)| = |F(\chi_E)| \leq \|F\| [\mu(E)]^{\frac{1}{p}} .$$

We claim that ν is a signed measure on $\langle X, \mathcal{B} \rangle$. Since $\nu(\emptyset) = 0$ and ν omits both values $\pm\infty$, it remains only to prove that ν is countably additive. If A and B are disjoint measurable sets, then $\chi_{A \cup B} = \chi_A + \chi_B$. Since F is linear we have

$$\nu(A \cup B) = F(\chi_{A\cup B}) = F(\chi_A + \chi_B)$$

$$= F(\chi_A) + F(\chi_B) = \nu(A) + \nu(B).$$

This proves that ν is finitely additive. To prove that ν is countably additive let $\langle E_n \rangle$ be a sequence of pairwise disjoint measurable sets. Put

$$E = \bigcup_{n=1}^{\infty} E_n \ , \quad S_n = \bigcup_{i=1}^{n} E_i \ , \quad G_n = E - S_n$$

Since $S_n \cap G_n = \emptyset$, and since ν is finitely additive we have

$$\nu(S_n) = \sum_{i=1}^{n} \nu(E_i) \quad \text{and}$$

$$\nu(E) = \nu(S_n) + \nu(G_n) = \sum_{i=1}^{n} \nu(E_i) + \nu(G_n) .$$

By Theorem 4, part (c), since μ is a measure, we

have

$$\lim_{n\to\infty} \mu(G_n) = 0 .$$

Hence $\nu(G_n) \to 0$ by (1), which implies that

$$\nu(E) = \sum_{n=1}^{\infty} \nu(E_i),$$

and so ν is a signed measure.

We have $\nu << \mu$. For if $\mu(E) = 0$ for some $E \varepsilon \mathcal{B}$, inequality (1) shows that $\nu(E) = 0$. Hence, by the Radon-Nikodym theorem there exists a function $g \varepsilon L^1(\mu)$ such that, for each $E \varepsilon \mathcal{B}$ we have

$$\nu(E) = \int_E g d\mu .$$

If $\phi = \sum_i c_i \chi_{A_i}$ is a simple function

$$(2) \qquad F(\phi) = \sum_i c_i F(\chi_{A_i}) = \sum_i c_i \nu(A_i) =$$

$$= \sum_i c_i \int_{A_i} g d\mu = \int_X \phi g d\mu .$$

We have

$$\left| \int_X g\phi d\mu \right| \leq \|F\| \|\phi\|_p$$

which, by Lemma 24, implies that $g \varepsilon L^q(\mu)$.

Now by Proposition 4 of Chapter 6, which remains true in the abstract setting of this chapter, the functional Φ defined on $L^p(\mu)$ by

$$\Phi(f) = \int_X fg\,d\mu$$

is bounded, and $\|\Phi\| = \|g\|_q$. Hence $F - \Phi$ is a bounded linear functional, and as (2) shows, it vanishes on the set S of simple functions. Therefore, since by Proposition 23 S is dense in $L^p(\mu)$, we conclude that $F - \Phi$ vanishes on $L^p(\mu)$. Thus for all $f \varepsilon L^p(\mu)$ we have

$$F(f) = \int_X fg\,d\mu$$

and $\|\Phi\| = \|F\| = \|g\|_q$.

It is easy to see that the function g is unique. For if g_1 is another function in $L^q(\mu)$ for which

$$F(f) = \int_X fg_1\,d\mu \quad \text{for all} \quad f \varepsilon L^p(\mu)$$

then

$$\int_X f(g_1 - g)\,d\mu = 0 \quad \text{for all} \quad f \varepsilon L^p(\mu)$$

which means that $g_1 - g$ gives the zero functional, and so $\|g_1 - g\|_q = 0$. This implies that $g = g_1$ a.e. $[\mu]$.

Step 2 We now prove the theorem in the general case. Since μ is σ-finite there is a sequence $\langle E_n \rangle$ of sets in $\mathcal{B}$, such that $X = \bigcup_{n=1}^{\infty} E_n$ and $\mu(E_n) < +\infty$ for each n. Set $X_n = E_1 \cup E_2 \cup \ldots \cup E_n$, $(n = 1, 2, \ldots)$. Then $\langle X_n \rangle$ is a nondecreasing sequence of measurable sets such that $X = \bigcup_{n=1}^{\infty} X_n$ and $\mu(X_n) < +\infty$ for each n. Denote by $\mathcal{B}_n$ the class of all elements in $\mathcal{B}$ contained in X_n, and by μ_n the restriction of μ to $\mathcal{B}_n$. Then for each n, $\langle X_n, \mathcal{B}_n, \mu_n \rangle$ is a measure space with $\mu_n(X_n) < +\infty$. Furthermore, we can identify $L^p(\mu_n)$ with the class of those elements of $L^p(\mu)$ which vanish outside X_n. Then, by the result of Step 1 applied to each $\langle X_n, \mathcal{B}_n, \mu_n \rangle$, there is a $g_n \in L^q$ which vanishes outside X_n such that

$$F(f) = \int_{X_n} fg_n d\mu_n = \int_X fg_n d\mu$$

for all f which vanish outside X_n. Moreover $\|f_n\|_q \le \|F\|$. Each such g_n is uniquely determined on X_n except on a set of measure zero. Since g_{n+1} has this property we may assume that $g_{n+1} = g_n$ on X_n. Define $g(x) = g_n(x)$ for $x \in X_n$. Then g is a well-defined measurable function and $\langle |g_n| \rangle$ is an increasing sequence such that

$$\lim_{n\to\infty} |g_n(x)| = |g(x)| .$$

Hence by the Monotone Convergence Theorem we obtain

$$\int_X |g|^q d\mu = \lim_{n\to\infty} \int_X |g_n|^q d\mu \leq \|F\|^q$$

and so $g \in L^q$.

Now let $f \in L^p$, and let $f_n = f$ on X_n, and $f_n = 0$ on $X - X_n$. Then

$$\lim_{n\to\infty} f_n(x) = f(x)$$

and

$$\lim_{n\to\infty} \|f_n - f\|_p = 0.$$

Since $f \in L^p(\mu)$ and $g \in L^q(\mu)$ we have that $fg \in L^1(\mu)$. Also $|f_n g| \leq |fg|$. By the Lebesgue Dominated Convergence Theorem we have

$$\int_X fg d\mu = \lim_{n\to\infty} \int_X f_n g d\mu$$

$$= \lim_{n\to\infty} \int_X f_n g_n d\mu = \lim_{n\to\infty} F(f_n) = F(f).$$

The uniqueness of g follows from the uniqueness of g_n on each X_n. The theorem is proved.

Remark *(a) Theorem 25 generalizes Theorem 5 of Chapter 6, for it is clear that the Lebesgue measure on the real line* $\mathbb{R}$ *is a* σ*-finite measure since* $\mathbb{R} = \bigcup_{n=1}^{\infty} [-n,+n]$.

(b) We only state here, without proof, that if $p = 1$ *the requirement that* μ *be* σ*-finite is necessary. If* $1 < p < \infty$, *then the* σ*-finiteness of* μ *is unnecessary.*

(c) As we proved in Chapter 8, section 4 the theorem is not true if $p = \infty$, *even in the simple case when* X *is a set of finite Lebesgue measure in* $\mathbb{R}$ *and* μ *is the Lebesgue measure in* $\mathbb{R}$.

10.7 PRODUCT MEASURES AND FUBINI'S THEOREM

Let f be a function of two real variables x, y, defined and continuous on the rectangle

$$A = \{(x,y) : a \leq x \leq b\ ,\ \ c \leq y \leq d\}$$

of the Euclidean plane. We may integrate the function f with respect to y, obtaining a function of x

$$\int_c^d f(x,y)dy$$

and then integrate this function with respect to x and obtain

$$\int_a^b \int_c^d f(x,y)dy\ dx \quad \text{written as} \quad \int_a^b dx \int_c^d f(x,y)dy.$$

In a similar manner we obtain by integrating first with respect to x and then with respect to y

$$\int_c^d dy \int_a^b f(x,y)dx.$$

The operation consisting of these two successive integrations is called an *iterated integral*. In contrast to the iterated integral, there is the operation of integrating the function f by a process appropriate to the Euclidean plane in which the function is defined. An integral of this kind is called a *double integral* and is denoted by

$$\iint_A f(x,y)dxdy.$$

The reader recalls, from his calculus courses, that the double integral is defined as a limit of approximating sums, each term of one of the sums being of the form $f(x',y')\cdot\Delta A$ where ΔA is the area of a subregion of A and (x',y') is a point of this subregion. We then have the theorem that the double integral and the two iterated integrals have the same value. This theorem is very important for it shows that the problem of calculating the double integral can be reduced to the problem of calculating integrals of functions of one real variable.

In this section we generalize the above ideas to abstract measure spaces.

Let $\langle X,\mathcal{A},\mu\rangle$ and $\langle Y,\mathcal{B},\nu\rangle$ be two measure spaces. What we would like to do is to define a measure (called the *product* of μ and ν) over some collection (σ-algebra) of sets in the Cartesian product $X \times Y$. Having defined this measure we shall consider double integrals over measurable subsets of $X \times Y$, as well as iterated integrals over subsets of the spaces X and Y. Finally we shall prove Fubini's theorem which enables us to interchange the order of integration, and to calculate double integrals by using iterated integrals.

Definitions *(a) If* $\langle X, \mathcal{A} \rangle$ *and* $\langle Y, \mathcal{B} \rangle$ *are measurable spaces a* measurable rectangle *is any set of the form* $A \times B$ *where* $A \in \mathcal{A}$ *and* $B \in \mathcal{B}$. *We denote by* $\mathcal{A} \times \mathcal{B}$ *the smallest* σ*-algebra of sets in* $X \times Y$ *which contains every measurable rectangle.*

(*Note*: The reader should notice that $\mathcal{A} \times \mathcal{B}$ is not the Cartesian product of two sets but a σ-algebra. Thus $\langle X \times Y, \mathcal{A} \times \mathcal{B} \rangle$ is a measurable space.)

(b) A subset S *of* $X \times Y$ *is said to be an* elementary set, *if and only if it is the union of a finite number of pairwise disjoint measurable rectangles.*

Definition *A* monotone class $\mathcal{C}$ *of sets is a collection of sets such that if*

$$A_i \in \mathcal{C}, \quad B_i \in \mathcal{C}, \quad A_i \subset A_{i+1}, \quad B_i \supset B_{i+1}$$

for $i = 1, 2, 3, \ldots,$ *then*

$$\left(\bigcup_{i=1}^{\infty} A_i\right) \in \mathcal{C} \quad \text{and} \quad \left(\bigcap_{i=1}^{\infty} B_i\right) \in \mathcal{C}.$$

(*Note*: We will prove [see Theor. 28] that the smallest monotone class that contains all elementary sets is precisely the class $\mathcal{A} \times \mathcal{B}$.)

Definition *Let X and Y be two nonempty sets, and let E be a nonempty subset of $X \times Y$. Define*

$$E_x = \{y \in Y: (x,y) \in E\},$$

$$E^y = \{x \in X: (x,y) \in E\}.$$

The sets E_x and E^y are called the **x-section** *and* **y-section** *of E.*

26. **Proposition** *Let $\langle X, \mathcal{A} \rangle$ and $\langle Y, \mathcal{B} \rangle$ be measurable spaces. Define $\mathcal{K}$ to be the class of all $E \in \mathcal{A} \times \mathcal{B}$ such that for every $x \in X$ (respectively $y \in Y$) we have $E_x \in \mathcal{B}$ (respectively $E^y \in \mathcal{A}$). Then $\mathcal{K} = \mathcal{A} \times \mathcal{B}$.*

Proof Since $\mathcal{K} \subset \mathcal{A} \times \mathcal{B}$ we need only to prove that $\mathcal{K} \supset \mathcal{A} \times \mathcal{B}$. If E is a measurable rectangle then $E = A \times B$ with $A \in \mathcal{A}$ and $B \in \mathcal{B}$. Clearly if $x \in A$ then $E_x = B$, and if $x \notin A$ then $E_x = \emptyset$. Hence every measurable rectangle belongs to $\mathcal{K}$. Next we show that $\mathcal{K}$ is a σ-algebra. Since $X \times Y$ is a measurable rectangle it belongs to $\mathcal{K}$. Also if $S \in \mathcal{K}$, since $\mathcal{B}$ is a σ-algebra we have $(\complement S)_x = \complement S_x$, so that $\complement S \in \mathcal{K}$. Finally, if $\langle S_i \rangle$ is a sequence of sets in $\mathcal{K}$ and if $S = \bigcup S_i$, then using again the hypothesis that $\mathcal{B}$ is a σ-algebra we have $S_x = \bigcup (E_i)_x$ and $S \in \mathcal{K}$ so that $\mathcal{K}$ is a σ-algebra. Since $\mathcal{K}$ is a σ-algebra containing all measurable rectangles, it follows from the definition of $\mathcal{A} \times \mathcal{B}$

that $\mathcal{H} \supset \mathcal{A} \times \mathcal{B}$. The proof for E^y is similar.

Remark Proposition 26 asserts that if $E \in \mathcal{A} \times \mathcal{B}$ then every x-section and every y-section of E are measurable.

27. Proposition *Let* $\langle X, \mathcal{A} \rangle$ *and* $\langle Y, \mathcal{B} \rangle$ *be measurable spaces, and let* $\mathcal{E}$ *denote the class of all elementary sets in* $X \times Y$. *Then*

(a) There exists a smallest monotone class $\mathcal{F}$ *in* $X \times Y$ *containing* $\mathcal{E}$.

(b) If $A_1 \times B_1$ *and* $A_2 \times B_2$ *are two measurable rectangles, then*

$$(A_1 \times B_1) \cap (A_2 \times B_2) = (A_1 \cap A_2) \times (B_1 \cap B_2) ,$$

$$(A_1 \times B_1) - (A_2 \times B_2)$$

$$= [(A_1 - A_2) \times B_1] \cup [(A_1 \cap A_2) \times (B_1 - B_2)] .$$

Hence the intersection of two measurable rectangles is a measurable rectangle and their difference is an elementary set.

The proof is left to the reader (see Prob. 24).

28. Theorem *Let* $\langle X, \mathcal{A} \rangle$, $\langle Y, \mathcal{B} \rangle$ *and* $\mathcal{E}$ *be as in Proposition 27. Then* $\mathcal{A} \times \mathcal{B}$ *is the smallest monotone class which contains all elementary sets in* $X \times Y$.

Proof By part (a) of Proposition 27 there exists a smallest monotone class $\mathcal{F}$ in $X \times Y$ such that $\mathcal{E} \subset \mathcal{F}$. Clearly $\mathcal{A} \times \mathcal{B}$ as a σ-algebra, is also a monotone class containing $\mathcal{E}$, and so $\mathcal{F} \subset \mathcal{A} \times \mathcal{B}$. All we need to prove is that $\mathcal{F}$ is a σ-algebra containing $\mathcal{E}$. We proceed by steps.

Step 1 If $E_1 \varepsilon \mathcal{E}$ and $E_2 \varepsilon \mathcal{E}$, then $E_1 \cap E_2 \varepsilon \mathcal{E}$ and $E_1 - E_2 \varepsilon \mathcal{E}$. This follows from part (b) of Proposition 27. Also since

$$E_1 \cup E_2 = (E_1 - E_2) \cup E_2 ,$$

and $(E_1 - E_2) \cap E_2 = \emptyset$ we have $E_1 \cup E_2 \varepsilon \mathcal{E}$.

Step 2 We prove that if $P \varepsilon \mathcal{F}$ and $Q \varepsilon \mathcal{F}$, then $P - Q \varepsilon \mathcal{F}$ and $P \cup Q \varepsilon \mathcal{F}$. Let P be any fixed subset of $X \times Y$, and let S_P be the class of all subsets $Q \subset X \times Y$ such that

$$P - Q \varepsilon \mathcal{F} \quad , \quad Q - P \varepsilon \mathcal{F} \quad \text{and} \quad P \cup Q \varepsilon \mathcal{F} .$$

By interchanging the roles of P and Q we observe that Q belongs to S_P iff P belongs to S_Q. Furthermore S_P is a monotone class. For if $\langle A_i \rangle$ is a sequence in S_P such that $A_i \subset A_{i+1}$, then for $i = 1,2,3,\ldots$ the sets $A_i - P$ belong to $\mathcal{F}$, and $A_i - P \subset A_{i+1} - P$. Since $\mathcal{F}$ is a monotone class the set $\bigcup_{i=1}^{\infty} (A_i - P) = \bigcup_{i=1}^{\infty} A_i - P$ also belongs to $\mathcal{F}$ so that by the definition of S_P, $\bigcup_{i=1}^{\infty} A_i \varepsilon S_P$. In a similar manner we prove that if $\langle B_i \rangle$ is a sequence of sets in S_P with the property that $B_i \supset B_{i+1}$ for $i = 1,2,3,\ldots$, then

$\bigcap_{i=1}^{\infty} B_i \in S_P$. Let P be a fixed element of $\mathscr{E}$. Then Step 1 shows that for all $Q \in \mathscr{F}$ we have $Q \in S_P$, so that $\mathscr{E} \subset S_P$. Since S_P is a monotone class it follows from the definition of $\mathscr{F}$ that $\mathscr{F} \subset S_P$. Next let Q be a fixed element of $\mathscr{F}$. We proved that if $P \in \mathscr{E}$, then $Q \in S_P$. We also proved that $Q \in S_P$ if and only if $P \in S_Q$. Thus $\mathscr{E} \subset S_Q$ so that $\mathscr{F} \subset S_Q$. This proves Step 2.

Step 3 The collection $\mathscr{F}$ is a σ-algebra of sets in $X \times Y$. To prove this we need to prove the three defining axioms of a σ-algebra. Since $X \times Y \in \mathscr{E}$ we have $X \times Y \in \mathscr{F}$. Also if $Q \in \mathscr{F}$, then $\complement Q \in \mathscr{F}$ since $X \times Y$ is in $\mathscr{F}$. Finally if $\langle P_i \rangle$ is a sequence in $\mathscr{F}$, then by Step 2, since the union of any two elements of $\mathscr{F}$ belongs to $\mathscr{F}$, we have that for each positive integer n the finite union $P_1 \cup P_2 \cup \ldots \cup P_n$ also belongs to $\mathscr{F}$. Put $Q_n = P_1 \cup P_2 \cup \ldots \cup P_n$. We have $Q_n \subset Q_{n+1}$ and

$$\bigcup_{i=1}^{\infty} P_i = \bigcup_{n=1}^{\infty} Q_n.$$

Since $\mathscr{F}$ is a monotone class it follows that $\bigcup_{i=1}^{\infty} P_i \in \mathscr{F}$. Hence $\mathscr{F}$ is a σ-algebra which contains $\mathscr{E}$. Since $\mathscr{A} \times \mathscr{B}$ is the smallest σ-algebra containing $\mathscr{E}$, and since $\mathscr{E} \subset \mathscr{F} \subset \mathscr{A} \times \mathscr{B}$ it follows that $\mathscr{F} = \mathscr{A} \times \mathscr{B}$. The theorem is proved.

Definition *Let* $\langle X, \mathscr{A} \rangle$ *and* $\langle Y, \mathscr{B} \rangle$ *be two measurable spaces, and let* f *be an extended real-valued function on* $X \times Y$. *With each* $x \in X$ *we associate a function* f_x *defined on* Y *by* $f_x(y) = f(x,y)$. *Similarly with each* $y \in Y$ *we associate a function* f^y *defined on* X *by* $f^y(x) = f(x,y)$.

29. Theorem *Let* $\langle X, \mathcal{A} \rangle$ *and* $\langle Y, \mathcal{B} \rangle$ *be measurable spaces. Let* f *be an extended real-valued* $\mathcal{A} \times \mathcal{B}$ *-measurable function on* $X \times Y$. *Then*

(a) *For each* $x \in X$, f_x *is* $\mathcal{B}$*-measurable.*

(b) *For each* $y \in Y$, f^y *is* $\mathcal{A}$*-measurable.*

Proof Let O be an open set in the extended real number system. Put $V = \{(x,y): f(x,y) \in O\}$, and observe that V is the inverse image under f of the open set O. Since f is by assumption $\mathcal{A} \times \mathcal{B}$-measurable it follows that V is a measurable set, that is $V \in \mathcal{A} \times \mathcal{B}$. Consider the x-section of V, $V_x = \{y \in Y: f_x(y) \in O\}$. By the remark made immediately after the proof of Proposition 26 we have that $V_x \in \mathcal{B}$. Since V_x is the inverse image under f_x of the open set O, it follows that f_x is $\mathcal{B}$-measurable. The proof for f^y is similar.

30. Theorem *Let* $\langle X, \mathcal{A}, \mu \rangle$ *and* $\langle Y, \mathcal{B}, \nu \rangle$ *be* σ*-finite measure spaces. Suppose* $E \in \mathcal{A} \times \mathcal{B}$. *Define the functions:*

$$\alpha : X \to \overline{\mathbb{R}} \text{ by } \alpha(x) = \nu E_x$$

and

$$\beta : Y \to \overline{\mathbb{R}} \text{ by } \beta(y) = \mu E^y,$$

where by $\overline{\mathbb{R}}$ *we denote as usual the extended real-number system. Then* α *is* $\mathcal{A}$*-measurable, and* β *is* $\mathcal{B}$*-measurable. Furthermore,*

$$(*) \qquad \int_X \alpha d\mu = \int_Y \beta d\nu.$$

Remark The functions α and β are well defined since for each $E \in \mathcal{A} \times \mathcal{B}$ the sets E_x and E^y are measurable.

Proof We proceed by steps.

Step 1 If E is a measurable rectangle, then (*) holds. To prove this let $E = A \times B$ where $A \in \mathcal{A}$ and $B \in \mathcal{B}$. Then since $E_x = B$ if $x \in A$, and $E_x = \emptyset$ if $x \notin A$, we have $\alpha(x) = \nu E_x = (\nu B)\chi_A(x)$. Similarly $\beta(y) = \mu E^y = (\mu A)\chi_B(y)$. Therefore

$$\int_X \alpha d\mu = \mu(A)\nu(B) = \int_Y \beta d\nu .$$

(Recall that $0\cdot(+\infty) = 0$.)

Step 2 If $\langle E_i \rangle$ is a sequence in $\mathcal{A} \times \mathcal{B}$ such that for $i = 1,2,\ldots,$ $E_i \subset E_{i+1}$, and such that (*) holds for each E_i, then (*) also holds for $E = \bigcup_{i=1}^{\infty} E_i$. To prove this let α_i and β_i be the functions associated with E_i in the manner in which α and β are associated to the set E in the statement of the theorem. Since μ and ν are countably additive set functions we have

$$\lim_{i\to\infty} \alpha_i(x) = \alpha(x) \quad \text{and} \quad \lim_{i\to\infty} \beta_i(y) = \beta(y).$$

Since α_i and β_i by assumption satisfy (*) we have for each i

$$\int_X \alpha_i d\mu = \int_Y \beta_i d\nu$$

and by the monotone convergence theorem we get

$$\lim_{i\to\infty} \int_X \alpha_i d\mu = \int_X \alpha d\mu = \int_Y \beta d\nu = \lim_{i\to\infty} \int_Y \beta_i d\nu.$$

This proves Step 2.

Step 3 We prove that if $\langle E_i \rangle$ is a sequence of pairwise disjoint sets in $\mathcal{A} \times \mathcal{B}$, such that for each E_i $(i = 1,2,3,\ldots)$ the conclusion of the theorem holds, then (*) also holds for $E = \bigcup_{i=1}^{\infty} E_i$. Put $Q_1 = E_1$ and for $n = 2,3,\ldots$ define $Q_n = E_1 \cup E_2 \cup \ldots \cup E_n$. Since the E_i are pairwise disjoint, we see that for each Q_n (*) holds. Since $\langle Q_n \rangle$ satisfies the assumptions of Step 2, and since $\bigcup E_i = \bigcup Q_n$ we have that (*) holds for $E = \bigcup E_i$.

Step 4 If $A \in \mathcal{A}$ and $B \in \mathcal{B}$, if $\nu(B) < +\infty$ and $\mu(A) < +\infty$, if $A \times B \supset E_1 \supset E_2 \supset \ldots$, and if for each E_i $(i = 1,2,3,\ldots)$ (*) holds, then (*) also holds for $E = \bigcap_{i=1}^{\infty} E_i$.

The proof of this statement is similar to the proof of Step 2 except that we use the dominated convergence theorem instead of the monotone convergence theorem. The use of this theorem is permissible since μA and νB are both finite. The rest of the proof is left to the reader (see Prob. 25).

Step 5 We now prove that (*) holds for every $E \in \mathcal{A} \times \mathcal{B}$. By assumption μ and ν are both σ-finite measures. This means that X is the union of countable many disjoint sets X_n with $\mu X_n < +\infty$, and that Y is the union of countable many disjoint sets Y_m with $\nu Y_m < +\infty$. If E is in $\mathcal{A} \times \mathcal{B}$, define $E_{mn} = E \cap (X_n \times Y_m)$ $(m,n = 1,2,3,\ldots)$. Let $\mathcal{F}$ be the class of all $E \in \mathcal{A} \times \mathcal{B}$ such that (*) holds for E_{mn}, and for all choices of m and n. It follows from Steps 2 and 4 that $\mathcal{F}$ is a monotone class. Steps 1 and 3 show that $\mathcal{E} \subset \mathcal{F}$. Since $\mathcal{F} \subset \mathcal{A} \times \mathcal{B}$ it follows from Theorem 28, that $\mathcal{F} = \mathcal{A} \times \mathcal{B}$. Now since for every $E \in \mathcal{A} \times \mathcal{B}$ the conclusion (*) holds for E_{mn} and for all choices of m and n, since the sets E_{mn} are pairwise disjoint, and since E is the union of the sets E_{mn}, it follows from Step 3 that (*) holds for E. The theorem is proved.

Definition *Let* $\langle X, \mathcal{A}, \mu\rangle$ *and* $\langle Y, \mathcal{B}, \nu\rangle$ *be* σ*-finite measure spaces. On* $\mathcal{A} \times \mathcal{B}$ *define the set function* $\mu \times \nu$ *by*

$$(\mu \times \nu)(E) = \int_X \nu(E_x)\,d\mu \qquad (E \in \mathcal{A} \times \mathcal{B}).$$

By Theorem 30, we also have

$$(\mu \times \nu)(E) = \int_Y \mu(E^y)\,d\nu.$$

The function $\mu \times \nu$ is called the *product* of the measures μ and ν.

31. Proposition *Let* $\langle X, \mathcal{A}, \mu\rangle$ *and* $\langle Y, \mathcal{B}, \nu\rangle$ *be as in Theorem 30. Then* $(X \times Y, \mathcal{A} \times \mathcal{B}, \mu \times \nu)$ *is a* σ*-finite measure space.*

Proof Clearly $\mu \times \nu$ is a nonnegative function and $\mu \times \nu(\emptyset) = 0$. To prove that $\mu \times \nu$ is countably additive let $\langle E_n\rangle$ be a pointwise disjoint countable family of sets in $\mathcal{A} \times \mathcal{B}$. By Theorem 7 we have

$$\mu \times \nu\left(\bigcup_{n=1}^{\infty} E_n\right) = \int_X \nu\left(\left[\bigcup_{n=1}^{\infty} E_n\right]_x\right) d\mu$$

$$= \int_X \sum_{n=1}^{\infty} \nu([E_n]_x)\,d\mu = \sum_{n=1}^{\infty} \int_X \nu([E_n]_x)\,d\mu$$

$$= \sum_{n=1}^{\infty} \mu \times \nu(E_n).$$

To prove that $\mu \times \nu$ is σ-finite it suffices to consider for each $E \in \mathcal{A} \times \mathcal{B}$ the sets E_{mn} as defined in the proof of Step 5 of Theorem 30. The rest of the proof is left to the reader. The theorem is proved.

Remark In proving Step 1 of Theorem 30 we saw that

if $A \times B$ is a measurable rectangle then $\mu \times \nu(A \times B) = \mu(A)\mu(B)$. We observe now that the latter condition determines the measure $\mu \times \nu$ uniquely. For if λ is any measure on $\mathcal{A} \times \mathcal{B}$ such that $\lambda(A \times B) = \mu(A)\mu(B)$ for all measurable rectangles $A \times B$, then $\mu \times \nu(E) = \lambda(E)$ for all elementary sets E. Since $\mu \times \nu$ is σ-finite, and since $\mathcal{A} \times \mathcal{B}$ is the smallest σ-algebra containing all elementary sets, it follows (see Prob. 27) that $\mu \times \nu(E) = \lambda(E)$ for all $E \in \mathcal{A} \times \mathcal{B}$. This observation shows that the equality

$$\mu \times \nu(A \times B) = \mu(A)\mu(B)$$

for every measurable rectangle determines uniquely the product measure, and it could be taken as the definition of $\mu \times \nu$.

We are now ready to state and prove the main theorem of this section. In its original form (published in 1943) the theorem of G. Fubini (1879-1943) was a theorem about expressing a multiple Lebesgue integral as an iteration of Lebesgue integrals in spaces of lower dimensionality. Subsequently the theorem was generalized to abstract measure spaces. Here we present the theorem in its abstract form.

32. The Fubini Theorem *Let* $\langle X, \mathcal{A}, \mu\rangle$ *and* $\langle Y, \mathcal{B}, \nu\rangle$ *be* σ*-finite measure spaces, and let* f *be a function defined on* $X \times Y$.

(a) If f *is a nonnegative extended real-valued* $\mathcal{A} \times \mathcal{B}$*-measurable function, and if*

$$p(x) = \int_Y f_x d\nu \quad \text{and} \quad q(y) = \int_X f^y d\mu \quad (x \in X,\ y \in Y),$$

then p is $\mathcal{A}$-measurable and q is $\mathcal{B}$-measurable, and

$$(1) \qquad \int_X p\,d\mu = \int_{X\times Y} f\,d(\mu \times \nu) = \int_Y q\,d\nu,$$

or by substituting p and q we have the usual form of (1)

$$(1') \qquad \int_X d\mu(x) \int_Y f(x,y)\,d\nu(y)$$

$$= \int_{X\times Y} f\,d(\mu \times \nu) = \int_Y d\nu(y) \int_X f(x,y)\,d\mu(x).$$

(b) If f is a complex-valued $\mathcal{A}\times\mathcal{B}$-measurable function and if $w(x) = \int_Y |f_x|\,d\nu$ and $\int_X w\,d\mu < +\infty$, then

$$\int_{X\times Y} |f|\,d(\mu \times \nu) < +\infty.$$

(c) If

$$\int_{X\times Y} |f|\,d(\mu \times \nu) < +\infty,$$

then f_x is an integrable function for almost all $x \in X$, and f^y is an integrable function for almost all $y \in Y$. Furthermore, the functions ϕ and ψ defined by

$$\phi(x) = \int_Y f_x\,d\nu \quad \text{a.e.,} \quad x \in X,$$

$$\psi(y) = \int_X f^y d\mu \quad a.e., \quad y \in Y,$$

are integrable and

$$\int_X \phi d\mu = \int_{X\times Y} fd(\mu \times \nu) = \int_Y \psi d\nu.$$

Proof *(a)* Since f_x is $\mathcal{B}$-measurable and f^y is $\mathcal{A}$-measurable the functions p and q are well defined. Next observe that if f is taken to be the characteristic function of a set $E \in \mathcal{A} \times \mathcal{B}$, then by Theorem 30 we have

$$\int_X pd\mu = \int_Y qd\nu = \mu \times \nu(E) = \int_{X\times Y} fd(\mu \times \nu)$$

which is the condition (1) in (a). It follows that (a) holds for all simple $\mathcal{A} \times \mathcal{B}$-measurable functions. In the general case let $\langle f_n\rangle$ be a sequence of nonnegative measurable simple functions with $f_n \le f_{n+1}$, and such that $\lim_{n\to\infty} f_n(x,y) = f(x,y)$ at every point $\langle x,y\rangle$ of $X \times Y$. The existence of this sequence is guaranteed by Theorem 3.

For each positive integer n, set

$$p_n(x) = \int_Y (f_n)_x d\nu.$$

We have

$$\int_X p_n d\mu = \int_{X\times Y} f_n d(\mu \times \nu) \qquad (n = 1,2,\ldots).$$

The sequence $\langle p_n \rangle$ is nondecreasing and so by the Monotone Convergence Theorem applied on the space $\langle Y, \mathcal{B}, \nu \rangle$ we have

$$\lim_{n\to\infty} p_n(x) = \int_Y f_x d\nu = p(x)$$

for all $x \varepsilon X$. Also, again by the Monotone Convergence Theorem, we have

$$\lim_{n\to\infty} \int_X p_n d\mu = \int_X p d\mu = \int_{X\times Y} f d(\mu \times \nu).$$

We obtain in a similar manner

$$\int_Y q d\nu = \int_{X\times Y} f d(\mu \times \nu).$$

Part (a) is proved.

Proof (b) Apply part (a) to $|f|$.

Proof (c) Without loss of generality we may assume that f is a real-valued function, for the complex case follows if we write $f = u + iv$ (see Chap. 3, p. 169). Let $f = f^+ - f^-$, where f^+ and f^- are respectively the positive and negative parts of f. Then part (a) applies to f^+ and f^-. Put

$$\phi_1(x) = \int_Y (f^+)_x d\nu, \quad \phi_2(x) = \int_Y (f^-)_x d\nu \qquad (x \varepsilon X).$$

Since $\int_{X\times Y} |f| d(\mu \times \nu) < +\infty$ and $f^+ \leq |f|$, and since (a) holds for f^+, we have that

$$\int_X |\phi_1| d\mu < +\infty.$$

Similarly we have $\int_X |\phi_2| d\mu < +\infty$. Since $f_x = (f^+)_x - (f^-)_x$ we have that $\int_Y |f_x| d\nu < +\infty$ for every x for which $\phi_1(x) < +\infty$ and $\phi_2(x) < +\infty$. But we have seen that ϕ_1 and ϕ_2 are integrable in the space $\langle X, \mathcal{A}, \mu \rangle$, which means that ϕ_1 and ϕ_2 have finite values almost everywhere on X. For every such value x we have $\phi_1(x) - \phi_2(x) = \phi(x)$. This shows that ϕ is integrable. We have

$$\int_X \phi_1 d\mu = \int_{X \times Y} f^+ d(\mu \times \nu)$$

and

$$\int_X \phi_2 d\mu = \int_{X \times Y} f^- d(\mu \times \nu).$$

If we subtract the second equation from the first we get

$$\int_X \phi d\mu = \int_{X \times Y} f d(\mu \times \nu).$$

We can prove in a similar manner that

$$\int_Y \psi d\nu = \int_{X \times Y} f d(\mu \times \nu).$$

The theorem is proved.

33. Corollary *Notations as in Theorem 32. If* f *is* $\mathcal{A}\times\mathcal{B}$*-measurable and if*

$$\int_X d\mu(x) \int_Y |f(x,y)|\, d\nu(y)$$

if finite, then the iterated integrals

$$\int_X d\mu(x) \int_Y f(x,y)\, d\nu(y) \;;\quad \int_Y d\nu(y) \int_X f(x,y)\, d\mu(x)$$

are both finite and equal.

The proof is left to the reader (see Prob. 28).

Before we close this section we wish to discuss very briefly the completion of a product measure.

If $\langle X, \mathcal{A}, \mu\rangle$ and $\langle Y, \mathcal{B}, \nu\rangle$ are measure spaces with μ and ν complete measures, then the following example shows that $\mu \times \nu$ need not be a complete measure.

Example

Let $X = Y =$ real line. $\mathcal{A} = \mathcal{B} = \mathcal{M} =$ Lebesgue measurable sets, and $\mu = \nu = m =$ Lebesgue measure. Let $A = \{x\}$ be any set in X consisting of a single point x, and let S be any nonmeasurable set of real numbers. Then $A \times S \subset A \times Y$ and $\mu \times \nu(A \times Y) = m \times m(A \times Y) = 0$. But by Proposition 26 we see that $A \times S$ is not a measurable set since the x-section of this set is S. This shows that $m \times m$ is not complete. However, we know that we can enlarge $\mathcal{A}\times\mathcal{B}$ to another σ-algebra $\overline{\mathcal{A}\times\mathcal{B}}$ (called sometimes the *completion* of $\mathcal{A}\times\mathcal{B}$) and extend the definition of μ to $\overline{\mathcal{A}\times\mathcal{B}}$ so as to obtain a complete measure. The question now arises whether or not the conclusions of Fubini's theorem hold if f is a $\overline{\mathcal{A} \times \mathcal{B}}$-measurable function on $X \times Y$. The answer to this question is given by the following theorem.

34. Theorem *Let* $\langle X, \mathcal{A}, \mu \rangle$ *and* $\langle Y, \mathcal{B}, \nu \rangle$ *be σ-finite complete measure spaces. Let* $\overline{\mathcal{A} \times \mathcal{B}}$ *be the σ-algebra which is the completion of* $\mathcal{A} \times \mathcal{B}$ *with respect to* $\mu \times \nu$. *Then all conclusions of Fubini's theorem hold except for the following changes:*

The function f_x *will not be measurable for all* x *in* X *but only for almost all* x *in* X, *so that the function* p *in (a) will be defined a.e.* $[\mu]$. *Similarly* f^y *will be measurable for almost all* $y \in Y$ *and* q *will be defined a.e.* $[\nu]$. For a proof of Theorem 34 the reader is referred to Rudin [24].

PROBLEMS

1. Prove Theorems 2 and 3.
2. Let f and g be measurable functions with range in $[-\infty, \infty]$. Show that the functions

$$f^+ = \max\{f, 0\} \quad \text{and} \quad f^- = \max\{-f, 0\}$$

are also measurable.
3. Show that the functions μ, as defined in Examples (a) and (b) immediately following Theorem 4, are measures.
4. Show that part (d) of Theorem 4 is false if the condition $\mu(E_1) < +\infty$ is not satisfied.

(Hint: Consider the counting measure on the set N of positive integers. Take $E_n = \{n, n+1, n+2, \ldots\}$.)

5. Let $\langle X, \mathcal{B}, \mu \rangle$ be a measure space and let μ be a complete measure. Let f and g be functions on X such that f is measurable and $f(x) = g(x)$ a.e. $[\mu]$. Show that g is measurable.
6. Let $\langle X, \mathcal{B}, \mu \rangle$ be a measure space and assume that μ is not a complete measure. Suppose also $f(x) = g(x)$ a.e. $[\mu]$. Show that g is measurable if $g(x)$ is a constant on the set $\{x: f(x) \neq g(x)\}$.
7. Let X be an uncountable set, let $\mathcal{B}$ be the collection of all sets $E \subset X$ such that either E or $\complement E$ is countable. Define $\mu(E) = 0$ if E is countable and $\mu(E) = 1$ if E is not countable. Prove that $\langle X, \mathcal{B}, \mu \rangle$ is a measure space.
8. On a measure space $\langle X, \mathcal{B}, \mu \rangle$ consider a sequence $\langle f_n \rangle$ of measurable real-valued functions. Let S be the set of all points x such that the limit $\lim_{n\to\infty} f_n(x)$ exists. Show that S is a measurable set.
9. Show parts (b) through (f) of Theorem 5.
10. Prove Theorems 6, 7, and 8.
11. Let $\langle X, \mathcal{B}, \mu \rangle$ be a measure space such that $\mu(X) < +\infty$. Let $\langle f_n \rangle$ be a sequence of bounded real-valued measurable functions on X such that

$$f_n \to f \quad \text{uniformly on } X.$$

Prove that

$$\lim_{n\to\infty} \int_X f_n d\mu = \int_X f d\mu$$

Is the above result true if the assumption $\mu(X) < +\infty$ is replaced by $\mu(X) = +\infty$?

12. Let $\langle X, \mathcal{B}, \mu \rangle$ be a measure space and let $f_n : X \to [0,+\infty]$ be measurable for $n = 1,2,3,\ldots$. Suppose that $f_n \geq f_{n+1}$ for $n = 1,2,3,\ldots$, $\lim_{n\to\infty} f_n(x) = f(x)$ for every $x \in X$, and f_1 is integrable on X. Prove that

$$\lim_{n\to\infty} \int_X f_n d\mu = \int_X f d\mu \ .$$

Show that the hypothesis that f_1 is integrable cannot be omitted.

13. Let $\langle X, \mathcal{B} \rangle$ be a measurable space and let $\mathcal{S}$ be the set of all finite signed measures on X (i.e., if $\mu \in \mathcal{S}$ then for every E in $\mathcal{B}$, $\mu(E)$ is finite). If μ and ν are in $\mathcal{S}$ let $\mu \prec \nu$ mean that $\mu(E) \leq \nu(E)$ for every $E \in \mathcal{B}$. Show that

(a) The relation $\prec$ is a partial order on $\mathcal{S}$, and that the zero measure, 0, belongs to $\mathcal{S}$.

(b) Show that if $\mu \in \mathcal{S}$, then μ^+ is the least upper bound of μ and 0 with respect to the partial order $\prec$, (i.e., $\mu^+ = \sup(\mu, 0)$). Show also that $\mu^- = -\inf(\mu, 0)$.

(c) Show that if μ and ν are in $\mathcal{S}$ then

$$\sup(\mu,\nu) = \mu + \sup(\nu - \mu, 0)$$

$$\sup(\mu,\nu) + \inf(\mu,\nu) = \mu + \nu.$$

14. Let $\langle X, \mathcal{B}, \mu\rangle$ be a measure space, and let f be an extended real-valued integrable function on X. Define

$$\nu : \mathcal{B} \to \mathbb{R} \text{ by } \nu(E) = \int_E f d\mu.$$

Show that ν is a signed measure, and that

$$\nu^+(E) = \int_E f^+ d\mu.$$

(Hint: Show that $\int_E f^+ d\mu \leq \nu^+(E)$, and $\nu^+(E) \geq \int_E f^+ d\mu$ separately.)

15. Let ν be a signed measure on a measurable space $\langle X, \mathcal{B}\rangle$, such that $\nu = \lambda_1 - \lambda_2$ where λ_1 and λ_2 are measures on $\langle X, \mathcal{B}\rangle$. Show that $\nu^+ \leq \lambda_1$ and $\nu^- \leq \lambda_2$.

16. Let $\langle X, \mathcal{B}, \mu\rangle$ be a measure space and let ν be a function defined on $\mathcal{B}$ by $\nu(E) = 0$ if $\mu(E) = 0$, and $\nu(E) = +\infty$ if $\mu(E) > 0$. Prove that $\langle X, \mathcal{B}, \nu\rangle$ is a measure space and that $\nu << \mu$. Find a nonnegative finite valued measurable function ϕ on X such that

$$\int_X f d\nu = \int_X f\phi d\mu$$

for all nonnegative extended real-valued measurable functions on X.

17. Prove the uniqueness of f in Theorem 21.

18. Let $\langle X, \mathcal{B} \rangle$ be a measurable space. A complex-valued function λ on $\mathcal{B}$ is called a *complex measure* iff

$$\lambda(\emptyset) = 0 \quad \text{and} \quad \lambda\left(\bigcup_{n=1}^{\infty} E_n\right) = \sum_{n=1}^{\infty} \lambda(E_n)$$

for all pairwise disjoint sequences $\langle E_n \rangle_{n=1}^{\infty}$ of elements of $\mathcal{B}$. Here also (see definition of signed measure, p. 501) it is implicit that the series $\sum_{n=1}^{\infty} \lambda(E_n)$ converges absolutely.

The total variation of λ is the function $|\lambda|$ defined on $\mathcal{B}$ by:

$$|\lambda|(E) = \sup \sum_{i=1}^{n} |\lambda(E_i)| \quad , \quad (E \in \mathcal{B}),$$

where the supremum is taken over all possible finite disjoint collections $\{E_1, E_2, \ldots, E_n\}$ of sets with $E_i \in \mathcal{B}$ and $E_i \subset E$ for $i = 1,2,\ldots,n$.

Let $\lambda(E) = R_e\lambda(E)$ and $\lambda_2(E) = \mathcal{I}m\lambda(E)$ be respectively the real and imaginary parts of $\lambda(E)$ for all $E \in \mathcal{B}$.

(a) Show that λ_1 and λ_2 are signed measures on $\langle X, \mathcal{B} \rangle$

(b) Show that for all $E \in \mathcal{B}$ we have

$$\lambda(E) = \lambda_1^+(E) - \lambda_1^-(E) + i\lambda_2^+(E) - i\lambda_2^-(E)$$

and

$$|\lambda|(E) \leq \lambda_1^+(E) + \lambda_1^-(E) + \lambda_2^+(E) + \lambda_2^-(E).$$

19. Let $\langle X, \mathcal{B} \rangle$ and λ be as in Problem 18. Show that

    (a) $\sup\{|\lambda(A)|: A \in \mathcal{B}, A \subset E\} \leq |\lambda|(E)$
    $\leq 4 \sup\{|\lambda(A)|: A \in \mathcal{B}, A \subset E\}$.

    (b) $|\lambda(E)| \leq |\lambda|(X) < +\infty$.

    (c) $\lambda_j^+(E) \leq |\lambda|(E)$ and $\lambda_j^-(E) \leq |\lambda|(E)$ for $j = 1,2$.

    Hence λ_1^+, λ_1^-, λ_2^+, λ_2^- and $|\lambda|$ are finite measures on $\langle X, \mathcal{B} \rangle$.

20. Let N be the set of all positive integers, and $\mathcal{B}$ the collection of all subsets of N. Let μ be the counting measure on N (see Sec. 10.2) and consider the measure space $\langle N, \mathcal{B}, \mu \rangle$. Characterize the elements of the space $L^p(\mu)$ with respect to $\langle N, \mathcal{B}, \mu \rangle$. Conclude that the space ℓ^p (all sequences $\langle x_n \rangle$ such that $\Sigma|x_n|^p < +\infty$) is complete.

21. Let $\langle X, \mathcal{B}, \mu \rangle$ be a measure space. A sequence $\langle f_n \rangle$ of measurable functions on X is said to *converge in measure* to the measurable function f iff for every $\varepsilon > 0$ there corresponds an M such that

$$\mu(\{x: |f_n(x) - f(x)| > \varepsilon\}) < \varepsilon$$

for all $n > M$. Assume $\mu(X) < +\infty$.

(a) Show that if $f_n(x) \to f(x)$ a.e. $[\mu]$ then f_n converges to f in measure.

(b) Show that if $1 \leq p < \infty$, $f_n \in L^p(\mu)$ for all n, and $\lim_{n\to\infty} \|f_n - f\|_p = 0$, then f_n converges to f in measure.

22. Suppose X consists only of two points a and b. Define $\mu(\{a\}) = 1$, $\mu(\{b\}) = \mu(X) = +\infty$ and $\mu(\emptyset) = 0$. Is it true for this μ that the dual space of $L^1(\mu)$ is $L^\infty(\mu)$?

23. Let $\langle X, \mathcal{B}, \mu\rangle$ be a measure space, $1 < p < \infty$, and $(1/p) + (1/q) = 1$. Show that $L^q(\mu)$ is the dual space of $L^p(\mu)$ even if μ is not a σ-finite measure.

24. Prove Proposition 27.

25. Complete the proof of Step 4 in the proof of Theorem 30.

26. With the notations of Proposition 31, prove that $\mu \times \nu$ is σ-finite.

27. With the notations of Proposition 31, prove that the equation

$$\mu \times \nu(A \times B) = \mu(A)\mu(B)$$

determines uniquely the measure $\mu \times \nu$.

28. Prove Corollary 33.

29. Let $X = Y = [0,1]$, and $\mu = \nu =$ Lebesgue measure on $[0,1]$ in Fubini's theorem. Let $\langle a_n\rangle$ be a sequence such that $0 = a_1 < a_2 < \ldots$, and

such that $\lim_{n\to\infty} a_n = 1$. For each positive integer n let g_n be a continuous function on $[0,1]$ which vanishes for $x \notin (a_n, a_{n+1})$, and such that

$$\int_0^1 g_n(t)dt = 1 \quad \text{for} \quad n = 1,2,\ldots .$$

Define

$$f(x,y) = \sum_{n=1}^{\infty} [g_n(x) - g_{n+1}(x)]g_n(y).$$

Show that

$$\int_0^1 dx \int_0^1 f(x,y)dy = 1 \neq 0 = \int_0^1 dy \int_0^1 f(x,y)dx.$$

30. Let $X = Y = [0,1]$, μ = Lebesgue measure on $[0,1]$, ν = counting measure on Y (see sec. 10.2 and Prob. 3) in Fubini's theorem. Define

$$f(x,y) = 1 \quad \text{if} \quad x = y$$
$$f(x,y) = 0 \quad \text{if} \quad x \neq y$$

for all x and y in $[0,1]$. Show that

$$\int_Y d\nu(y) \int_X f(x,y)d\mu(x) = 0 \neq 1$$

$$= \int_X d\mu(x) \int_Y f(x,y)d\nu(y) .$$

31. Explain why the conclusion of Fubini's theorem does not hold in the examples considered in Problems 29 and 30. (For another example see Gelbaum and Olmsted [10] p. 124.)

BIBLIOGRAPHY   INDEX

Bibliography

1. *T. M. Apostol:* "Mathematical Analysis," Addison-Wesley Publishing Company, Inc., Reading, Mass., 1957.
2. *N. Artémiadis:* "A Remark on metric spaces," Koninkl, Nederl., Academie van Wetenshappen-Amsterdam, Proc. Series, 68, 1965.
3. *G. Bachman and L. Narici:* "Functional Analysis," Academic Press, Inc., New York, 1966.
4. *R. P. Boas, Jr.:* "A primer of real functions," John Wiley and Sons, Inc., 1960.
5. *N. Bourbaki:* "Théorie des Ensembles," Herman, Paris, 1958.
6. *N. Bourbaki:* "Topologie Générale," Herman, Paris, 1958.
7. *C. Carathéodory:* "Vorlesungen über reele Functionen," 2d ed., Chelsea Publishing Company, Inc., New York, 1948.
8. *G. Choquet:* "Topology," Academic Press, Inc., New York, 1966.
9. *J. Dieudonné:* "Foundations of Modern Analysis,"

Academic Press, Inc., New York, 1958.

10. *B. R. Gelbaum and J. M. H. Olmsted:* "Counter-examples in Analysis," Holden-Day, Inc., San Francisco, Calif. 1964.

11. *C. Goffman:* "Real Functions," Rinehart, New York, 1955.

12. *D. W. Hall and G. L. Spencer:* "Elementary Topology," John Wiley and Sons, Inc., New York, 1955.

13. *P. R. Halmos:* "Measure Theory," D. Van Nostrand Company, Inc., Princeton, N.J., 1960.

14. *P. R. Halmos:* "Naive Set Theory," D. Van Nostrand Company, Inc., Princeton, N.J., 1960.

15. *E. Hewitt and K. Stromberg:* "Real and Abstract Analysis," Springer-Verlag New York, Inc., New York, 1965.

16. *J. L. Kelley:* "General Topology," D. Van Nostrand Company, Inc., Princeton, N.J., 1955.

17. *A. Kolmogoroff and S. Fomin:* "Elements of the Theory of Functions and Functional Analysis," Graylock Press, Rochester, N.Y., vol. I, 1957.

18. *C. Kuratowski:* "Topologie I," "Monografje Matematyczne," vol. 3, Warsaw, 1933.

19. *E. J. McShane and R. Botts:* "Real Analysis," D. Van Nostrand Company, Inc., Princeton, N.J., 1959.

20. *I. P. Natanson:* "Theory of Functions of a Real Variable," Frederick Ungar Publishing Co. New York, vol. I, 1964; vol. II, 1960.

21. *J. F. Randolph:* "Basic Real and Abstract Analysis," Academic Press, Inc., New York, 1968.

22. *H. L. Royden:* "Real Analysis," The Macmillan Company, New York, 1963.

23. *W. Rudin:* "Principles of Mathematical Analysis," 2d ed., McGraw-Hill Book Company, New York, 1964.

24. *W. Rudin:* "Real and Complex Analysis," McGraw-Hill Book Company, New York, 1966.

25. *G. F. Simmons:* "Introduction to Topology and Modern Analysis," McGraw-Hill Book Company, New York, 1963.

26. *A. E. Taylor:* "General Theory of Functions and Integration," Blaisdell Publishing Company, Waltham, Mass., 1965.

27. *G. Valiron:* "Théorie des Fonctions," Masson et Cie, Editeurs, Paris, 1948.

28. *A. Wilansky:* "Functional Analysis," Blaisdell Publishing Company, Waltham, Mass., 1964.

29. *A. Zygmund:* "Trigonometric Series," 2d ed., Cambridge University Press, New York, 1959.

Index of Symbols

Index